THE FOUNDERS OF WESTERN THOUGHT – THE PRESOCRATICS

BOSTON STUDIES IN THE PHILOSOPHY OF SCIENCE

VOLUME 257
For further volumes:
http://www.springer.com/series/5710

THE FOUNDERS OF WESTERN THOUGHT – THE PRESOCRATICS

A Diachronic Parallelism Between Presocratic Thought and Philosophy and the Natural Sciences

CONSTANTINE J. VAMVACAS

Translated from the original Greek version
into English by Professor Robert Crist of the
University of Athens, Greece.

Springer

Originally published
as Oi Themeliotes tis Dytikis Skepsis-Prosokratikoi
by the Crete University Press
Herakleion
Greece 2001

ISBN 978-1-4020-9790-4 e-ISBN 978-1-4020-9791-1

DOI 10.1007/978-1-4020-9791-1

Library of Congress Control Number: 2009920271

Printed on acid-free paper

9 8 7 6 5 4 3 2 1

springer.com

To
EVA
and
KATERINA

The question was raised long ago, is still and always will be,
and always baffle us – What is Being (to be)?
Aristotle

May you become yourself, through learning.
Pindar

Preface

There can be little doubt that the Greek tradition of philosophical criticism had its main source in Ionia ... It thus leads the tradition which created the rational or scientifi attitude, and with it our Western civilization, the only civilization, which is based upon science (though, of course, not upon science alone).

Karl Popper, *Back to the Presocratics*

Harvard University physicist and historian of Science, Gerald Holton, coined the term "Ionian Enchantment", an expression that links the idea back in the 6th century B.C. to the ancient Ionians along the eastern Aegean coast, while capturing its fascination. Approximately within a seventy-f ve year period (600–525 B.C.) -a split second in the history of humanity- the three Milesian thinkers, Thales, Anaximander and Anaximenes, without plain evidence, but with an unequalled power of critical abstraction and intuition, had achieved a true intellectual revolution; they founded and bequeathed to future generations a new, unprecedented way of theorizing the world; it could be summarized in four statements: beneath the apparent disorder and multiplicity of the cosmos, there exists order, unity and stability; unity derives from the fundamental primary substratum from which the cosmos originated; this, and, consequently, the cosmic reality, is one, and is based not on supernatural, but on physical causes; they are such that man can investigate them rationally. These four statements are neither self-evident nor self-explanatory. The attempt was to orient the mind in a rational, critical spirit, and this attitude would prove decisive and seminal for the entire development of Western thought.

Virtually all of the Presocratic thinkers are from Ionia. Thales started the presocratic period, followed by Anaximander, Anaximenes, Pythagoras (mathematization of nature), Xenophanes ("absolute" and "apparent truth"), Heraclitus (change and "becoming"), Parmenides (change and "being"), Empedocles (four "elements", two opposite "forces"), Anaxagoras ("mind" and "matter"). The atomists Leucippus and Democritus would round out the great age of the Presocratics. By the mid-fift century B.C., there had already been a turn in Greek thought from "nature" to "man". The anthropological view of the cosmos introduced by Socrates was exhaustively explored in the work of Plato and Aristotle. Attic philosophy is, however, not only a new creation; it is also a recapitulation. All the themes that were treated during the Presocratic age are examined anew and ordered within a wider and fuller unity, creating the unparalleled edific of Platonic and Aristotelian philosophy. The widely accepted assumption that Greek philosophy and science starts with

Socrates is, thus, unfounded. "What would Plato have been, as well as Socrates", asks Martin Heidegger, "without Parmenides"?

The present book is a *balanced interdisciplinary philosophic-scientifi* presentation of the presocratic tradition. Two are its main goals: to make the founding and evolution of Western thought accessible to the reflect ve man of our day, and to shed greater light also on the scientifi dimension of the Presocratics' work, showing–probably for the firs time systematically–its timeless value. The Presocratics were the firs to interpret the universe *critically*, through a unique combination of *rational thought*, *intuition* and *observation*, excluding any intervention of divine and supernatural powers. At the same time they appealed to man's *conscience*, elevating man to a *free* and *responsible* position. They posed the fundamental questions about "truth", "being", "becoming", laid the foundations for the tradition of *critical*–not dogmatic–investigation, without which science would not exist; and invented 'mental tools' for approaching the physical world, concepts which prevail in today's natural sciences, such as: universality of the physical laws; unity underlying seeming plurality; hidden ultimate reality underlying phenomena; interrelation and interconnection of all beings; man-observer as an inseparable part of the natural world; merging of opposite concepts; shift in emphasis from matter to process; inherent dynamic balance; mathematization of nature; atomic theory; intelligent-design theory; attribution of quality to quantity; symmetry; isotropy; measure; order; evolutionary process and natural selection. Karl R. Popper has described this impetus as the "Presocratic Enlightenment": "As to the Presocratics, I assert that there is the most perfect possible continuity of thought between their theories and the later developments in physics; whether they are called philosophers, or pre-scientists, or scientists matters very little, I think. . .Few philosophers or scientists are aware of the influenc exerted by some of the oldest ideas of Greek philosophy and Greek science upon our most advanced scientifi theories: upon classical physics and chemistry, relativity, quantum theory, genetics, and even molecular biology".

The spirit of the Presocratics is above all *holistic*. Thus, *philosophy* (metaphysics, ethics, psychology, sociology) and *science* (physics, chemistry, cosmology, biology) are two areas that have as yet remained inviolably *joined and unifie* . Immediately after the Presocratics, *wholeness* disintegrates. The fragmentation into separate sciences and specializations would become the exclusive context of thinking and investigation of things–a phenomenon that leads, finall , to the perception that this detached state of affairs corresponds to ultimate reality itself. Martin Heidegger summarized the situation in one sentence: "What the word *cosmos* expresses for us today is the unregulated confusion of a technical mechanism of information which arose in the face of unfragmented *nature* and usurped her place, while its functioning remained accessible and controllable only through calculation".

There is, however, today an increasing keen longing for unified-all-embracin knowledge, as inherited by the Presocratic tradition. For the firs time since the Presocratic age, there is an effort in the Western world toward a "reconciliation" and "reunification of the natural sciences, arts and humanities. Harvard University professor emeritus, Edward O. Wilson, in his visionary book *Consilience: The Unity of Knowledge*, notes: "I had experienced the Ionian Enchantment. That recently coined expression I borrowed from the physicist and historian Gerald Holton. It means a belief in the unity of the sciences-a conviction far deeper than a mere working proposition that the world is orderly and can be explained by a small number of natural laws. Its roots go back to Thales of Miletus, in Ionia, in the sixth century BC". Many of the new directions and movements in the last half of the 20th century are in their essence transdisciplinary, leading to corrosion of the boundaries between the disciplines by unifying their subjects. R. Sanders Williams, the medical school dean of Duke University, at the convocation for new graduate and professional students on August 21, 2003, cautions them from becoming isolated in their particular discipline: "essential as this may be, the mastery of specialized knowledge and technique should not be all that define your experience at the University". Only fluen y across the boundaries will provide a clear view of the world as it really is, of who we are and of why we are here. This longing is wonderfully expressed by one of the most prominent physicists of the 20th century, Werner Heisenberg: "The most significan areas of pure science are those in which there is no longer any concern for applications, in which pure intellect, chiefl , traces the hidden harmonies of the universe. This most inner region, on which science and art can scarcely be distinguished from one another any longer, is perhaps for mankind today the sole place that reveals the truth with utter clarity. . .through which unfolds a glimpse of portions of the universe about which one can speak only in parables".

All the above views of contemporary scientists revive anew the spirit of the Presocratics. To be sure, a 'return' to them is impossible. What is feasible, however, is to *assimilate their thought* in an attempt to *transcend* today's situation: "What is needed is to learn afresh, to observe, and to discover for ourselves the meaning of wholeness. . .Such insight implies an original and creative act of perception into all aspects of life, mental, and physical. . .Then the whole fiel of *measure* will come into *harmony*, as fragmentation within it comes to an end. . .When such a harmony prevails, man can then not only have insight into the meaning of *wholeness* but, what is much more significant he can realize the *truth* of this insight in every phase and aspect of his life". These words of D. Bohm, a distinguished 20th century physicist and philosopher, best reflec the everlasting spirit of the Presocratics.

Athens, Greece — Constantine J. Vamvacas

Contents

Introduction 1

The Juncture 3
Nature 4
Society 6
Religion 8
Myth and Epic Poetry 9
Hesiod – The First Cosmogonies 10
Personality – Lyric Poetry – Art 12
Foreign Influenc 15

Introduction to the Presocratics 19

Thales of Miletus (ca. 625–546 B.C) 29
Personality – Life 29
Cosmology – Mathematics 30
First *'arche'*: Water 30
Soul 32
The Divine 32
Overview 32

Anaximander of Miletus (ca. 610–546 B.C) 35
The Extant Fragment 35
Personality – Life 35
First *'arche'*: The Boundless 36
The Divine 37
The Opposites 37
Justice – Necessity 38
Cosmogony 39
Cosmology 40
Meteorology 41
Overview 41

Anaximenes of Miletus (ca. 585–525 B.C) ... 45
Introduction ... 45
First *'arche'*: Air ... 45
Soul ... 47
The Divine ... 47
Cosmogony – Cosmology – Meteorology ... 48
Overview ... 48
Epilogue: The Milesian *'physicoi'* ... 49

Pythagoras of Samos (ca. 570–496 B.C) ... 53
Personality ... 53
Life ... 55
The Way of Life ... 55
'Acusmatici' – 'Mathematici' ... 57
'Acusmata' – 'Symbola' ... 57
Immortality of the Soul ... 58
Homogeneity ... 59
Purif cation ... 59
Harmony of Soul ... 59
Ethics ... 60
Society ... 60
Education ... 61
Theology ... 61
Philosophy ... 62
Mathematics ... 63
Arithmetic ... 64
Geometry ... 65
Incommensurability ... 66
Geometrical Algebra ... 67
Music ... 68
Harmony ... 70
The opposites ... 71
Number ... 71
Cosmogony ... 72
Cosmology ... 74
Overview ... 75
  Physical Sciences ... 77
  Philosophy – Holism ... 81

Xenophanes of Colophon (ca. 570–470, B.C) ... 85
Life ... 85
Personality ... 85
Cosmology ... 86
Biology ... 86

Physics 87
Theology 88
Anthropocentric Theory 89
Progress 89
Ethics 91
Community 92
Knowledge 92
The Tragic Element 93
Overview 95

Heracltus of Ephesus (ca. 540–480 B.C) 101
Introduction 101
Life 101
Writings 102
Union of Opposites 104
Perpetual Strife of Opposites 107
Becoming 107
Fire 109
Logos 110
Truth 112
Logos: Common-Private 112
Logos: Law-Measure 113
Logos: One-All 114
Wisdom 115
Knowledge 115
The Soul 116
Death 117
Cosmology-Physics 118
Ethics – Community – Law 119
The Divine 121
The Tragic Element 122
Overview 124
Epilogue 133

Parmenides of Elea (ca. 515–450 B.C) 135
Being 135
Personality 136
Life 137
Writings 137
'Prologue' 138
Being–Non-being 138
Attributes of Being 141
The Opinions of Men 142
The Cosmology of Seeming 143

The Path of 'Truth' – The Track of 'Opinion' 145
What is 'Being'? 145
'Being' and 'Not Being' 146
'Being' and 'Truth' 147
'Being' and 'opinions' 148
'Being' and 'Apprehending and Saying' 148
The Eleatic School 150
Melissus of Samos 150
Zeno of Elea 151
The Paradoxes of Zeno 151
Plurality 151
Space 152
The Millet Seed 153
First paradox: 'The dichotomy' (dividing in two) 153
Second paradox: 'Achilles and the tortoise' 154
Third paradox: 'The flyin arrow is stationary' 154
Fourth paradox: 'The Stadium' 155
Heraclitus – Parmenides 157
Influenc 159
Overview 162
Cosmology 162
Ontology 164

Empedocles of Acragas (ca. 494–434 B.C) 167
Personality 167
Writings 168
The Natural World 169
The 'Roots' 170
Becoming 171
Forces 171
The 'Cosmic Cycle' 172
Cosmogony – Cosmology 173
Creation of Life 175
Knowledge 176
'Purif cations' 177
Introduction 178
The Primal Stage of Bliss 179
The Fall 179
Guilt 179
Purif cation 180
Daemons 180
The Relationship of the Two Poems 181
Overview 182
Sense Experience 182
Experiment 183

Practical Applications ... 183
Determinism ... 184
The Universality of Natural Laws ... 184
Forces ... 185
Light ... 186
Chemistry ... 186
Cosmogony ... 189

Anaxagoras of Clazomenae (ca. 500–428 B.C) ... 197
Personality ... 197
Life ... 197
Writings ... 198
Matter ... 198
Mind ... 199
Cosmogony ... 201
Cosmology ... 201
Creation of Life ... 202
Physiology ... 202
Overview ... 203

Democritus of Abdera (ca. 460–360 B.C) ... 209
Personality – Life ... 209
Writings ... 210
Leucippus ... 210
The Roots of Atomic Theory ... 211
Atoms and Atomic Compounds ... 214
Motion ... 216
'Necessity' and 'Chance' ... 218
Cosmogony ... 219
Cosmology ... 220
Natural Phenomena ... 221
Biology – Medicine ... 221
Soul – Life – Death ... 222
Sensation ... 222
Thought ... 225
Knowledge ... 225
Other Sciences ... 226
Theology ... 227
Ethical Thought ... 228
Overview ... 233

Epilogue ... 249

Appendix: A Brief Historical Retrospect: On the Origins of the Greek People 255

Bibliography 261

Index of Names 271

Index of Subjects 277

THE TIME: Chronological Table 291

THE SPACE: The Greek Civilization during the Presocratic Period 293

Introduction

> *What we call intellectual cultivation is f rst of all the need to rediscover, inherit, and augment that, which created the nobility of the world.*
>
> André Malreaux

Paul Valery remarked somewhere that only the person who possesses a passion for the *future* can conceive the substance and value of the *past*. What is the value and meaning of today's thinking man becoming involved with the developments of the Greek spirit? Is it the wonder or is it the respect for the unalloyed clarity of a deep spirit capable of penetrating with a unique combination of *intuitive* and *rational* power the profoundest problems, the solutions to which will thereafter challenge mankind? M. Heidegger observed that the origin of all philosophic inquiries is to be found in the Presocratics. And what of the answers they provided? Did they simply establish the *beginnings* of European rationality? Or were they inspired by a *timeless* critical and logical vision that continues today to cast light on contemporary human issues of freedom, the environment, philosophy, and the natural sciences, together with its most advanced theories regarding symmetry, chaos, force, interaction, basic particles, the creation of the universe, dualism, holism, et al.?

Indeed we are fille with wonder, excitement, respect, a sense of timelessness; at the same time, however, we feel a thirst for self-knowledge, a passion to understand *how our thought itself* took shape, forming our perception of the cosmos and life. We yearn to become acquainted with a mode of thought less empirical than our own and because of that *less biased*. Prior to the firs Greek thinkers, neither philosophic schools nor scientifi views existed. Setting out from the abstract world of myth and poetry, the Presocratics would create for the firs time on European ground those *intellectual foundations* upon which Western thought would thereafter expand. Here precisely lies the complex difficult of examining the genesis of the Greek spirit. The difficult is both *objective* (language, texts) and *subjective* -the angle (present or past) from which this approach is attempted.

Within today's crystallized intellectual structures, *language* comprises a comfortable means of interpersonal communication and understanding. In the case of the study of the origins of European thought, however, language and logic cease to be rigid, predetermined intellectual tools and become themselves subject to examination in *relation* to the developmental stages of the Greek thought. Evidently difficult in comprehending the firs Greek thinkers is therefore to be expected, given that not only their *ideas* but also their very *words* are under investigation.

The problem becomes more intense when we turn to the primary sources as they were organized in the memorable work of Hermann Diels. The extant authentic

C.J. Vamvacas, *The Founders of Western Thought – The Presocratics*, Boston Studies in the Philosophy of Science 257, DOI 10.1007/978-1-4020-9791-1_1,

texts are few. These are the fragments contained in post-Socratic allusions by other authors, especially by Plato, Aristotle, and his pupil Theophrastus, expressed in the light of the problems, criticism, and philosophic systems of these thinkers themselves. The later Testimonia tradition is based to a great extent on the work of Theophrastus and other indirect information. Thus, it is evident that misinterpretations and arbitrary conclusions are likely to arise. Contemporary researchers have cast new light on Presocratic thought, but most of them have been unable to avoid interpretations colored by their own point of view.

Which view is the correct one? Inevitably, every interpretation is to some degree subjective and one-sided. Only the careful interrelation of these views and illumination of concepts from different perspectives can bring about a systematic, spherical (though as yet faint) picture of early ancient thought. For there is a factor which we must not neglect: Then the road of thought was *single* and *unifie* . Only later did it diverge into various philosophical and scientifi paths that led to today's specialization and thereby to one-sided views of the world. With a certain nostalgia physicist E. Schrödinger was to observe: "I am of the opinion that for this reason the philosophy of the Greeks is so attractive to us today, for nowhere in the world, before or after, did there arise such a progressive and harmonious structure of knowledge and thought, *without* the fatal *fragmentation* which has been at work for centuries and has become unbearable in our time".[1]

With awe and great care we will attempt to unearth the development of Presocratic thought, which was at once so many-sided and bold. This was a mode of thinking that, striving to open new horizons, created for the firs time philosophic and scientifi conceptions -an achievement which would be perceived by many as a true *intellectual revolution*.

"The world in which we live", writes C. Malevitsis, "is a mysterious and wondrous showing forth. This is the dizzying phenomenon of the revelation of being. In the face of this we are so small that we cannot perceive its unlimited significance Or perhaps we do not as yet possess that *perceptiveness* whereby we will become able to conceive it".[2] Presocratic thinking offers us precisely this *perceptiveness*.

---

[1] E. Schrödinger, *Die Natur und die Griechen*, 28.

[2] C. Malevitsis, *Philosophia kai Thriskeia (Philosophy and Religion)*, 15.

The Juncture

Mankind's historical pivotal point appears to be situated around 500 B.C.,
within conceptual developments that unfold between 800 and 200 B.C.
Here is found the deepest juncture of history.
Man as we know him today took shape.
This epoch will be named in short,
The 'Pivotal Period' (Die Achsenzeit).

Karl Jaspers

The Greek spirit did not spring suddenly into being. There was a slow, gradual development from a-logical, mythological conceptions toward rational thinking, which would constitute the basis of European philosophy, science, and ethics. For centuries the boundary between the *mythological* and the *rational* would remain blurred. The historical moment at which the Greek spirit began to realize that *truth* could be traced by means of *critical thought itself* is the 7th century B.C., This is the decisive, seminal moment for the subsequent intellectual development of humanity. During that same historic epoch, in an inexplicable and enigmatic way, various fundamental developments happened to be taking place in as yet mutually isolated cultures: China (Laotse, Confucius), India (Machavira, Buddha), Israel (the great prophets Jeremiah, Ezekiel), Iran (possibly Zoroaster), and Greece (the Presocratic philosophers, Socrates, Plato, Aristotle). It is the singular historical juncture in the world that K. Jaspers named "the pivotal period of human history".[1]

Whereas these fundamental intellectual developments coincide chronologically and possess the common basic purpose of seeking the essence of the world through thought and contemplation, they spring from *different points of departure* and proceed in different directions: In the immense kingdom of *China* one find a dominant *practical concern* for proper human relations within a well-governed state. In *India* the *religious longing* of man poses the crucial issue regarding the deeper meaning of life. In *Greece*, conversely, where neither a unifie state nor a religious hierarchy exists, "it is through wonder that men now begin and originally began to philosophize".[2] It was the *wonder* and *questioning* the Greek felt toward the harmony of nature that led him to posit for the firs time the vital question of the *origin*

[1] K. Jaspers, *Vom Ursprung und Ziel der Geschichte*, 14.

[2] Aristotle, *The Metaphysics*, 982b11.

C.J. Vamvacas, *The Founders of Western Thought – The Presocratics*, Boston Studies in the Philosophy of Science 257, DOI 10.1007/978-1-4020-9791-1_2,

and the *essence* of the world.[3] This was a decisive inquiry that created the rational foundations of Western philosophy and science, also leading, however, to the fina *intellectual dividing line* between Europe and the East, with all that this might entail.

W. Nestle was to write in the introduction to his work, *Die Vorsokratiker*: "The very fact that the Greeks turned to philosophy already sets them apart from others. The *way*, however, in which they did so raises them above the other nations of antiquity".[4] Here we must take care to observe a fundamental point: indubitably, it was the rational spirit of the Greeks that established the foundation of philosophy and the physical sciences – an event that was decisive in the development of European civilization. To attribute *precedence*, however, to that civilization as opposed to great Eastern cultures like those of China and India constitutes a one-sided conception, exclusively from a European point of view. The Greeks *founded* Western civilization, simultaneously *alienating* it, however, from Eastern cultures. This decisive *polarization* led to the formation of the Western spirit, at the same time depriving it of other human potentialities, which will remain for the Western world undeveloped and unexploited.

In what way did the Greeks philosophize? Why did they philosophize thus and not otherwise? What were the factors that determined the direction of Greek Presocratic thought and unquestionably firs stamped the European way of thinking philosophically and scientifically These are crucial questions that must not receive one-sided answers. To comprehend the genesis of Greek Presocratic thought, we must return to the origins and firs steps of intellectual development in Greece. The natural environment, social structures, myth, religion, impact from Eastern cultures, and other basic factors contributed to the formation of the personality of the Greek, as manifested in the Homeric epic as well as in the lyric and elegiac poetry of the time, which we will briefl examine below.

Nature

Greek nature, and probably she alone
find herself not at the 'moment' of Hegelian antithesis,
but at the 'moment' of Hegelian synthesis *with man.*
Konstantinos Tsatsos

The natural environment, *in conjunction* with the way it is viewed, undoubtedly played a significan role in the formation of the character and destiny of the ancient Greek. The Greek landscape is made up of a continuous series of definite yet not extreme, contrasts. Extended mountain ranges divide the land, creating numerous small, enclosed valleys and plateaus, as well as innumerable bays and inlets. The

---

[3] Twenty-f ve centuries later, the now elderly Goethe would write: "The greatest height man can reach is wonder (*das Staunen*)". (Letter to his friend, Eckermann, 2–18-1829).

[4] W. Nestle, *Die Vorsokratiker*, 7.

shores of Asia Minor from which civilization set forth, as well as those of Epirus and the gleaming isles of the Aegean, are characterized by a seamless linking of land and sea, under a deep blue, luminous sky. The Greek lives in a supremely mild, delectable climate, bathed by an unparalleled abundance of light. Compared with other countries, "Greece possesses the best and most gentle climate",[5] confirm world-traveler Herodotus. The pure light and embroidered beaches that divide the land's edge from sea's threshold are factors that decisively influence the formation of the Greek character. The endless interpenetration and symbiosis between land and sea offered him an archetypal pair of opposites, and the roaring waves on the sea's surface, in contrast to her mute, invisible, and motionless depths, might well have created in him the firs unconscious hint of the antithesis between 'seeming' and 'being,' 'becoming' and 'being'.

The challenge of survival in a land that is not very fertile would reinforce his virtuosity and inventiveness. The natural world around him was a harmonious and approachable sphere that challenged him toward mastery and understanding. These achievements brought him satisfaction and self-confidence The vastness of the sea and sky that spread before him evoked wonder, curiosity and the desire for greater discoveries and expansion. The sea would constitute an ideal means of communication with neighboring peoples. This contact would reveal different civilizations, religions, manners and customs, knowledge, and ways of life, and would provide opportunities for comparison, competition, interaction, conflicts doubts. All this, for the restless and critical spirit of the Greek, would constitute fertile soil for his mental development. "The [Greek] sea", Odysseus Elytis, Nobel laureate in Literature (1979), observed, "bears no relation to the vast fl wing meadow which spreads out before other sea-side countries. This element, which elsewhere divides people, here unites them".[6]

The Greek *viewed* the spectacle of the surrounding natural cosmos with a sense of religious wonder and questioning. This kind of *theorizing* is unique in the world and would be decisive in the birth and development of Western philosophy and science. To theorize, *theoro* [derived from *thea* (spectacle) + *orao* (see; look) > *theoros* (spectator) > *theoria* (viewing; contemplation; consideration)] signifie at once 'look at', 'observe', 'perceive', 'contemplate', consider, and later: 'interpret of facts of a certain category'. R. Bultmann observes significantly "At the base of this notion of vision is found the special Greek conception of being, of reality, as shaped material. Those things are true which one can savor the texture of, describe their form, their relationships and their structure ... To view a thing is at the same time a touching of the form which define the object as an entity".[7] The ancient Greek identifie the beautiful – in formal plasticity, in symmetrical lines, in the harmonious whole. This was the ground of his creation of unsurpassable monuments of sculpture and architecture. From the time of Homer, he glowingly described nature that for

---

5 Herodotus, *Historiae*, III, 106.

6 O. Elytis, *Anichta Chartia* (*Open Papers*), 601.

7 R. Bultmann, *Zur Geschichte der Lichtsymbolik im Altertum*, 17.

him constituted a harmonious, well-ordered (*cosmios*) totality. The Greek gods are visible; they possess a human shape; they move among us, see, and are seen. For the ancient Greek it was a deep expression of reverence to immortalize the gods' human form in sculpture. Referring to this unique way of *theorein* (observing) by the ancient Greek, W. Otto notes: "What is dominant and supreme is not power expressed in action but *being* (das Sein) which is revealed *through form*".[8]

What is the ancient Greek perception of nature? The word *physis* (nature) derives from the root *phy-*, meaning 'birth', 'growth', 'evolution', but also 'essence', 'existence', 'the natural constitution of a person or thing'. Opinion differs as to which meaning is the oldest. Thus, the interest of the ancient Greek in nature turns toward two basic understandings: on the one hand toward the process of *genesis* and *growth* of beings, and on the other hand toward the true *essence* and *constitution* of beings. The last constitutes a clearly *dynamic* Greek view of the universe, which demonstrates how unjust has been the notion in some quarters that Greek thought was static.[9]

The ancient Greek perceived nature as an ordered whole which he would represent in the richly meaningful term *cosmos*, meaning firs (Homer) 'ornament', 'decoration', but also suggesting the existing order in the universe and in society as well. Man conceives himself to be an inseparable, organic part of nature. Man is not reduced to being "the dust of the ground"[10] of the Old Testament, fallen through original sin, and distanced from God, from whom he received the terrible pronouncement, "cursed shall be the ground through you".[11] On the contrary, the ancient Greek lived and moved amidst nature "overfl wing with gods",[11A22] believing that "the world is most beautiful, being the creation of God".[11A1(35)] A miracle is not needed to perceive God, since for the ancient Greek nature herself, within which he lives, is divine. This Greek conception of harmonious nature and the human being's inseparable place, as man, within her, will certainly later enable the Greek to dare to inquire into the world, gradually shaking off the 'supernatural' of Greek mythology.

Society

Democracy is better than tyranny.
Periandrus the Corinthian (7th century B.C.)

The twilight of the brilliant Mycenaean civilization (12th century B.C.) was followed by centuries of extensive relocations involving the descent of less advanced Hellenic tribes and the establishment of settlements reaching the islands of the

[8] W. Otto, *Die Götter Griechenlands*, 10.

[9] Only three centuries later in the works of Aristotle the original meaning of the world *physis* as *development* will be abandoned, to establish its second meaning as the true essence and form of a thing.

[10] Old Testament, *Genesis*, 2, 7.

[11] Old Testament, *Genesis*, 3, 17.

Aegean, the shores of Asia Minor, an -later- the coasts of Sicily and Southern Italy. In the 8th century B.C., these shifts led to the creation of many small 'city-states', civilized communities of an ethnic character with political autonomy. Aristotle was to give the following definitio of them: "A community made up of a great number of towns forms the perfect city, which, finall attained the limit of virtually complete sufficien y, and thus while it was created for the sake of life, exists for the good life".[12] The Greek landscape indubitably contributed to the formation of these small autonomous 'cities', because of the configuratio of the landscape, which *divides* the Greek land into numerous small plateaus, valleys and gulfs enclosed by mountain ranges, and the Aegean Sea with its countless islands.

The existence of many independent 'city-states' did not prevent the simultaneous creation of a common *Greek* consciousness. The commerce and colonization that brought the Greeks into contact with the neighboring 'barbarians' (the word 'barbarian' meaning peoples of other tongues and races without the derogatory sense it later took on) fostered their awareness of their shared Greek cultural heritage – their common manners and customs, the same gods, social-political structure, and language; the *Greek (Hellenic) language* is characterized by unparalleled lucidity, precision, suppleness, and aesthetic dexterity. The appellation *Hellene* (Greek) would prevail as a mark of this ethnic identity which was expressed and reinforced both in the international political sphere through military alliances, religious affiliations Amphictyonic Leagues, bi-lateral agreements, guilds, and community federations, and on the cultural level in epic poetry, geometric art, pan-Hellenic athletic meets (e.g. Olympic games) and musical events, and shared centers of worship.

During this period, ca. 800 B.C., the Greeks constructed the Greek alphabet that comprises a significan and historically unaltered contribution of the Greeks to European civilization. The older form of Greek writing was the Mycenaean B Script, which seems to have fallen into oblivion around 1200 B.C. along with the collapse of the Mycenaean culture. During the Geometric period the Greeks created a new Greek alphabet. Using Phoenician writing as a model, they proceeded to a fundamental innovation: for the firs time they introduced vowels into the Northern Semitic alphabet (which was made up exclusively of consonants and a few semivowels), thus transforming this script into an elegant *phonetic* alphabet in which each letter represents a specifi phoneme sound. Such a simple but ingenious modificatio is justifiabl considered a discovery of universal historic importance. Henceforth all of the wealth and tradition of Western civilization would be inscribed in Greek, or in slightly different forms of the Greek alphabet (Latin, Slavic). The establishment of the Greek alphabet undoubtedly signifie a decisive development for future relationships and deeper communication among the Greeks, contributing at the same time to the founding of a common, pan-Hellenic consciousness.

Nature, religion, community were at firs unbreakably linked concepts whereby the same words were used on different planes of thought. Thus, for example, the word *cosmos* that, as we have seen, expressed the harmony of nature, was also used to characterize a government of justice and law. The idea in physics and geometry

12 Aristotle, *The Politics*, 1252b28.

of *symmetry* and *homogeneity* that will appear in the firs Greek cosmological theories is reflecte in the 'city-state' both structurally and geographically. The laws established by the ancient Greek as the basis of community life will also be sought by him in nature, as he dares *fi st* to set forth a deterministic physical cosmological system, free of supernatural and divine forces.

During the course of these radical socio-political changes of the 7th and 6th centuries, B.C., the Greek realized that as opposed to Justice, injustice is linked to arbitrary actions, conflict arrogance, to *hybris* (insolence; wanton violence). This realization led to the formulation of general ethical principles that reflec the spirit of the age as it is also expressed in the relevant aphorisms of the *Seven Wise Men*.

Religion

For the Greeks nature itself was inherently divine,
since their gods were also not outside or above nature.
Friedrich W.L. Schelling

The gods of the Greeks are firs of all *powers* and secondly *persons*. However, the Olympian gods are not omnipotent. Behind and above them lies *Moira* (fate), *Eimarmeni* (destiny), a significant early abstract concept. Religious life belonged to the social life of the city; and irreverence -as would be seen in the tragic example of Socrates- constituted a crime against the *polis* itself. Since society was the guardian of the sacred, the *priesthood* simply comprised a formal authority without the power and influenc that it held in other cultures like those of Egypt and India. Consequently, the priesthood did not inhibit or influenc the growth of the rational spirit.

The gods were seen as an inseparable part of the natural order in which they participate. Their interventions are neither uncontrolled nor illogical. To Greeks the divine was expressed within the cosmic order; they did not require, nor did they seek, miracles or prophets to infer divinity. Given this view, it is evident that the moment the Greek will attempt to interpret the world *rationally*, his religious feeling toward the Olympian gods will be radically shaken.

There was no dogma and consequently no proselytizing. For the same reason-unlike people of other faiths- they could not conceive the creation of the universe *by* god because they believed that the divine was already a part of the cosmos. This conception would enable the Greek, undistracted by religious assumptions, later to discover exclusively *natural* causes for the beginning of the universe.

This emancipation of the spirit from religion was also unquestionably aided by the fact that *knowledge* was never considered irreverent, for "it cannot be that god is jealous". [13] According to the old Testament knowledge comprises original sin: "And the Lord God laid this command upon the man: 'From every tree in the garden you

[13] Aristotle, *The Metaphysics*, 983a2.

are free to eat; but from the tree of knowledge of good and evil you must not eat; for the day that you eat of it you shall certainly die".[14] On the contrary, the Greek believed that through *knowledge* "he had a share of the divine attributes".[15]

It is evident that with the continual intervention of the gods, the Greek of the Homeric period did not feel independent, and thus he did not consider himself responsible for his actions and feelings. He attributed everything to the gods, and he even lacked the realization that he himself could be the cause of his decisions and feelings. In spite of this, he never felt overwhelmed or humiliated by the gods. He viewed their interventions and actions more with amazement than with awe. Furthermore, as B. Snell observes, "the Greek divinity -in contrast to the Hebrew, the Hindu, and the Chinese- invites man to imitate it. For this reason the Greeks always ran the risk of becoming complacent and arrogant, exceeding human limitations. This vanity, which the Greeks called *hybris*, was inherited by the European mind in spite of its Christianization".[16]

Already, however, in the Homeric epics, one can detect in various elements parodying the gods, a weakening of 'faith' in them. The gradual weakening of their trust in the Olympian gods would lead gradually to the laying down of two roads of abstraction: the *rational philosophy* of the Ionians and the *mysticism* of the Orphic teachings, which would answer the profoundest religious needs of the Greeks. This new mystical religion originating in Thrace would extend during the 8th to 6th centuries B.C. to the major areas of Greece, and in spite of initial resistance would finall merge with local religious practices without replacing them. Its content is so radically different from the Greek spirit of the Homeric age that it is justifiabl assumed that its provenance was eastern, most likely India.

The mystic tendencies of Orphism aside, we see from the early archaic period forward the inborn tendency of the Greek toward a global natural and social determinism and order. During the Presocratic period Orphism was not to exercise a great influenc on the formation of Greek thought. As W.K.C. Guthrie was to observe epigrammatically: "Orphism was too philosophic for the masses, too mythological for the intellectual pride of youthful philosophy. To fin a wider response, it had to wait until the distinctive greatness and distinctive limitations of the classical age had broken down".[17]

Myth and Epic Poetry

No other people has demonstrated such a balance between imagination and reason as did the Greek.
Wilhelm Nestle

---

[14] Genesis 2, 16–17.

[15] Plato, *Protagoras*, 322a.

[16] B. Snell, *Die Entdeckung des Geistes*, 55.

[17] W.K.C. Guthrie, *Orpheus and Greek Religion*, 238.

Myth comprises the initial attempt to comprehend reality through images and stories, which, however, conceal the truth. From the moment the question, "What is truth?" will be posed, will begin the stirring effort of logical thought to uncover and reveal the truth encoded in myth. Greek myth inherently contains the seeds of reason because it is characterized by causal relationships. In pre-Homeric Ionia -a center for colonists from various areas of the Greek mainland, who brought with them the mythical traditions of their birthplaces- began the gradual shaping and classificatio of the diverse legendary material that finall would comprise the great mythical cycles of the *Argonauts*, of *Thebes*, of *Argos*, of *Troy*. In the same systematic manner, the clear structure of the Olympian pantheon would crystallize from the firs theogonic and cosmogonic myths.

Already evident within these materials is the inborn tendency of the Greek toward *order*, a tendency that two centuries later would lead the Ionian thinkers to the realization of the firs natural philosophy. "No people progressed as did the Greek, entirely on his own initiative and internal necessity, on the road from myth to reason. Nowhere can distinguish so clearly the dialectic of mythical consciousness, which derived from man recognizing himself as the *creator* of the mythical forms that till then had been viewed as a world outside and above himself".[18] In these words W. Nestle would underline the uniqueness of the Greek spirit.

The main means of the handing down of myth is the heroic epic. Already cultivated from the beginning of the Geometric period (1000 B.C.) as an oral composition, epic poetry achieved its most perfect expression and development in Homer, who from all indications lived during the 8th century, B.C. In the *The Iliad* and *The Odyssey*, the firs tendencies toward *demythificatio* are already clearly drawn. And when comparing the *Iliad* with the later *Odyssey*, we fin a shift in the ideal of the hero from that of *somatic* to that of *intellectual* prowess. Homer, however, never exposes this intellectual capacity, in detail.

The impact of the Homeric epic on the cultural development of the Western world was enormous. Although it was primordial, the Homeric perception of man and the world comprised the firs step in European thought. The ensuing one would be Hesiod.

Hesiod – The First Cosmogonies

Every true poet is a prophet and a pioneer.
Hesiod is the herald of lyrical art *and* Presocratic philosophy.
Evangelos Roussos

Ascending the narrow road to the left of the small crossing a few kilometers outside Thebes on the way to Delphi, we soon arrive at a charming little village on the deep green slope of Mt. Elikon. It is Askra. Here Hesiod was born around of the end of the 8th century B.C. His life was difficul and laborious. He tended sheep on the mountainside, and here in the midst of nature he conceived his *Theogony*. As he

[18] W. Nestle, *Vom Mythos zum Logos*, 7, 20.

tells us, one moment the nine muses appeared to him: "and gave a scepter to me, a shoot / Of blooming laurel, wonderful to see, / And breathed a sacred voice into my mouth / With which to celebrate the things to come / And things which were before".[19] He would be the *fi st* European poet to present his work *eponymously*. In contrast to the Homeric epics, which reflec the aristocratic life of Ionia, his verses are frugal and modest, derived from tough rural life.

The rational thought of Hesiod lies at the opposite pole from the mythic concepts of Homer, though they are very close in time: In his two great didactic works, *Theogony* and *Works and Days*, Hesiod is the firs Greek thinker who does not entertain but illuminates. He does not narrate 'myths' but communicates 'truths'. He does not line up divinities but places them in a consistent, complete, ordered system. He poses the question as to the beginning of the world, and he transmutes theogony into cosmogony. He does not extol heroism but highlights diligence and justice as the highest good. Considered by many to be the firs Greek philosopher, his mind had a decisive influenc on later speculative thought.

Hesiod consciously and emphatically opposes for the firs time to the 'many lies' of myths the 'truth' that the gods entrusted to him through the muses: "we know enough to make up lies / Which are convincing, but we also have / The skill . . . to speak the *truth*".[20] This pronouncement, which comprises a major juncture in the history of thought, marks at once the birth of philosophical and scientifi thinking – the threshold of the long road that leads thenceforth to the quest and exploration of verity.

Hesiod's cosmogony may not have been unique, but certainly it was the one in the Geometric age that was most mature and complete. From the various other theogonies of the so-called 'theologians'[21] -Orpheus, Musaius, Akousilaus, Epimenides, Pherekydes- only fragments have survived. Both the authenticity and the chronology of the Orphic theogony are difficul to establish. Perhaps we had best speak simply of a group of parallel versions of the Orphic theogony which, set down long afterward chiefl by neo-platonic writers, express the same general spirit but with many points of difference – for example, in regard to the origin of things.

"First of all (*protista*) *Chaos* came to be, and then broad-bosomed Earth (*Gaia*), . . . and Love (*Eros*), the most beautiful among the immortal gods".[22] In these lines Hesiod begins his *Theogony*. At the same time he answers the two major questions he raises, as to the *beginning* (*protista*) and the *all* (*panton*), and he clarifie in embryonic form four basic concepts which will occupy philosophy and science ever since: regarding 'time' (*protista*), 'space' (*chaos*), 'matter' (*gaia*), and 'force' (*eros*).

Hesiod would pose for the firs time the question of the beginning of the world in a manner more scientifi than theological. Responding to the indirect question of *who* created the world, the Old Testament answers: "In the beginning God created

---

19 Hesiod, *Theogony*, 30–32.

20 Hesiod, *Theogony*, 27–28.

21 Aristotle, *The Metaphysics*, 983b29, 1000a9, 1075b26.

22 Hesiod, *Theogony*, 116–117, 120.

the heavens and the earth".[23] Hesiod -and all later presocratic thinkers as well- was not concerned with the *who* of the Creator, but with the *what* that took place. "Chaos was created firs of all, and then / Broad-bosomed Earth... /and Eros", Hesiod would write in his *Theogony*. "At the beginning, there was an explosion," Steven Weinberg was to write in his book, *The First Three Minutes*.[24] While the answer of today's science appears different as regards subject matter, it responds to the very question firs posed by Hesiod, referring not to a person-creator but to *what* happened. His response then was *matter* (earth) and *force* (Eros); today the answer continues to be *matter* (free elementary particles) and *force* (weak, strong, electromagnetic and gravitational forces, originally *unifie*).

With the same rational approach that he conceived the beginning, he would also grasp the *all*, the Universe, in which he would consistently and systematically place the genealogical tree of the gods, thus becoming the firs Greek theologian. His various gods comprise not only objects of worship but also organic parts of cosmic coherence and order. The *all* is conceived as something ordered, consistent, harmonious – an idea later to be expressed in the word *cosmos*.

Hesiod deeply influence ancient Greek thought. He posed for the firs time three fundamental questions – as to 'truth', the 'beginning', and the 'coherence of the *cosmos*'. His answers predictably still reside within the mythic realm of personification Demythificatio and the transition from *analogy of personal beings* (theology) to the causal *relations between things* (physical sciences) will take place in stages with the maturation of the Greek mind. Nevertheless, the ground-breaking questions he raised for the firs time mark a decisive dividing line in the history of philosophical thought and in the beginning of the Greek philosophic quest for 'truth' and the unveiling of 'being'. With Hesiod, the Geometric period closes. "He epitomizes its experience", E. Roussos was to state epigrammatically; "he summarizes its ideas, and he blazes the trail toward the Archaic period. He transmutes the old and prophesies the new. Every true poet is a prophet and a pioneer. Hesiod is the herald of *lyrical art* and *Presocratic philosophy*".[25]

Personality – Lyric Poetry – Art

The fairest thing in all the world... to me
Is the heart's beloved.
Sappho (ca.600, B.C.)

On July 4, 1964, on the hill of the Pnyx, just below the Acropolis of Athens, the Japanese physicist Hideki Yukawa, Nobel Prize Laureate in physics, gave a lecture in the conference 'Athens Meeting 1964' on 'Intuition and Abstraction in Scientifi

[23] *Genesis* 1, 1.

[24] Weinberg, S., *The First Three Minutes*, 2.

[25] *Historia tou Hellinikou Ethnous (History of the Greek Nation)*, II, 179.

Thinking'. Yukawa posed the question as to why science achieved its present level, as a result of its foundations in Greece and not in any other land. The Chinese, too, with Laotse and Chuangtse had developed a sound philosophy of nature and life, yet for some reason it did not lead to science. What was the characteristic trait of the Greek mind that constituted the foundation of philosophy and science? The answer, for Hideki Yukawa, was: "The important point would appear to be the *balance* or *cooperation* between *intuition* and *abstraction*... Speaking in this way, I can not help tracing the history of science back to Greece. There, not only were intuition and abstraction in complete harmony and balance with each other, but there was also no such thing as the estrangement of science from philosophy, literature and the arts. All these cultural activities were close to the *mind* and *heart* of human beings".[26] In the character of the Greeks there existed neither excessive imagination to the detriment of the mind, as most likely in India, nor suppression of the feelings by reason, as probably in China. In accordance with their innate genius, *reason* and *intuition*, *instinct* and *logic*, co-existed in fertile balance. It is precisely this unique harmony which sets them apart from other great peoples. "It was the combination of passion and intellect", B. Russell was to state, "that made them great, while they were great. Neither alone would have transformed the world for all future time as they transformed it".[27]

During the two centuries following Hesiod, fundamental social and political developments, growth of commerce, encounters with other peoples, and the widening of geographical horizons would lead to a gradual awakening of the personality of the Greek that would be expressed in all facets of life: in freedom of thought, in personal responsibility, in lyric poetry, in the plastic arts, and in human relationships, and in civic life. Perhaps no other people behaved less like a herd or a mass than did the Greek. Already the fact that the Greek political structure -in contrast to the vast governmental establishments of other eastern countries- was the small 'city-state' in which the individual was not nullifie by insignificance suggests how crucial it was for him to operate as an influentia social unit. This was also a vital pre-condition for the later establishment of Democracy. In the art works of the great eastern civilizations, the artist remained in the background, anonymous. In Greece, conversely, already Hesiod had found a way to refer to himself in the third person as the creator of the *Theogony*; and the later artists, poets, thinkers, political leaders presented their works under their own name with pride and a high sense of responsibility. In this fertile ground of free thought and intellectual competition, the seeds of philosophy and science would soon bear fruit.

The awakening of the personality of the Greek appears nowhere more clearly than in the early lyric poetry, which developed mainly in the Ionian and Aeolian communities. The *epic* -the nostalgic narrative of the mythical heroic age that belonged definitel to the past- no longer expressed the restless spirit of the Greek who during the 7th and 6th centuries, B.C., was encountering great socio-political

---

[26] H. Yukawa, *Intuition and Abstraction in Scientifi Thinking*, 58, 65.

[27] B. Russell, *History of Western Philosophy*, 41.

turmoil. Now the Greek was deeply feeling for the firs time the need to express his emotions, his thoughts, his personal experiences, his very being. Consequently there was a historical necessity to turn from the literary genre of the *epic* to that of *lyric poetry* and finall to an independence expressed in Attic *tragedy,* reflectin the human journey from dependence upon the gods to emancipation of the personality.

Lyric poetry, like the Homeric epic, presented itself at the outset in a highly perfected form. The *Iambus* (Archilochus, Simonides, Hipponax) satirized, ridiculed, and mercilessly blasted persons and situations, highlighting the nothingness of human existence. The *elegy* (Callinus, Tyrtaeus, Mimnermus) urging resoluteness in both battle and political strife, elevated love within the transcience and mutability of human life. The *lyric monody* (Terpandrus, Alcaeus, Sappho) to the musical accompaniment of the lyre sang of pain, friendship, erotic passion, the beauty and delight of life. *Lyric choric* poetry (Alcman, Stesichorus) -celebrative, gnomic, often exhortative- was based on myth and hymns to gods, heroes and living persons. From the latter *tragedy* will later arise.

Through these matchless monuments of the oldest period of lyric poetry the personality of the Greek arose and expressed itself. For the firs time there arises a disparity between outward and inward values, between traditional and personal convictions. In Homer, *external* appearance coincides with the *internal* endowments of the man. Now Archilochus derides the imposing office of noble bearing and prefers the one who is short and bandy-legged, provided his "feet are on the ground and his heart is full".[28]

The individual develops his own personal subjective values: "The fairest thing in the entire world . . . to me / Is the heart's beloved",[29] Sappho would sing. Feeling and sensuous experience take on value. Now, for the firs time appears an abstract concept of the *soul* – not as a bodily organ but as the bearer of an inward power, not as an entity that functions only after death, when it deserts the Homeric body, but as a vital vehicle of an esoteric intensity.

"The Greeks", Bruno Snell notes, "developed a counter-awareness. (To glory and might) they opposed the simplicity, frugality, and naturalness which constitute the essential characteristics of the Greek people . . . . They remained true to Solon's perception that a super-human power punishes wrong, even though they felt the new obligation to support the right as *their personal concern.* Also, in this regard, they valued prudence and meaning of measure".[30] For, as Solon wrote, if these virtues are not present, people "are fain to make ruin of their great city by their thoughtlessness".[31] In Athens toward the end of the 6th century, B.C., Cleisthenes would establish the foundations of the firs democratic political system. The same era saw the extension of Greek colonies along the whole Mediterranean coast as well as

---

[28] Archilochus, 58.

[29] Sappho, 5, 3–4.

[30] B. Snell, *Die Entdeckung des Geistes*, 213–214.

[31] Solon, *Elegy*, 3.5.

on the Black Sea, an event which was to affect the later development of European civilization.

The Persian wars deeply marked the Greek character. Around the middle of the 6th century B.C., the Greeks became aware of the impending threat of the Persian empire. This realization would lead to the firs rallying of all the Greek peoples and their eventual victorious united struggle against the Persians. The 5th century, B.C., dawned beneath the heavy shadow of Persia's plans for expansion, which foresaw the imposition of a vast empire. United for the firs time, the Greeks countered the numerical superiority of the Persians with moral resolve and tactical prowess. They would figh for freedom itself and prove victorious. The victories at Marathon and Salamis assured the continuity and development of Greek culture and European civilization as a whole. The contribution of the Greeks in curbing the Asian thrust determined the future of Europe.

The Persian wars marked not only the later historic direction of the Western world but also the rise of thought and art from the *archaic* to the *classical* period. Heraclitus's statement, "War is father of all things"[12B53] expresses thus not only a philosophic conviction but also *personal experience*. The founding of democracy and the birth of Attic Tragedy comprise peak manifestations of the self-awareness and self-knowledge developed by the Greeks. The serene smile of the archaic *kouros* was to vanish, the center of gravity of his body was to shift and lend the firs *motion*, the firs step toward the communal, as expressed in one of the firs statues created after the battle of Salamis, 'The Boy of Critias'[32] (ca.485, B.C.). At the same moment the center of gravity of Presocratic thought also turned toward the problem of *change* and *motion*. The balance of symmetrical facets in archaic art would give way to dynamic, harmonious *syntheses of opposing forces*, a conception which was to fin its most perfect expression in that singular creation of the classical age, the Parthenon. At the same time, Heraclitus would make the unity of all things depend on the *balanced reaction of opposites*. Shortly thereafter, the Athenian Anaxagoras would place 'mind' in front. The same is true in classical -as opposed to archaic- art where the rational and the intellectual prevail. Thus, English art historian A. Ashmole was to observe: "The idea dominates, and one feels that the creation of a work of art has now become a totally conscious intellectual process that is understood and controlled. That is classical art".[33]

Foreign Influenc

The gift for abstraction, with its limitless possibilities
and (we must add) its inherent danger,
was the peculiar property of the Greeks.
W.K.C. Guthrie

---

[32] Acropolis Museum, Athens.

[33] Cited in *Historia tou Hellinikou Ethnous* (*History of the Greek Nation*), III.2, 271.

"If we extensively research the literature of the peoples of the eastern Mediterranean", C. Malevitsis observed, "aiming to record the 'categories' of motifs that are common in works so different as the Epic of Gilgamesh, the texts of Ougarit, the Old and New Testament, the Homeric epics, the texts of the Presocratics, the Tragedies, et al., we will be astounded by the golden thread we will fin running from the one text to the other through common motifs. We must develop an awareness of the common Mediterranean Myth so as better to comprehend the sources of our culture and intellectual life".[34] At the time of the historic awakening of the Greek spirit, peak civilizations already existed in neighboring areas: Phoenicians, Hittites, Egyptians, Summerians, Babylonians, Assyrians, Persians, Israelites. To be sure, the sea facilitated communication with these neighboring peoples. Quick to arise, therefore, is the question as to what degree these eastern cultures shaped Greek thought. The issue is complex, and there have long been conflictin interpretations. "It is true", G. Lloyd notes, "that the source of a Greek idea is sometimes represented not as an Egyptian, Babylonian or Hittite myth itself, but some *archetype* from which all the variant versions are derived... [Thus,] there is the problem of distinguishing fortuitous from non-fortuitous parallels or similarities".[35] Today we can support in essence the position that certainly the Greek absorbed valuable *knowledge* from neighboring peoples, but *philosophy* and *science* are of exclusively Greek construction. "Let us take into consideration that whatever the Greeks took from the barbarians they completed and perfected",[36] Plato was to say.[37]

Astronomy originated in Egypt and Babylonia and developed in Babylonia long before it did in Greece, where it began with Thales in the 6th century, B.C. Certainly, the Greeks took much from these peoples. Thales's prediction of the eclipse of the sun on May 28, 584, B.C., is supposedly based on Babylonian tables that Thales could have learned about on travels to Egypt. The *polos* and the *gnomon* -instruments for determining the time of day and the solstice and equinox- originated in Babylonia, but Anaximander perfected them through much more precise geometric methods.[38] Following Anaximander, Greek astronomy was mainly *geometric,* while Babylonian astronomy, in contrast, was basically *arithmetic*.[39] From the fact that Pythagoras introduced *numbers* into astronomy -"the whole universe is harmony and number"[40] -one can deduce that he was influence in this respect by his travels in Mesopotamia and Egypt. Certain factors in regard to the *periodicity* of

---

[34] C. Malevitsis, *Philosophia kai Thriskeia, (Philosophy and Religion),* 144.

[35] G.E.R. Lloyd, *Methods and Problems in Greek Science,* 285.

[36] '*Barbarians*' is not used derogatorily here; it simply means 'foreigner', i.e. 'one who does not speak Greek'.

[37] Plato, *Epinomis,* 987d.

[38] B.L. van der Waerden, *Erwachende Wissenschaft: Die Anfänge der Astronomie,* II, 134, 254.

[39] B.L. van der Waerden, *Erwachende Wissenschaft: ägyptische, babylonische und griechische Mathematik,* I, 117.

[40] Aristotle, *The Metaphysics*, 986a2.

astronomical phenomena could also have originated in the east.[41] In contrast, Hesiod's agricultural calendar can be considered as the firs stage of Greek astronomy.[42] Likewise to be considered early Greek astronomical discoveries are, among others, those referring to the moon as a heavenly body illuminated by the sun, the sun and moon having an earthly composition, the earth being suspended in air, and the earth being spherical,[43] as well as the true causes of solar and lunar eclipses.

Certain ancient sources (Herodotus, Aristotle, Proclus) indicate that the Greeks borrowed elements of *arithmetic* and *geometry* from the Egyptians. Recent studies, however, have weakened that assertion: "In the older Egyptian texts we do not fin geometry in the Greek sense, as practiced by Thales", the reputable mathematician B. van der Waerden would note. "The Egyptians of the Middle Kingdom of course knew arithmetical methods of assessing area and volume, but a geometry of constructions and proofs we do not fin in their texts".[44]

Concepts of the *transmigration* and *immortality of the soul* in Orphic and Pythagorean beliefs are similar to elements of eastern (Persian?) doctrines, suggesting an influenc of the latter. Persian (Zoroastrian) influence could, in turn, have affected monotheistic and certain ethical precepts of Xenophanes and the followers of Orphism and Pythagorism. Some scholars suspect the parallel influenc of a northern shamanist current that came to Greece from the depths of central Asia and Siberia by way of Scythia and Thrace, involving legends about mythical or historical figure like Orpheus, Musaeus, Aristaeus, Epimenides, Ermotimus, Pythagoras, and Empedocles. These legends are characterized by mystical, spiritualist, shamanist conceptions (e.g., body and soul as opposites).

It is commonly believed that the Greeks of the early archaic age received *knowledge* from the East, the extent of which is difficul to determine, but not *scientifi* systems, which were non-existent in eastern countries. And they were non-existent because knowledge served *practical* purposes. Precisely this is the cause of one of the most basic and crucial differences between Greek thought and that of the great neighboring civilizations. These peoples developed mythological systems and accumulated a vast amount of observational data that, however, served purely utilitarian ends. The arithmetical calculations of the Egyptians related mainly to *accounting* problems of quantitative measurement and distribution of agricultural products, salaries, and the like. Their geometry answered the necessity of surveying and reestablishing of agricultural land boundaries following floods In Babylonia likewise, the recording of astronomical phenomena served mainly *astrological* purposes. B. van der Waerden was to sum it up thus: "In antiquity and the Middle Ages one of the main reasons for the emphasis on astronomy was its application in astrology. Yet that is only one side to the story . . . The Greeks developed their

[41] B.L. van der Waerden, *Erwachende Wissenschaft: Die Anfänge der Astronomie*, II, 223.

[42] Hesiod, *Works and Days*, 383–387, 564–567, 614–616, 619–622, 663–665.

[43] "There is . . . no evidence that [Pythagoras] borrowed the theory [that the earth is spherical] from any non-Greek source". T.L. Heath, *Aristarchus of Samos: The Ancient Copernicus*, 48.

[44] B.L. van der Waerden, *Erwachende Wissenschaft: Die Anfänge der Astronomie*, II, 127.

scientifi astronomy ... *not* for astrological use but for the sake of a special interest in *astronomy* itself".[45]

"Philosophy is a distinctive enterprise, marked out by, and indeed define in terms of, its own aims and methods", G. Lloyd notes. "The adaptation of the theses and ideas from Greek and non-Greek myths and religion should certainly not be ruled out. But what Greek philosophy took over from such sources, it generally recast in a new, dialectical, framework".[46] "The Greek philosophy of nature, as it took shape for the firs time in Miletus and later in other Greek cities, is an entirely new intellectual creation which does not exist in any previous civilization". All of the concepts that comprise the system of philosophy and physics were conceived during that period of Greek philosophy. The concepts of *matter*, *force*, *number*, *size*, *movement*, *becoming*, *being*, *continuity*, *discreteness*, *space*, *time*, and the *atom* are distinct creations of the firs Greek philosophy. The Greek mind was not motivated by the practical concerns of eastern peoples but by the quest for *truth* within a broader conception of *cosmos*.[47]

The Greeks "philosophized," wrote Aristotle, "to avoid ignorance, and thus it is evident that they pursued science for the sake of knowledge, not for any practical utility".[48] The presentation of the work of the Presocratics, without ignoring possible likely influence from great neighboring civilizations, will show that their thought transcends whatever elements, practical or not, which they gleaned from other peoples; their conceptions sprang from unprecedented critical, dialectical, and deductive methods which they were the firs to discover, thus founding Western philosophy and science. "Only where a *powerful* spirit encountered pressure from the east was foreign material transformed into philosophy", U. Hölscher -one of the most profound researchers in the area- was to note. "Philosophy is not eastern; on the contrary, the discovery of the East constituted one of the functional parts of Ionian science itself".[49]

---

45 B.L. van der Waerden, *Erwachende Wissenschaft: Die Anfänge der Astronomie*, II, 6.

46 G.E.R. Lloyd, *Methods and Problems in Greek Science*, 290

47 However, an 'Afro-centric' theory has recently appeared in the United States. Ignoring the work and distorting the conclusions of distinguished classical philologists and philosophers, and arbitrarily assuming that virtually all of the other ancient civilizations of the Eastern Mediterranean are represented by Egypt, this position maintains that Egypt constitutes the source of all Greek achievements and that to Egyptian not Greek culture is owed the development of philosophy and science. These writers' arguments have not convinced the academic world, since errors have been found and it is obvious that these books clearly serve afro-centric, racist political and social interests in the United States. The essay collection of M. Lefkowitz and MacLean Rogers Guy, *Black Athena Revisited,* as well as M. Lefkowitz's, *Not Out of Africa, How Afrocentrism Became an Excuse to Teach Myth as History*, counter with objectivity, careful research, and full documentation.

48 Aristotle, *The Metaphysics*, 982b20.

49 U. Hölscher, *Anfängliches Fragen*, 69, 82.

Introduction to the Presocratics

The error of the common mind lies in transforming the apparent into the obvious; while for the inquiring mind nothing is obvious.

Christos Malevitsis

"The Greeks knew the proper season to start philosophy", Nietzsche was to note; "that is, not later, in bitterness – something wrongly supposed by those who derive philosophy from melancholy. Instead, they began in happiness, in mature adolescence, in the burning exuberance of bold, victorious manhood. That the Greeks philosophized in such a period teaches us about what philosophy is and what it means...".[1] In *Ionia* philosophy and science were born. Virtually all of the Presocratic philosophers are from Ionia. "These Ionians", wrote the widely-traveled Herodotus, "set their cities in places more favored by skies and seasons than any country known to us".[2] These were the thinkers who for the firs time posed in rational and critical terms the fundamental questions that henceforth would engage the human mind.

The period of Presocratic philosophy extends from the beginning of the 6th century, B.C., to around the end of the 5th century, B.C., which corresponds -in accordance with its conventional name- with the period Socrates appeared. As we established above, Presocratic philosophy is not the manifestation of a sudden awakening of the Greek spirit. It is the culminating result of a long development and maturation of the Greek mind.

Intuition and *reason* tend toward a unique balance, and thought begins to turn away from purely practical ends toward theoretical principles and concepts. Hesiod has already illuminated the pathway that will be blazed by Greek thought. The time is now ripe for the Greek spirit to make the crucial step toward philosophy and its still unalienable partner, science.

This step will be taken by the Presocratics on the basis of four axioms that will henceforth be decisive in the development of the mind. As a point of departure, they may be summarized thus:

- Beneath the apparent disorder and multiplicity of the cosmos, there existsorder, unity and stability.

[1] F. Nietzsche, *Die Philosophie im tragischen Zeitalter der Griechen*, 263.

[2] Herodotus, *Historiae*, I, 142.

C.J. Vamvacas, *The Founders of Western Thought – The Presocratics*, Boston Studies in the Philosophy of Science 257, DOI 10.1007/978-1-4020-9791-1_3,

- That stability derives from the fundamental primary substance from which the cosmos originated.
- This primary substance, and consequently the cosmic reality, is one, and is based not on supernatural, but on physical, causes.
- These physical causes are such that man can investigate them rationally.

These statements are neither self-evident nor self-explanatory. Of course, their background is that long maturation of the Greek mind. But the crucial fina step was the exclusive achievement of the Presocratic thinkers. Now, for the firs time the human mind focuses on the truth, seeking rationally and critically the inherent order and stability in nature.

However, the question as to whether the Presocratics were also the founders of *science* demands deeper analysis. The history of contemporary science begins with the Renaissance. While Galileo Galilei (1564–1642) was the founder of contemporary experimental science, Francis Bacon (1561–1626), though not himself a scientist, is considered the originator of the new inductive method: "The ancients", he was to write, "had a particular form of investigation and discovery, and their writings show it. But it was of such a nature, that they immediately fl w from a few instances and particulars to the most *general* conclusions or the principles of the sciences, and then by their intermediate propositions deduced their *inferior* conclusions, and tried them by the test of the immovable and settled truth of the first" [3] According to Bacon, this method is faulty. Instead of attempting to reveal the truth *deductively* on the basis of general syllogistic rules, one should begin *inductively* with particular observations and statements, proceeding gradually to general conclusions. In his great work, *Novum Organum* -the title consciously chosen as opposed to Aristotle's *Organon*- he was to demonstrate the scientifi method of *induction*, maintaining that the true science "constructs . . . axioms from the senses and particulars, by ascending continually and gradually, till it finall arrives at the most general axioms, which is the true but unattempted way".[4]

Bacon himself, however -who was a profoundly learned in ancient Greek literature- would exclude from his polemics against Platonic, and especially, Aristotelian dogmatism, the work of the Presocratics: "The more ancient Greeks, as Empedocles, Anaxagoras, Leucippus, Democritus, Parmenides, Heraclitus, Xenophanes, Philolaus, and the rest (for I omit Pythagoras as being superstitious), did not (that we are aware) open schools, but betook themselves to the investigation of truth with greater silence and with more severity and simplicity, that is, with less affectation and ostentation. Hence in our opinion they acted more advisedly, (and) however their works may have been eclipsed in course of time . . . (they) exhibit some sprinkling of natural philosophy, the nature of things, and experiment".[5]

---

[3] F. Bacon, *Novum Organum*, 243.

[4] F. Bacon, *Novum Organum*, 244.

[5] F. Bacon, *Novum Organum*, 327, 333, 338.

Is Bacon in principle correct? Is the inductive method of proceeding from particular observations to general theories the *sole* scientifi procedure? Does the fact that the Presocratics reasoned *deductively*, beginning with universal general principles and not particular, special observations, mean that they were lacking in scientifi thinking? The very history of contemporary science refutes that claim. Russell was to observe: "Bacon's inductive method is faulty through insufficien emphasis on hypothesis. He hoped that mere orderly arrangement of data would make the right hypothesis obvious, but this is seldom the case".[6] The answer to Bacon's anguished question as to how to reach objective truth does not depend upon the -by nature impossible- stripping from observation any trace of theory or conjecture, but on critical assumptions and critical experimental method. There is nothing unscientifi about guessing, observes R. Feynman, Nobel laureate in Physics, 1965. "It is only uncertain. It would have been unscientifi *not* to guess. It has to be done because the extrapolations are the only things that have any real value".[7] Conjecture is what *precedes* and guides observation. "For reasons of principle, it is entirely incorrect", Einstein stresses in a conversation with Heisenberg, "for one to wish to found a theory exclusively on the weight of observational data. For in reality precisely the opposite occurs. From the outset, it is the *theory* which determines what one can observe".[8] And he adds: "The method of the theorist is based on the use of general presuppositions, so-called Principles, from which consequences can be deducted .... The researcher must somehow 'strain his ears' (ablauschen) to those general principles in nature, tracing certain general features into larger webs of empirical facts that can be sharply defined However, as long as the principles that can be used as a base for deduction have not been found, the single empirical fact in the firs place is useless; even more so, the theorist is in no position to start anything with isolated, empirically derived general laws".[9] Consequently, the objection that the Presocratics do not exemplify scientifi thought because they do not set out from empirical observations does not stand up.

Nevertheless, one can append another objection: that the Presocratics failed to submit their theories to experimental control, as we understand it today. The objectivity of a theory is based not only on critical thought and/or intuition but also on careful examination of experiments that will lead to the acceptance or rejection of the theory. This objection appears, at least on theoretical grounds, to be valid. Why were the Presocratics not aiming at the experimental confirmatio of their theories? Was it from carelessness and lack of discernment, or was there a deeper reason? We must not forget that we fin ourselves in the 6th–5th centuries, B.C. Technology is rudimentary; equipment for experiments and instruments for measurement are as yet virtually non-existent; great practical difficultie impede the easy exchange of ideas and information between various geographical areas; a vocabulary for the

---

[6] B. Russell, *History of Western Philosophy*, 529.

[7] R. Feynman, *The Meaning of it All*, 25.

[8] W. Heisenberg, *Der Teil und das Ganze*, 80.

[9] A. Einstein, *Mein Weltbild*, 110–111.

expression of philosophical and scientifi terms does not as yet exist. For the firs time the Presocratics themselves have to invent a basic set of terms to serve their purposes. Physics and all of the sciences are as yet indissolubly linked with philosophy. A philosophical theory, even today, is not proved by experiments. If we transport ourselves back to that era, we must realize that it would be difficul for them to distinguish between a *physical* hypothesis, subject to experimental proof, and an *a priori philosophical* theory. In general terms, that was the framework within which the Presocratics functioned.

There may also be a profounder reason for the lack of experimental verificatio of their theories. This is the very *stance* taken by the Presocratics toward nature. "An important feature of experiment", S. Sambursky was to observe, "is the *isolation* of a natural phenomenon in its distinctness, in order to study it more exhaustively and systematically. This isolation is *artificia* , for natural phenomena are always an integral part of a whole web of interrelated processes. . .In this sense, we can regard experiment as *unnatural*",[10] to the degree that a process is isolated from its natural environment and submitted to conditions selected and controlled by man. For the ancient Greek who viewed nature in its entirety as an unbreakable unity, this *artifi cial* division of a harmonious totality and the isolation and change of the parameters of a part would have been inconceivable. This would not only be improper for the Presocratics but it could also have constituted an act of *hybris* (insolence) as an intervention into the phenomena of the harmonious *cosmos.*

They would probably believe, moreover, that to the extent that they would deal with partial experimental data, the boundaries of their thought would be narrowed. The goal of the Presocratics is ambitious. Their aim is to *comprehend* not to describe nature. They are so acute and visionary as to perceive that any involvement with partial, experimental details would necessarily mean diversion from their sole end, which is to *apperceive* the cosmos in its *wholeness*. Twenty f ve centuries later, Heisenberg was to note with a certain bitterness and nostalgia how right they were: "The more we explore the fiel which opens before us in physics, chemistry and astronomy, the more we tend to replace the phrase 'explanation of nature' with the more modest 'description of nature,' and it becomes ever more evident that this progress has nothing to do with direct knowledge but with analytic understanding".[11]

In any case, even if the Presocratics had wished to check their theories experimentally, most of them would be so general that it would be impossible at this early stage to design -and even more difficult carry through relevant experiments in support their theories. What experiment could verify a hypothesis concerning the origin of the universe, or Democritus's atomic structure of matter? Could it be, then, that they acted unscientificall in the sense that they hastened to announce their theories without firs securing tangible proof of their validity? The answer is 'no'. The history of science during the last three hundred years is replete with exam-

---

[10] S. Sambursky, *Das physikalische Weltbild der Antike*, 608, 610.

[11] W. Heisenberg, *Wandlungen in den Grundlagen der Naturwissenschaft*, 17.

ples of distinguished scientists who formulated their theories long before empirical verification It suffice to cite in the twentieth century Einstein's General Theory of Relativity and Pauli's neutrino hypothesis. Of course, these great physicists were not 'un-scientific in announcing purely theoretical hypotheses. Einstein was to note: "An instance can present itself in which clearly formulated principles lead to conclusions that are found entirely or virtually beyond the bounds of experimental facts presently available. In such a case, many years of empirical research work may be needed to verify the extent to which the theoretical principles represent the truth".[12]

A fina question remains: Are the Presocratics' physical theories correct? Although -as was the case with most of them- they were later disproved, does that deny them their scientifi qualifications The answer is negative. "But are not Anaximander's theories false, and therefore non-scientific" Popper asks. "They are false, but so are many theories, based upon countless experiments, which modern science accepted till recently, and whose scientifi character nobody would have dreamed of denying, even though they are now believed to be false.... A false theory may be as great an achievement as a true one. And many false theories have been more helpful in our search for truth than some less interesting theories which are still accepted".[13] Was it not the -retrospectively- *faulty* classical laws of black-body radiation of W. Wien and J. Raleigh – J. Jeans, which led Planck to the critical revision and formulation of the quantum theory? Was it not the *faulty* atomic hypothesis of Rutherford upon which Bohr based the development of his atomic theory? As Popper notes in his study *Back to the Presocratics,* "there is the most perfect possible continuity of thought between [the Presocratics'] theories and the later developments in physics. Whether they are called philosophers, or pre-scientists, or scientists, matters very little".

The true theory of knowledge, according to Popper, is based on "a practice which arose in Ionia and which is incorporated in modern science (though there are many scientists who still believe in the Baconian myth of induction): the theory that knowledge proceeds by way of *conjectures* and *refutations*".[14] No theory can claim absolute truth. This would be stressed with amazing perceptiveness by Xenophanes, Heraclitus and Democritus, a fact that also shows how unjustifie was the claim that Presocratic thought is dogmatic: "Certain truth has no man seen, nor will there ever be a man who knows about gods and about everything of which I speak; for even if he should fully succeed in saying what is true, even so he himself would not know it, but in all things there is opinion"[21B34], Xenophanes was to observe. For, as Heraclitus was to say, "nature loves concealment".[22B123] And Democritus was to confirm "In reality we know nothing, for truth is hidden in the depths".[68B117]

The conclusion can be summed up in the words of S. Sambursky: "that these natural philosophers are the *intellectual forefathers* of our epoch can be doubted by no one who compares today's science with the achievements of Greek science, which

---

[12] A. Einstein, *Mein Weltbild*, 112.

[13] K. Popper, *Back to the Presocratics*, 12.

[14] K. Popper, *Back to the Presocratics,* 12.

are a portion of our heritage – their methodology, their imagination and inspiration, their enormous associative powers, and their capacity to arrive at logical conclusions".[15] The permanence and stability which the Presocratics perceived behind the surface plurality of natural phenomena, and their determination to probe it with a critical rational mind was the cornerstone of the structure upon which science to this day has been based and developed.

The above general outline of the Presocratics' thinking aside, there are significan differences between them. Each has his own personality and his unique intellectual radiance. Thus it is difficul to classify them in groups. The following can serve as an initial chronological orientation to their individual work.

On the eastern edge of the Greek settlements, in Ionian Miletus, one the largest and most prosperous Greek cities of the 6th century, B.C., *Thales*, *Anaximander*, and *Anaximenes* were to shape the firs philosophic discourse. Inspired by wonder and curiosity toward their natural surroundings, they would search beyond the plurality of phenomena the single, fundamental substratum, which also would be the primary source from which the universe originated. For Thales the *proti arche*[16] (the 'firs principle and element') was *hydor* (water); for Anaximander, the *apeiron* (the boundless); for Anaximenes, *aer* (the air). A common view of the three Milesians is that the *proti arche* means not only the world's permanent constitution but also its capacity to *change* and *evolve*. The primary substratum is both self-moving and eternal.

During the same period, *Pythagoras of Samos* settled in Croton, a Greek colony in Southern Italy, founding there the school, which took his name. Now both the motive and philosophical direction changes. Beyond knowledge, the *Pythagoreans* would aspire to endow consciousness with a deeper religious content. The orientation toward the *proti arche* (the firs principle) is not toward *matter* but instead to the *form* and the *relationships* between beings. These relationships are for the firs time expressed *quantitatively* by mathematical means, by numbers, and they are considered changeless and eternal, in contrast to the perishable, mutable material world. Thus, there appears a decisive shift of the spirit from the quest for natural causes toward more abstract mathematical concepts: "Number is the essence of all things"[58B8] is the doctrine, and "immortal. . .is the soul".[14.8a]

Not long thereafter, *Xenophanes of Colophon,* fleein from Persian rule, left Ionia for Southern Italy. A poet who was a profound thinker and deeply religious as well, he would preach for the firs time the 'one god' within the framework of negative theology, denouncing the anthropomorphic attributes which had prevailed up to that time. In the intellectual realm, he is justly considered the firs reformer. He would introduce an anthropocentric theory, inviting men for the firs time as autonomous,

---

[15] S. Sambursky, *Das physikalische Weltbild der Antike*, 619.

[16] '*Arche*': there are different senses of this term depending on author and context: 'origin'; 'beginning' or 'starting point'; 'originating cause' or 'the original state out of which the manifold world was developed'; 'primary element' or 'substratum'; 'firs principle' or 'permanent constitution'; also 'sovereignty' 'magistracy'.

responsible beings, void of any divine intervention, their only power, the *logos* (reason), to advance toward sustained progress, toward the continuous 'better'. At the same time he would humbly posit the *limits* of human progress and knowledge from the point of view of the gnoseological, ethical, social and philosophical context. He would explicitly stress the distinction between 'ultimate reality' and 'seeming', between 'absolute' and 'apparent truth', positing one of the fundamental problems of philosophy, which would immediately thereafter deeply concern Heraclitus and Parmenides.

Toward the end of the 6th century, B.C., the theory of the three Milesian *physicoi* (natural philosophers) regarding a 'primary substratum' that is stable and immutable creates the firs questions about the relationship between the 'senses' and 'reality', between 'being' and 'becoming'. The answer to these two basic questions, as to *knowledge* and *change*, will be given in two diametrically opposed ways by the peerless thinkers Heraclitus and Parmenides. From that time, and up to today, no European philosophical system could ignore their work.

Heraclitus of Ephesus would see in ceaseless cyclical *change* and *movement* the sole reality. He refuses to accept matter as the static substratum of truth, and simple sensuous perception as well, if that is not controlled by *logos* (reason). The problem of self-knowledge and self-consciousness takes on importance, and it is certainly not coincidental that Attic tragedy firs appears in the same period. *Matter* is not the ultimate reality; it is *processes* that are real, deriving from a perpetual dialectical conflic and synthesis between opposite tendencies in which there are moments of balance or apparent stability, while the alternating dominance of the one over the other leads to perpetual flu and cyclical birth and destruction of all things. "War is father of all things".[22B53] Eternal is the *logos* alone, which controls perpetual movement and change, as well as fire which particularly expresses this change. "This world-order (*cosmos*)", Heraclitus was to pronounce, "the same for all, none of the gods nor of men has made, but it was always and is and shall be: an ever-living fire which is being kindled in measures and extinguished in measures".[22B30]

Parmenides of Elea, student of Xenophanes and teacher of Zeno, set out independently from the same basic principles of Heraclitus -that the *cosmos* is unified and perceiving by the mind, not the senses, illuminates the road to truth. But the road to truth that appeared to Parmenides is diametrically opposed to that of Heraclitus. The *cosmos*, according to Parmenides is one, homogeneous, indivisible, *unmoving*, and *unchangeable*. For the firs time truth is explored according to ontological criteria, and for the firs time is posited a *conceivable cosmos* as a *real* world, independent of the senses. This Parmenidean conception is based on the metaphysical argument that what 'is' exists, while what 'is not' does not exist. 'Nothing' does not exist; thus 'void' does not exist. Since void does not exist by which matter would be separated, the *cosmos* is *single* and *compact*, and this characteristic of course permits *no movement* since there is no void space toward which something would move. In the same way, *change* cannot exist. Change means that something that 'is' is transmuted to something that 'is not' (for it would mean a break in its ontological continuity for it to be changed into something else which, in turn, also 'is' – which exists). Nor, however, could it be changed into something which 'is not', for the latter by definitio does not exist, while we are referring to existing things. "*Being*", Parmenides would conclude, "neither was, in the past, nor shall be, since it is now in its entirety one, continuous . . . motionless . . . without beginning, or end".[28B8]

Both the Heraclitean dynamic conception of the *cosmos* as a perpetual and immaterial dialectical process of change and movement, and the diametrically opposed Parmenidean ontological conception of being as singular, eternally unchangeable and motionless, are in grave conflic with *everyday experience*. Man is asked to forsake his trust in the sensory perception of an actual world -an attitude, which is difficul to adopt. The remaining Presocratic thinkers, Empedocles, Anaxagoras, and Democritus, will consequently strive, under the dense shadow of the Parmenidean antinomy, to provide new interpretations of *movement* and *change* in the tangible material world. They will necessarily abandon the *monistic* theories of the Milesian *physicoi* concerning a *single* primary substratum, which from the powerful blow by Parmenides are no longer in a position to explain either the *plurality* of the physical, empirical world, nor its continuous *change*. Instead, they will develop new pluralistic theories that will also contain for the firs time *kinetic principles.*

Empedocles of Acragas, in the middle of the 5th century, B.C, would hold that there are four entities or elements, the 'roots': earth, water, air, f re. Given this pluralistic primal condition, the multiplicity of the *cosmos* would derive from the various combinations with different proportions of the four 'roots'. The combinations come about under the influenc of two external kinetic principles, 'Love' (attraction) and 'Strife' (repulsion) that cause change, decay and birth.

In contrast to Empedocles, whose theories also contained some mystical elements, *Anaxagoras of Clazomenae,* of approximately the same period, will theorize the *homoeomeroi* (having parts like its other and the whole) or 'seeds' as the component entities, and for the firs time will distinguish *mind*, as opposed to *matter*, as the primal moving force that directs and controls the universe: "the mind controls all things".[59B12] For the firs time a *dualism* of *mind-matter* will be introduced that will influenc philosophic movements thereafter.

The *atomists*, *Leucippus* and *Democritus of Abdera*, will take a different, more scientific direction. They realized that the empirical dead-end which they had inherited derived from Parmenides's concept that 'being' exists, while 'non-being' does not exist. In order to escape from that logical impasse, they arrived at the inspired idea of abandoning the latter statement of Parmenides and thereby to maintain that void, although 'non-being', *exists*. "[They say] the elements that exist are the *Full* and the *Void*. . .calling the one 'being' and the other 'not-being'. Of these they identify the full or solid with 'being' and the void or rare with 'not-being' ".[67A6] The characteristic movements and relative positions and distances between the solid, small particles in void create the natural world that we experience through the senses. These smallest basic bodies are viewed as inseparable and indivisible, and are termed *atoms*[67A14] (*a-tomos:* indivisible). Democritus's atomic theory, which deterministically explains the cause of *motion*, as well as *plurality* and *change* in the world, will play a decisive role in the evolution of scientifi thought to this day.

In the upcoming discussion of the Presocratics' work, we shall often encounter principles, statements, and conclusions, which appear to be in qualitative agreement with contemporary scientifi theories. At this point, we must clearly outline the *confine* of Presocratic thinking:

- The Presocratics firs posed the fundamental philosophical and scientifi *questions* that will thenceforth concern Western thought; for vivid philosophical thought, *the question* takes priority over the answer.
- They were the initial founders of the tradition of *critical* -as opposed to dogmatic inquiry- without which science would not exist. "To my knowledge", Popper was to observe, "the *critical* or *rationalist* tradition was invented [by the Presocratics] *only once*. It was lost after two or three centuries, perhaps owing to the rise of the Aristotelian doctrine of *episteme*. ... It was rediscovered and consciously revived in the Renaissance especially by Galileo Galilei".[17]
- They firs developed 'conceptual tools' and qualitative *methods* of approaching and understanding the physical world that have been used creatively up to our time – as, *measure*, *rhythm*, *symmetry*, *analogy*, *order*, *cosmos*, *universal determinism*, *continuous discrete*, *reduction of qualitative to quantitative differences*, *merging of opposing concepts*, *dynamic balance* (Heraclitus), *mathematization of nature* (Pythagoras), etc.
- Finally, they arrived at answers guided by an impressive combination of *reason, intuition,* and *observation*, but not (with certain exceptions) through experiments.

The last point is the source of the basic difference between Presocratic thought and today's natural sciences. During the last three hundred years *experiment and measurement* have played a *decisive* role in the acceptance or rejection of a theory. Thus, whatever agreements exist -and, as we shall see, there are many- Presocratic concepts and contemporary scientifi theories derive from different approaches and interpretations, and it would be a *serious mistake* to believe that the latter constitute simply a continuation and experimental verificatio of the former.

How, then, can the frequent surprising parallels be explained? Do they occur by pure chance? Or do they occur because these thinkers possessed -in addition to their rational critical capacities- a seminal intuition and perceptiveness that led them in the correct scientifi direction? Or do the parallels occur because their 'mode of thought' was grafted on all the later development of Western thought, so that the latter traces the footsteps of the former? Simple answers are risky. Emphasizing sheer coincidence of views may end up in wrong conclusions, whereas ignoring similarities may lead to oversimplificatio and total misunderstanding of the Presocratics' work. Thus, the answer to the question had best be left to the reader's considered judgment....

[17] K. Popper, *Back to the Presocratics*, 23.

Thales of Miletus (ca. 625–546 B.C.)

Thales – the un-mythical.
Friedrich Nietzsche

Personality – Life

Thales is considered the firs Greek *physikos* (natural philosopher), the "firs . . . to have revealed the investigation of nature to the Greeks".[11B1] Aristotle named him "the founder of this philosophy".[11A12] He called him the "firs wise man",[11A1(22)] the "wisest"[11A1(32)] of the Seven Wise Men. Both during his lifetime and after death, he was to enjoy panhellenic respect and admiration. It is not certain, whether he was the author of certain studies (*Nautical Star Guide* and *On the Solstice and the Equinox*). Even if they may have been his, they have not survived. He was a wise man, sunk in his thoughts and theories, but at the same time a practical person, extremely ingenious, experienced and self-confident with strong interests in public affairs.[1] His life and work confir this unique combination of theory and practice, abstract thought and effective solutions for concrete practical problems.

To Thales is attributed the profound saying, 'Know thyself',[11A1(40)] among many aphorisms of acute perceptivity and good sense. To the question, "How can we live in the best and most just way"? he answered, "By not doing what we criticize in others".[11A1(36)] "Who is happy?" "It is he who is healthy in body, magnanimous, and well-educated".[11A1(37)] Worthy of note are the three attributes, which Thales deems necessary for human happiness: not wealth; not power; not glory. Instead: health, a 'kind heart' and education. Neither the bad man, nor the uncultured man can gain felicity. It is worth while to contemplate a moment this attitude of Thales toward life. When he was asked, "What is difficult" he answered, "To know yourself". As for "What is easy"? he responded, "To advise another".[11A1(36)] He would say, in turn, "The oldest of all beings is god, because he is unborn; the most beautiful thing is the cosmos, because it is the creation of god. The largest thing is space, because it contains everything. The swiftest thing is the mind, because it outspeeds everything.

[1] Nietzsche was to observe with a dose of irony, "The old Greek philosophy is a philosophy of pure political leaders. How miserable the situation is with our politicians of today! In this, moreover, the Presocratics differ from the Postsocratics". (F. Nietzsche, *Wissenschaft und Weisheit im Kampfe*, 345).

C.J. Vamvacas, *The Founders of Western Thought – The Presocratics*, Boston Studies in the Philosophy of Science 257, DOI 10.1007/978-1-4020-9791-1_4,

The strongest thing is need, because it exceeds everything. The wisest thing is time, because it reveals everything".[11A1(35)]

Cosmology – Mathematics

To the *physikos* (natural philosopher) Thales are attributed a wide range of astronomical and mathematical discoveries; among others: (a) He was able to predict an eclipse of the sun (28 May 585, B.C.). (b) He provided the correct interpretation of a solar eclipse, considering it to be caused by the shadow thrown by the moon, lighted by the sun, upon the earth. (c) He was the firs to maintain that the sun and other stars are composed of the same 'earthly' stuff, and that "the moon is lighted by the sun".[11A17b] (c) He calculated the equinoxes and their variation; and for navigation, he replaced the constellation of the Great Bear with that of the Little Bear.

Although he must have gained some of the above information and measurements from contact with the other great eastern cultures -the Babylonian and the Phoenician- he is still correctly considered the firs Greek *astronomer*.

It is equally correct to name Thales the firs Greek *mathematician* and the founder of *theoretical geometry*. Following his journeys in Egypt, he introduced geometry in Greece, not simply as a body of practical knowledge, but for the firs time in the form of scientifi theorems confirme by his own inductive reasoning, such as: (a) The diameter bisects the circle. (b) The angle inscribed in a semicircle is a right angle. (c) The angles at the base of an isosceles triangle are equal. (d) The opposite angles of two intersecting straight lines are equal. (e) A triangle is determined given its base and the angles relative to its base. (f) Based on the properties of like triangles, he determined the height of the pyramids, comparing the length of their shadow with that of staff of known height, just as he calculated from shore the distance of a ship at sea.

Thales both observed and experimented. Experimenting with amber he determined its capacity to attract other bodies magnetically. His correct observation that the annual summer flood of the Nile delta coincide with annual northern winds, would lead him to conclude that the latter caused the flooding "blocking the Nile's fl w into the sea".[11A16] What is of importance here is not that a correct observation is followed by a faulty conclusion; this occurs in all sciences till today. What is of importance is that for the firs time he would seek the *causation* of this phenomenon and would place it in the domain of the *physical* -not the *mythical*, as did the Egyptians.

First *'arche'*: Water

"Most of the earliest philosophers thought that the *arche* (the originating cause; the primary element) which were in the nature of matter were the only principles of all things.... On the number and nature of such *arche* they do not all agree.

Thales, who led the way in this kind of philosophy, says that the *arche* is water".[11A12] In these words Aristotle was to present the fundamental point of departure in the history of human rational thought attempted by Thales. It was a decisive turning point, for it led for the firs time:

- to the demythificatio and rational explanation of natural phenomena;
- to the search for a natural unity in the seeming multiplicity of the universe;
- to the discovery of a deterministic causation in nature.

The word *arche* was known at that time and could mean: (a) beginning, (b) original state out of which the manifold world was developed, (c) firs principle or substratum. For Thales, this *arche* was *water*. How did he arrive at this conclusion? There are various possibilities involving both *mythical* and *rational* explanations. Homer refers to the "Ocean as the father from all things".[2] Besides, Thales knew the great riverside civilizations of Babylonia and Egypt. Aristotle was to say that Thales "declared that the earth rests on water",[11A12] that is, the created (earth) was held upon the creator (water). These mythic elements may have played a role in the formulation of Thales *arche*, but they were not the sole influence His *critical mind* must have been decisive in his fina selection of *water* as the primary substratum of all things.

The *arche* had to satisfy for Thales certain *presuppositions*, which would be confirme by *observation* and, possibly, by *experiment*:

(a) It must supply and support *life*, since for Thales nature is a living organism: "the world is besouled".[11A3] In this early period of philosophy the dichotomy between animate beings and inanimate matter did not exist.
(b) Being the *sole* origin of nature, it necessarily also had to include the primary cause of *change* and *motion*.

On the basis of these criteria Thales would logically arrive at water as *arche*. In regard to the firs point, he would fin in *water* the necessary presupposition for the existence of *life*: "his supposition may have arisen from the observation that the nourishment of all creatures is moist, . . . and that the semen of all things have a moist nature and water is for moist things the origin of their nature.".[11A12] Water is the principle vehicle of life and consequently of nature as a whole. This logical conclusion, which Thales would arrive at through observation, is wholly confirme by contemporary molecular biology: "Hydrogen bonding among water molecules is of crucial importance for life on earth because all life requires an aqueous environment".[3]

On the second point, Thales would observe that water is both *easily transformed* and *mobile*. It was the only substance that could appear -in his times- in the three physical states, solid (ice), liquid (water), and gas (steam). Ice melts into water,

---

[2] Homer, *The Iliad*, 14, 246.

[3] J. Darnell, H. Lodish, and D. Baltimore, *Molecular Cell Biology*, 27.

water evaporates into steam, and again, conversely, the latter condenses into water, and in winter water is solidifie into ice. In addition to its transformability, water is the most mobile of visible substances, since it immediately takes the shape of the space that contains it. Rivers and seas, moreover, are characterized by perpetual motion. "Through elemental moisture there penetrates a divine power that moves it".[11A23] Thus, it is not by chance, nor only through mythological elements that Thales found water to be the firs *arche*.

Soul

For Thales, soul was both the fountainhead of *life* and of *motion* and *change*: "Thales was the firs to declare that soul was of a perpetually self-moving nature".[11A22a] He still considered all of nature to be besouled, with an inherent property of movement and change.

The Divine

Indeed, being the primary *arche* "always preserved"[11A12] and besouled, it can also be no other than divine: "And some said that [soul] is blended in the universe; probably for this reason Thales too believed all things are full of gods".[11A22] Thales said that "the mind of the world is god, and that all things are besouled, and full of daimons [gods]".[11A23] He demythifie phenomena as he sought for a natural explanation for everything. Mythic, divine *personification* were abolished, and nature maintained an impersonal but sacred hypostasis, a single, divine *arche* expressed within the powers of life, of change, and of movement. This constitutes an ontological view of the divine and its actions, which belong to the cosmic realm of beings.

Overview

Thales found himself before two paths leading to the understanding of nature. The one was toward the traditional, theistic personificatio of natural phenomena and led to *theology*. The second proceeded to demythificatio and objective, natural inquiry, leading to *science*. Being the firs to choose decisively and persistently the second path -with a boldness astounding for his time- Thales would unquestionably become the founder of scientifi thought. What made him immortal, as historian of science W.P.D. Wightman observed, was that "it was Thales who firs attempted to explain the variety of nature as the modification of something *in* nature".[4]

The *demythificatio* attempted by Thales, and the emancipation of nature from anthropomorphic, theistic, arbitrary approaches, the firs location of the *arche within*

[4] W.P.D. Wightman, *Growth of Scientifi Ideas*, 10.

the *material* world, his belief in a *lawful order* of the universe, his persistence in a causal investigation of natural phenomena, and the transcendence he attempted within empiricism so as to attain general *theoretical knowledge*, indicate beyond doubt Thales's *scientifi* thought. That which also makes him the firs *philosopher* -and not a natural philosopher only- is the idea of the *unity*, the 'one principle' of the world, within its phenomenological complexity. This position, which is stated for the firs time in the history of western thought by Thales, is not only scientifi but also *ontological*. His thought is not only focused on things themselves, but strives to probe deep into the *unifie substratum*, which underlies them.

The search for the *oneness* of the world is ageless, and we fin it in all cultures. It constitutes the guiding thought in science as in philosophy. Thales would be the firs to remove the mythic, irrational elements in this inquiry, and to identify *unity* through a logical, critical method in *one* element within a deterministic *cosmos*. Thus, he would bring about the decisive, historical division between the *mythic* and the *scientifi* eras, choosing the path of reason upon which Western man would tread thereafter.

Anaximander of Miletus (ca. 610–546 B.C.)

The effort to understand the universe is one of the very few things that lifts human life a little above the level of farce, and gives it some of the grace of tragedy.

Steven Weinberg

The Extant Fragment

"Anaximander named the arche of existing things the apeiron (the boundless, the unlimited). Things perish into those out of which they have their being according to necessity; for they make just recompense to one another for their injustice according to the assessment of time".[12B1]

This passage of Anaximander's constitutes the *fi st* text in the history of Western philosophy, the "most ancient saying[1] of the Western mind",[2] as Heidegger would put it. At the same time it is the *fi st* extant passage written not in the tongue of poetry but in colloquial *prose*, the language appropriate to science and philosophy.

Innumerable thinkers, philologists and philosophers have bent over these forty words, striving to fathom and explain their deepest meaning. The result is a succession of interpretive studies, pragmatical treatises, and visionary essays, which frequently clash and erase one another in relation to virtually every word of the text. Jaspers was to succinctly explain the reason for this conflict "All the interpretations show: In the thought of Anaximander exist a plethora of latent meanings, whose possibilities extend further than any given specifi approach, which either wanders far afiel or falls short of the import".[3] Viewed in this light, Anaximander's fragment takes on an even deeper meaning and constitutes inexhaustible stimulation for the intellectual quester today.

Personality – Life

Anaximander was to be the firs thinker to offer a comprehensive cosmological system, beginning with cosmogony and ending in man. The titles of his *prose* works are recorded as *On Nature*, *Circuit of the Earth*, *On the Fixed Stars*, *Celestial Globe*,

---

[1] We do not know whether the passage is a saying, or a portion of a larger text.

[2] M. Heidegger, *Der Spruch des Anaximander*, 317.

[3] K. Jaspers, *Aus dem Ursprung denkende Metaphysiker*, 21.

C.J. Vamvacas, *The Founders of Western Thought – The Presocratics*, Boston Studies in the Philosophy of Science 257, DOI 10.1007/978-1-4020-9791-1_5,

and some other works.[12A2] None of these have survived -with the exception of the above fragment. Like his fellow countryman and teacher Thales, he was to play an active part in the political life of Miletus. He was to be the firs cartographer of the then known world: "He firs dared draw the inhabited world on a tablet".[12A6] Although the Egyptians too used maps of limited areas, the creation of a *world map* undoubtedly constitutes "a grand gesture which symbolizes the desire to investigate and seek an *overall* vision of the world",[4] as O. Gigon was to state.

First *'arche'*: The Boundless

"He said that the *arche* and primary element of existing things was the *apeiron* (the boundless, the unlimited), being the firs to introduce this name [of] the *arche*".[12A9] It is not clear whether the meaning here is that Anaximander firs introduced the term *arche*, or was firs to consider the *apeiron* as *arche*. Both versions have been supported, and the firs appears to have prevailed. At the same time both versions might well apply – that is, that he is the firs who not only used the word *arche* but also introduced the term *apeiron* as his concept of the *arche*. As for the meaning which Anaximander conveys with the word *arche*, that could be -without ruling out 'originating cause'- the 'firs principle', that which preexisted. Thus, the *apeiron* itself constitutes the very 'origin'. The term *apeiron* -which Anaximander conceived abstractly by nominalizing the neuter of the adjective *apeiros* (boundless)- is polysemantic. Anaximander certainly was familiar with the *chaos* of Hesiod. His *boundless,* however, is not the same concept. The Hesiodic *chaos* is empty and neutral, constituting the situation in which cosmogony comes about. The Anaximanderian *apeiron* is replete and dynamic, constituting the source but also the receiver of the perishable and finite

- *Temporally*, it is "unborn and imperishable", [12A15] "immortal, and indestructible", [12B3] in other words, without start and without finish eternal.
- *Spatially*, it is unbounded, which means: (1) without beginning and end, "necessarily being without limit";[12A15] (2) without external boundaries (as the surface of a sphere has no boundaries), infinite (3) immeasurable, vast, impenetrable. It is seen "to encompass all things".[12A15]
- *Quantitatively*, it is inexhaustible, unending, "in order that becoming might not fail".[12A14] According to Anaximander in order for balance to exist in the world, opposite the 'one' boundless must exist 'innumerable' finites thus he concludes that there are 'innumerable worlds'.[12A14] The power of this critical concept is astonishing, for it cannot be based on any known mythological or empirical indication.
- *Qualitatively*, the *apeiron* is considered indefinable indeterminate, without internal boundary lines, undifferentiated. It cannot be considered a mixture or an

[4] O. Gigon, *Der Ursprung der griechischen Philosophie*, 90.

amalgam but something homogeneous, shapeless, without qualities. It would contain all of the potentialities of the visible world, none of which is as yet differentiated, developed, or materialized. As the *fi st principle*, the boundless cannot be identifie with any of the elements (e.g., water), because then that element would negate all balance in the world and would dominate over all the other elements. Accordingly, the boundless exists before and beyond the differentiation between the material and spiritual world. The genius of Anaximander appears precisely in the fact that his mind *conceived* the *arche as* the 'boundless', yet the latter in itself is mentally *inconceivable.*

The Divine

The 'boundless' is divine. "Of the boundless there is no beginning . . . but this seems to be the beginning of the other things. . .It is also unborn and indestructible".[12A15] Considering the 'boundless' to be without beginning but as such the eternal, immortal, and imperishable originating cause of all things, Anaximander identifie it with the divine: "and this is the divine; for it is immortal and indestructible".[12A15] Anaximander elevates the Homeric and Hesiodic sense of the divine from an entity born and anthropomorphic to the eternal and impersonal hypostasis of the 'boundless', lending it at the same time a dynamic property that speaks to the deepest religious feelings of man: "it encompasses all things and steers all".[12A15] Anaximander's *divine* as captain-guide-protector expresses a profound religious import. He moreover considers the 'boundless' to exercise ontological rule over worldly beings. It is apparent that such a stance toward cosmological phenomena derives by extension from the civil society of the established judicial-political order.

Anaximander conceives the *apeiron* and consequently the *divine* in a chain of abstractions that lead to *negative* attributes: *a-peiron* (un-limited), *an-archon* (without beginning), 'im-mortal', 'in-destructible', 'un-born', 'in-corrubtible'. We will again encounter this apophatic or negative definitio of the divine in the neoplatonic philosophy of Plotinus.

The Opposites

Anaximander theorizes that from the *apeiron* "the opposites are separated out by eternal motion . . . opposites being hot, cold, dry, wet, and the others".[12A9] At the time of Anaximander, the concepts 'hot', 'cold', etc. were not considered simple attributions that could be isolated from the noun they modify. It is believed that there did not as yet exist a clear distinction between the *nominal* and *adjectival* function, between qualities (hot, cold, wet, dry etc.) and things. Only much later, at the time of Plato and Aristotle would the absolute distinction between *material* bodies and *properties* be completed.

The opposites constitute components of Anaximander's cosmogony, and from them are created the various embodied parts of the universe such as the earth (dry), the water (wet), the air (cold), the fir (hot). The opposites themselves are found to be in continual struggle, the one with the other, and the dominance of the one brings about the abolition of the other (hot-cold, dry-wet, light-darkness). The same is true for the entities that derive from the opposites; these too are found to be in a perpetual process of birth and destruction: "things perish into those out of which they have their being".[12B1] This conception is not new. What is new is the framework within which Anaximander places this process; namely, the *struggle of opposites* in which the birth of one inevitably coincides with the demise of its complementary opposite, only for the latter to return later and obliterate the former. In other words, there is a perpetual creation-destruction-recreation of the opposites and of their derivative things and animate beings. This takes place 'according to necessity', because 'they make just recompense', the 'one to the other', for their 'injustice', 'according to the assessment of time'.

Justice – Necessity

The entire process of genesis and obliteration comes to be seen as a process of injustice and punishment. Within the perpetual struggle of the opposites, the dominance of the one over the other constitutes an injustice, leading inevitably to overbearing pride that entails *hybris* (insolence, presumption), causing guilt, compensation for which must be given to the wronged party -that is, the opposite in the pair ('to one another'),[5] which in turn will dominate and later make its own recompense.

This ceaseless alternation between dominance and defeat occurs according to dire 'necessity' and 'justice'. Impersonal 'necessity' replaces for the firs time the former uncontrolled powers and arbitrary acts of the gods, expressing both, natural determinism and causality, as well as impartial 'justice'. The imposition of justice is aimed at the 'equal share' of the balance of the opposite tendencies and comes about 'according to the assessment of time'. Time is viewed here as a temporal framework, the fi ed, fina time-limit within which worlds, beings, things, come into being and pass away.[6]

---

[5] When the words *one another* as yet were missing from the fragment, older commentators (Nietzsche, Ritter, Diels, Rohde, and others) saw the rule of justice to be imposed on things by the *apeiron*, as a penalty for the injustice of their existence itself, which was considered to be 'a fall from the primordial unity of things' into their imperfect, separate hypostasis. This ontological interpretation ceased to hold from the time when it was established by the phrase 'one another' that the redress and recompense do not refer to the 'apeiron' but to the 'pairs of opposites'.

[6] Time in archaic and classic thought, beyond the cosmogonical dimension attributed to it by Pherekydes ("*Zeus and Chronos -time and earth- are eternal*")[7B1] plays a decisive role as a supreme power. We have already met Thales's saying, "*Time most wise, as it reveals everything*". Solon was to write, "*Time, the just*" (Fr.24.31); Pindar was to say, "*Judgment of all things is imposed by blessed Time*" (Bowra, Fr.14); "*Of manly right, Time most excellent savior*" (Bowra, Fr.145);

The meanings and associations that we can draw from Anaximander's fragment are inexhaustible. His thought develops on many levels that for him are as yet inseparable and unifie – the ethical, the aesthetic, the social, the scientific the philosophical, the ontological, the religious. The entire evolution in nature and in life is determined by a single strict *ethical* necessity ('according to necessity, for they make just recompense to one another for their injustice') that reflect the social life and legal system of the *polis* (city-state) of that time. That same necessity also constitutes the primary physical law that governs the universe.

Cosmogony

Bertrand Russell was to observe that "wherever Anaximander is original he is scientifi and rational".[7] His cosmogony and cosmology were developed in that spirit: From the *apeiron* at a given moment the pair of opposites, hot-cold, were 'separated out'. Aristotle speaks of a 'vortex'[8] – something like whirling nebulae causing the further separation of the opposite elemental masses. Thus, there "was formed over the air surrounding the earth a sort of fier sphere, like the bark around the tree. When this sphere burst and closed in various circles, the sun, the moon, and the stars were created".[12A10] This leads to the formation of fier circles around the earth, covered over by the dark clouds that hide them, except for openings shooting fire thus becoming visible as the sun, the moon and stars. When these openings are partially or wholly blocked, we have the phenomena of the phases of the moon and the eclipses.

Life firs forms in the watery element under the influenc of the sun – that is, the sun's rays and heat: "living creatures come into being from moisture evaporated by the sun".[12A11(6)] The descent of man cannot be the same as that of the other animals, "because after a short time after their birth the other animals can care for themselves and only man requires extended maternal support; for that reason he [the *fi st* man] would not survive, if this was his original form".[12A10] From this observation arises the logical conclusion that "man was originally similar to another creature, namely the fish"[12A11(6)] Anaximander is here describing the origin and evolution of mankind. According to another reference, "man was born at firs in

---

"*Time, the father of all*" (Olympic Odes, 2, 17); "*Time, who exonerates only just causes*" (*Olympic Odes, 10, 65)*. The great tragic dramatists frequently allude to time. Aeschylus, *Prometheus Bound, 981:* "*Maturing time is the teacher in all things.*" Sophocles, *Oedipus Rex*, 614: "*Time only brings to light the honorable man*". Sophocles, *Ajax*, 645: "*Long and immeasurable time brings forth all things that are obscure and when they have come to light hides them again*".

[7] B. Russell, *History of Western Philosophy*, 47.

[8] Aristotle, *On the Heavens*, 295a13.

the fis and having been nourished like sharks[9] and proved themselves capable of supporting themselves, emerged and walked on land".[12A30]

Even so, we recognize the concept of an evolution process through time, because otherwise man would still today be born in the fish Thus, Anaximander would be the firs to attempt a *logical* interpretation of man's descent and the firs to state -in a manner both imaginative and original- the idea of gradual *evolution* and *adaptation* to the environment. This theoretical conception takes on even greater value since it was based on observation of the diet of marine mammals.

Cosmology

Through the rational thinking, Anaximander would develop his cosmology. Its originality springs from its *geometrization* of the physical universe. He introduces also numbers and numerical magnitudes among the heavenly bodies. For the firs time the heavens assume the shape of a *sphere* instead of a domed roof; the earth is *suspended* at the center, motionless, equidistant from all points of the spherical heavens. It remains motionless 'by necessity', because as the *symmetrical* center of *isotropic* space, 'it is not befitting that it move in any direction whatsoever "either up or down[10] or to the side, remaining steady at the center on account of its equal distance from everything".[12A11(3)] Anaximander applies the principle that the *precondition* for any movement is the existence of differences. This principle of *sufficien cause* will be clearly stated a century later by Leucippus in the saying, "nothing occurs at random, but everything for a reason and by necessity",[67A2] and more recently by G.W. Leibniz as the 'principle of sufficien reason'.

The circular orbits of the stars, moon, and sun around the immobile earth have circumferences that are, respectively, 9 [3X3X1], 18 [3X3X2], and 27 [3X3X3] times greater than that of the earth. Anaximander would be the firs Greek who introduced mathematics into *astronomy*. The emphasis on the importance of the number three clearly indicates that behind these firs mathematical relationships there are as yet strong mythic and religious, but also aesthetic, elements.

---

[9] We do not know by what criteria Anaximander theorized the shark to be the distant ancestor of men. It is a fact, however, that today's science found that one of the most complex systems of an organism, the immune system, shows striking similarities between that of the shark (which has been existing for as many as 450 million years) and the Homo Sapiens (who has been around for approximately half a million years). "Sharks have a spleen, which, as in humans, is a rich source of B cells.... The similarities extend to cellular immunity. Like humans, sharks have a thymus, in which T cells mature and from which they are released. Sharks also have T-cell receptors ... [and] as in humans, diversity in these receptors arises from the same kind of generic mechanisms that give rise to antibody diversity.... The protein structures of shark and human antibodies are very similar ... Sharks may be our only remaining link to the distant origins of T and B cell immunity". (G.W. Litman, *Sharks and the Origin of Vertebrate Immunity*).

[10] To transcend the notion of an absolute 'up' and 'down' -in contradiction to all experience-demands a power of intuition and abstraction that one cannot overlook.

Meteorology

As one would expect, Anaximander's 'meteorology', though still incomplete, is also purely naturalistic, excluding any intervention or influenc of mythical gods or powers. Wind is the result of purely *natural causes*: "wind is a fl w of air occurring when the fines and most moist elements in it are set in motion or liquefie by the sun".[12A24] "Rain comes from the vapor that arises from the earth due to the sun, lightning from the wind breaks the clouds".[12A11(6)] Thunder, whirlwinds and typhoons are due to the wind.

Overview

The great intellectual leap attempted by Anaximander was to introduce for the firs time in the history of philosophy and science the concept of the 'infinite' of 'opposites', of 'necessity', of 'symmetry' in time and in space, of 'dynamic balance', of 'relationship', of 'evolution'.

He firs conceived the creation of the universe not as a process which begins from something concrete and already known, but as an evolution that springs from a unified undefined undifferentiated, boundless 'originating cause' and ends in the plurality of defined individual beings and things. Two and a half thousand years later philosophy was to defin evolution precisely as the "change from an incoherent homogeneity to a coherent heterogeneity".[11] The question as to whether the unique primary substratum is one of the existing, known ones, or *different* and *above* them, also appears in contemporary sub-atomic physics. In the search for a unifie theory, from which will arise mathematically all the elementary particles and their properties, must physics turn -as it has in recent decades- toward the known elementary particles and determine that some of them are 'fundamental', or must it seek an entirely *different*, general, "primary substratum" (which could be matter or energy) as the source of all the rest? W. Heisenberg observes: "None of the various elementary particles can essentially be differentiated from the rest as an especially 'fundamental' elementary particle. This last concept corresponds exactly to Anaximander's doctrine, and I myself am convinced that in contemporary physics this conception is the correct one".[12]

There are further correspondences between Anaximander's cosmogony and today's theoretical physics. Anaximander would speak of 'infinit worlds', successive or coexistent. Today's science precludes neither of these two possibilities. Innumerable *successive* worlds come about when the as yet average density of the matter of the universe proves to be greater than three protons per cubic meter, whereby we have an "unending cycle of expansion and contraction".[13] Each expansion creates

---

[11] H. Spencer, *First Principles of a New System of Philosophy*, 360.

[12] W. Heisenberg, *Physik und Philosophie*, 43.

[13] S. Weinberg, *The First Three Minutes*, 143.

a new universe, which will begin with the contraction and disappearance of the previous one, in perpetuity. Also multiple *coexistent* worlds are not excluded according to a relatively recent, still unproved, theory of quantum mechanics concerning 'many worlds' or the 'bifurcating universe'[14] suggested in 1957 by Physicist Hugh Everett III.

Transformation and change are based on the *principle of reciprocity,* which according to Anaximander rules in relationships, both as an ethical and social norm as well as a physical law. With the principle of the equality between action and reaction, classical physics confirm that "forces always possess a reciprocal character".[15]

"Anaximander is the firs in the West to conceive of the world as a community of justice, as an *order* of things," W. Jäger was to state with K. Jaspers.[16] This conception would lead him to the firs formulation of a causal succession of natural processes 'according to necessity', within 'the assessment of time'. Just as wrong conducts lead unavoidably to punishment, so also the cyclical, alternating rule of opposites in nature expresses the inevitable relationship of cause and effect. 'According to necessity' constitutes the firs abstract conception -still indissolubly linked with the ethical law of justice and punishment – which will form the nucleus of the physical principles of *determinism* and *causality.*

Anaximander introduces for the firs time the concept of *symmetry* within time and space. *Temporally* he conceives a periodic, cyclical, ascendancy of the elements which is based on the symmetrical effect of action-reaction. *Spatially*, he theorizes a universe characterized by *geometrical balance.* This conception leads by exclusively *rational criteria* to the conclusion -radical for his time- that the earth which is found at the center of the universe must be located symmetrically the same distance not only from the spherical dome of the heavens both above, in the visible hemisphere, but also from the invisible hemispherical dome below. He rejects Thales's position that the earth must be supported (upon water) so as not to fall, by critically applying the principle of *isotropic symmetry* – considering, that is, like *contemporary* classical physics, that space is *isotropic*, i.e., all its directions are equivalent. Thus, he concludes that any support is superfluous because the earth does not move upward, there is also no reason for it to move downward. Accordingly, he is the firs to declare that "the earth hangs freely, held up by nothing, but steady on account of its equal distance from everything".[12A11(3)] The idea of an earth hovering in space would be so revolutionary that none of the succeeding thinkers would accept it. "Their rejection probably constitutes the oldest recorded conflic between mathematical science and common sense",[17] C. Kahn was to write. K. Popper considers this conceptual innovation of Anaximander's to be "one of the

---

[14] B.S. DeWitt and N. Graham, *The Many-Worlds Interpretation of Quantum Mechanics.*

[15] E.N. Oeconomou, *I Physiki simera (Physics Today)*, 39.

[16] K. Jaspers, *Aus dem Ursprung denkende Metaphysiker*, 20.

[17] C. Kahn, *Anaximander and the Origins of Greek Philosophy*, 80.

boldest, most revolutionary and most portentous ideas in the whole history of human thought".[18]

The notion of *symmetry* which is introduced for the firs time by Anaximander as a fundamental physical principle continues to constitute today, in contemporary physics -within, of course, a more abstract framework- one of the cornerstones aiding in not only explaining but *predicting* natural phenomena, just as Anaximander had attempted two and a half millennia ago. A recent example can be found in the unifie fiel theory of physics. In the sixties, *gauge symmetry* aiming at unifying nuclear weak interaction and electromagnetism allowed S. Glashow, A. Weinberg and A. Salam to make some concrete *predictions* about the existence -in addition to the photon- of the three intermediate vector bosons W^+, W^-, Z^0. The particles W^+, W^-, Z^0 are the carriers of the weak force but they were *unknown* at that time; they simply appeared in the basic equations of the unifie theory in a *symmetrical* way. At firs only a few people believed in the existence of these weak force carriers. However, over the next ten years or so this prediction could be experimentally confirmed Referring to this great scientifi break-through (for which S. Glashow, A. Weinberg and A. Salam were awarded the Nobel Prize for Physics in 1979), physicist I. Iliopoulos would note: "With enthusiasm we observe that a *theory* based purely on intellectual and aesthetic foundations showed us the way to great experimental discoveries"[19]. This *intellectual* and *aesthetic* scientifi foundation was proposed and successfully applied *fi st* -in the theory concerning the earth *suspended in vacuum*- by Anaximander.

All the above perceptions of Anaximander's genius led for the firs time to the conception of the universe as *self-regulated* and *self-controlled* by a dynamic relation of balance and causative conditions. This constitutes an overwhelming pronouncement of *liberation* from the bonds of uncertainty, which arose from what were, up to that time, unspoken and arbitrary interventions of mythic deities. Fear and doubt surrendered their position to a *rational* and *balanced* cosmos in which inviolable law regulates relationships and changes.

If in Thales we recognize keen observations which will for the firs time fragmentarily cast light on the foundations of a demythifie universe, in Anaximander we encounter an unsurpassed intellectual temperament which with the *visionary* power of genius but also with bold *critical* thought creates the firs consistent and integrated cosmo-theory known in the West. "Anaximander's achievement is valuable in itself, like a work of art. Besides, his achievement made other achievements possible, among them those of great scientists (like Aristarchus, Copernicus, Kepler, Galilei)",[20] K. Popper was to state; and K. Jaspers would add: "We sense how our own Western thought itself appears here both initially and powerfully in a grand pattern".[21]

---

[18] K. Popper, *Back to the Presocratics*, 9.

[19] Cited in E.N. Oeconomou, *I Physiki simera* (*Physics Today*), 216.

[20] K. Popper, *Back to the Presocratics*, 12.

[21] K. Jaspers, *Aus dem Ursprung dendende Metaphysiker*, 22.

Anaximenes of Miletus (ca. 585–525 B.C.)

> *A few single themes – unspoken assumptions and intuitively held prejudices that originate outside science – underlie all scientifi thought.*
>
> Gerald Holton

Introduction

Anaximenes, fellow citizen and student of Anaximander, was the firs to promote a *physical* interpretation of the *change* of things, attributing it to the degree of *rarefaction* or *condensation* of the primary substance, air. His empirical thought is found at the opposite pole from Anaximander's abstract mind. Thus, Anaximenes would follow a course more systematic and inductive that would lead him to two basic conclusions. These might lack the intuitive power of his mentor, but would prove extremely fertile for the later development of scientifi thought. First, he establishes the continuous and uninterrupted evolution of an infinit primary substance (air) toward the plurality of the world, while its very material essence (air) remains unaltered, and second, he is the firs to attribute all *qualitative* changes to *quantitative* differentiations.

First *'arche'*: Air

Anaximander consciously separates and contrasts the indefinite undifferentiated, imperceptible, infinite 'First Principle' or 'Substratum', the *apeiron* (the boundless) with the plurality of the definite qualitatively differentiated, tangible world. This view presents an inherent difficulty the *transition* from the *supra sensorial* to the *material* world. Anaximenes would not accept his teachers theory, which is characterized by *discontinuity.* Concentrating his attention precisely on this weak point regarding the *process* of genesis and change, he would conceive a less imaginative but by the same token more empirical and equally ingenious cosmological theory that, for the firs time, would insure unbreakable continuity and cohesion. This conception is founded on *two* basic presuppositions, which evade the difficul ties encountered by his fellow citizens Thales and Anaximander.

The *fi st assumption* states that the originating cause retains the properties of the *apeiron* but it is no longer the *apeiron* (the boundless). It has a real state of material existence and "he calls it air".[13A5] From it proceeds whatever "comes to be, or has done so in the past, or shall be".[13A7(1)] The thought of Anaximenes thus

C.J. Vamvacas, *The Founders of Western Thought – The Presocratics*, Boston Studies in the Philosophy of Science 257, DOI 10.1007/978-1-4020-9791-1_6,

develops *from the outset* within a purely *material* context and thus can establish -in contrast to the ungraspable, indefinit *arche* of Anaximander- a cohesive, *physical* interpretation of the world's entire evolution.

This constitutes precisely his *second position*: That genesis and change depend upon a unified *mechanical* process of *condensation* or *rarefaction* of the primary substance, air, "that it differs in rarity and density according to the different substances. Rarefied it becomes fire condensed, it becomes firs wind, then cloud, and when condensed still further water, then earch and stones. Everything else is made of this. (Anaximenes) also postulated eternal motion, which is indeed the cause of the change".[13A5]

According to Anaximenes, all the elements consist of air itself and differ from one another only in the degree of condensation or rarefaction. Accordingly, he is the firs to reduce *qualitative* change to a natural, material *quantitative* property – the density (of the air). The foundation of *qualitative* difference upon a *quantitative* base constitutes the firs step toward, and the firs suggestion of, what is accepted by today's science: that the existence of different substances and materials is due exclusively to the *quantity* and arrangement of the atoms which constitute their parts; that the existence of different elements is due to the *quantity* of protons and neutrons which constitute their nuclei; and the difference, again, between protons and neutrons is due to the *number* of the subatomic particles which constitute them. "If, in fact", notes physicist E. Schrödinger, "instead of air, Anaximenes had said 'dissociated hydrogen gas' (something that one could truly not have expected from him), he would not have been a great distance from our view today".[1]

Anaximenes would consider the opposites themselves to be the consequence of rarefaction-condensation. The dry and the hot are produced by rarefaction, while the wet and the cold are produced by condensation. He would base many of his theories on observation. The above conclusion derived from the observation that when we blow air on our hand with *wide open* mouth, the air is *hot*, because -according to Anaximenes- it is rarefie and expanded, while if we blow with *half-closed* lips, it is cold, because it is condensed and compressed. Even though we know today that what takes place is precisely the opposite -that is, air compression leads to heat and air expansion to freezing- the fact that for the firs time Anaximenes related the concept of *temperature* to the physical process *compression-expansion* is indeed impressive. In general terms, Anaximenes's concept that the *physical properties* can be reduced to *quantitative* differences would prove extremely fertile in the development of the natural sciences.

'Air' could have been chosen as the firs principle or substratum for many reasons. Air exists everywhere; it fill and contains everything. It occupies all of space and is characterized by a subtle mobility that makes it "prone to change".[2] This mobility satisfie a basic presupposition considered self-evident by the f rst Ionian 'physiologists', since the *distinction* between *inert matter* and *moving force* was

---

[1] E. Schrödinger, *Die Natur der Griechen*, 107.

[2] Simplicius, *Aristotelis De Caelo Commentaria*, 615.20.

not yet understood. The primary substance, air, had to be *autokinetic* -as indeed it appears to be- so that it could effect various changes: "He considers motion eternal and says that this is what causes change".[13A5]

Soul

Pneuma[3] (from *pneo*, breath), signifie 'blast of air', or 'wind'. But at the same time it also means 'breathed air', the 'breath of life', 'breathing', 'life' itself – the 'spirit', the 'soul' (*psyche*, from the verb *psyho,* breath, blow). Anaximenes was to use the words *pneuma* and *air* as synonyms,[13B2] thus attributing to air a much broader significanc – that of the cosmic *life-engendering* breath, a conception that probably derived from old orphic and eastern beliefs. Here Anaximenes attempts, as usual, a bold analogy: "Just as our soul. . .being air holds us together, so *pneuma* (breath) and air encompass [and guard] the whole world".[13B2] This saying is considered to be the only extant statement from Anaximenes's prose study. Although opinion is divided concerning the precise meaning of this aphorism, most commentators agree on one point: Anaximenes is the firs western thinker who introduces a parallel between *microcosm* and *macrocosm*, applying by *analogy* what is *known* about man to what is *unknown* about the universe. One can also fin here, however, an *identificatio* of the soul, as the breath of life, with the *fi st principle* (air) of nature.

The Old Testament, too presents the *pneuma* both in the founding of the world ("And the Spirit [*pneuma*] of God moved upon the face of the waters"[4]) and in the creation of man ("And the Lord God. . .breathed into his nostrils the breath of life [*pneuma* as engendering breath]: and man became a living soul"[5]). Anaximenes cosmogony, however, differs radically: the *physicos* (natural philosopher) Anaximenes recognizes no *creator* of the universe, nor does the spirit (*pneuma*) of a creator *guide* man, as it does through the entire course of God's work as guardian of the people of Israel. Thus, in the minds of the firs Greek thinkers, man stands *tragically* alone.

The Divine

The air itself, which is perpetually moving, boundless, eternal and life-engendering, can be seen as being like the divine[13A10]. In this sense divine may refer to "the powers that permeate the elements or the bodies"[13A1] or to the various Olympic gods, born of the boundless air: "infinit air was the originating cause from which. . .gods and the divine came into being."[13A7(1)] In this manner, he retains but at the same time also lowers the conventional deities to the simple creations of the 'firs principle'.

---

[3] H.G. Liddell, R. Scott, *A Greek-English Lexicon.*

[4] Old Testament, *Genesis*, 1, 2.

[5] Old Testament, *Genesis*, 2, 7.

Cosmogony – Cosmology – Meteorology

Anaximenes's cosmogony and cosmology are not inspired by the pioneering ideas of his teacher. He abandons Anaximander's bold vision of the cylindrical earth suspended in space, affirmin that the fla earth is held up by "riding upon the air".[13A20] Exploring further his concept of the opposite pairs, *rarefaction – condensation*, Anaximenes would develop a theory of cosmic evolution and natural processes that would apply to all cosmogonical, cosmological, and meteorological phenomena: "everything is produced by a kind of condensation (of the air) and again rarefaction".[13A6]

Overview

Given the limited vocabulary even in his own age, Anaximenes was not the firs to exploit analogy so that -like the earlier Milesian 'natural philosophers'- he could make certain observations of his more understandable and vivid: "The sun is fla like a leaf",[13A15] "the earth is table-shaped",[13A20] etc. He was the first however, who would use *analogy* based on similarities and correlations so as to extend his theory from known conditions and arrive at conclusions about cosmic phenomena as yet not understood ("Just as when you dip a hot piece of iron in water, it cools with a great noise, thunder causes noise when air attempts to pierce through clouds"[6]), or a parallel between man and the universe ("Just as our soul being air holds us together, so *pneuma* and air encompass the whole world",[13B2] etc.

Anaximenes's use of *analogy* is worth noting mainly for two reasons. On the one hand, it indirectly expresses his conviction that the same *universal* natural laws apply by *analogy* both on earth and in the heavens. This belief is by no means self-evident. Not until two thousand years later would it be proved mathematically by Newton. On the other hand, he introduces the method of drawing conclusions *by analogy* from familiar, established facts to puzzling phenomena, thereby explaining the uncertain through the certain. This method has been used ever since by the physical sciences to develop certain theories. Thus, in the recent past, the chemical properties of certain elements as yet to be found and isolated were determined *by analogy* to known, closely related elements. The quantisation (Quantelung) of gravity has not been achieved to this day, though it was foreseen by Heisenberg and Pauli from the 1930s 'according to precise analogy'[7] to the applied procedure in their quantum fiel theory.

The systematic, empirical thought of Anaximenes was to influenc later thinkers and the general development of philosophy as well. Natural laws and properties take on a *universal* character and apply from the depths of the earth to the boundaries of space, a concept that leads to the founding of a unifie and unique natural

---

[6] Seneca, *Naturales Questiones*, II, 17.

[7] W. Heisenberg, *Wandlungen in den Grundlagen der Naturwissenschft*, xxv.

science. Anaximenes establishes unbroken *continuity* in the physical evolution of the universe.

The synonymous words *aer* and *pneuma* would also become terms in ancient Greek medicine. "Air is of the highest ruler in every respect", notes the author of *Peri physon 3* of the *Corpus Hippocraticum.* "Air is essential for life but is also a carrier of illnesses. Air empowers both objects and living organisms".[8] The role of the Anaximenean *pneuma* would expand and would be studied, in turn, by Diogenes of Apollonia, so that it would occupy a key position in stoic philosophy – that of the cohesive physical fiel that is the carrier of all material properties.

The reference of *qualitative* differentiation to *quantitative* change, firs conceived by Anaximenes, would mark the subsequent course of science. "In contemporary science," Sambursky notes, "the process of mathematization went very far in this direction, stripping the natural world and all of its phenomena of *qualities* and replacing them with *quantities* - that is, with number and measure. Though the path from the theoretical teachings of Anaximenes to the extremely abstract calculations of today's physicist and mathematician is very, very long, the principle is the same".[9] The path that was blazed by Anaximenes would, however, also mark the beginning of a *relinquishment,* which is characterized by Heisenberg as "deeply significan for all the following period – the relinquishment of the 'direct' understanding of qualities".[10]

Epilogue: The Milesian *'physicoi'*

The circle of the Milesian philosophic tradition closes with Anaximenes. In 494 B.C. Miletus would be razed by the Persian army, a horrible punishment for the Milesians's attempt to rebel against the dynasty. This would also mark the end of this intellectual center in which the firs philosophic and scientifi discourse was developed.

Though Anaximenes was the student of Anaximander, and Anaximander, the student of Thales, each of the three Milesian *physicoi* would develop -each one separately- a *distinct* cosmological system without rejecting the precursor's perception of the universe and without creating contention among themselves. Between teacher and student no absolute theory and dogmatic continuity are introduced, but each theory constitutes an invitation and a challenge for critical examination by those who follow. This means that from the outset there existed a spirit, if not of encouragement, at least of *tolerance* toward a *critical* theorizing of the world. "There is the historical fact that the Ionian school was the firs in which pupils criticized their masters, in one generation after the other. There can be little doubt that the Greek tradition of philosophical criticism had its main source in Ionia", Popper observes.

---

[8] A.D. Papanikolaou, *Aer: Anaximenes und Corpus Hippocraticum*, 319–326.

[9] S. Sambursky, *Das physikalische Weltbild der Antike*, 25.

[10] W. Heisenberg, *Wandlungen in den Grunlagen der Naturwissenschaft*, 12.

"To my knowledge the critical or rationalistic tradition was invented [by the Presocratics] only once. It was lost after two or three centuries, perhaps owing to the rise of the Aristotelian doctrine of *episteme*.... It was rediscovered and consciously revived in the Renaissance, especially by Galileo Galilei."[11]

Having arrived at the end of the 6th century, B.C., and looking backward, we realize with amazement and wonder that within *around seventy-fiv years* (600–525, B.C.) -a split second in the history of humanity- the three Milesian thinkers, without plain evidence, with an unparalleled power of critical abstraction and intuition, had achieved a true intellectual revolution that would prove decisive and seminal for the entire development of Western thought. The veil of myth is suddenly raised, permitting for the firs time a rational critical examination of the cosmos. Uncontrolled interventions of anthropomorphic divinities are abolished and give way to purely natural phenomena and causations. The Milesian mind founded and bequeathed to future generations a new, unprecedented way of theorizing the world that can be very generally summarized in the following points:

- Attribution of the apparent plurality and disorder of the world to a single, stable firs principle, *arche*.
- Acceptance of a deterministic, exclusively physical causation that possesses a universal character.
- Conviction that natural phenomena and natural laws are subject to the critical, rational inquiry of the human mind.
- Reduction of quality to quantity.
- Quantitative understanding of the world through number and measure.
- Use of geometrical and mechanical models in the formulation of theories.
- Use of symmetry and analogy to arrive at conclusions.
- Non-existence of a world creator.
- The firs *arche* is eternal and contains the dynamic of evolution.
- Nature is material, alive, besouled, divine, autokinetic.

The last points also mark the limits of Milesian thought. The cause of motion is not explained. The autokinetic property appears to be wanting. The lack of a creator is disturbing. Man is define as a natural entity like any other within nature. The concept of change continues to be problematic. The inherent nature of matter and spirit, matter and life, is in doubt. The relationship between the senses and reality raises questions.

The value of a philosophic system derives not only from the solutions it offers but also from the questions it posits. Both the logical interpretation of the universe attempted by the Milesians and the inevitable questions left unanswered would constitute fertile ground for the sequel in Presocratic thought, as expressed in the mystical-mathematical concepts of Pythagoras, the religious orientation of Xenophanes, the ontological dimension of Parmenides, the dynamic conception of

[11] K. Popper, *The World of Parmenides*, 23.

Heraclitus, the synthesis of Empedocles, the teleological principle of Anaxagoras, the atomic theory of Democritus.

In Ionia, the Milesians, with an unbounded faith in their powers of reason, attempted to provide the firs systematic, rational explanation of natural phenomena. Their thought thus marked the path of the *natural sciences*. In contrast, in southern Italy philosophy would be founded in the sense we know it today. Pythagoras would firs attempt a synthesis of the rational and the irrational, seeking the deeper meaning of life, while Parmenides would firs posit his ontological question 'what does *being* mean?' "The sixth century [B.C.]", Arthur Koestler would observe in his familiar graceful way, "evokes the image of an orchestra expectantly tuning up, each player absorbed in his own instrument only, deaf to the caterwaulings of the others. Then there is a dramatic silence, the conductor enters the stage, raps three times with his baton, and harmony emerges from the chaos. The maestro is Pythagoras of Samos".[12]

[12] A. Koestler, *The Sleepwalkers*, 25.

Pythagoras of Samos (ca. 570–496 B.C.)

I believe that it is the destiny of the West continually to bring these two basic tendencies into contact with one another-the critical rational on one side, striving to understand, and the mystical irrational on the other, seeking redemptive, integral experience.
In the soul of man these two states will forever reside,
the one perpetually bearing within itself,
as the seed of its antithesis, the other.

Wolfgang Pauli

Personality

Who was Pythagoras? This question might at firs seem naive, yet it is totally justified While so much has been said over the centuries about his teachings, nothing appears to be certain concerning the *individual himself.* A differentiation between what *he himself* taught, and Pythagorism during the period of its development, remains a matter of pure conjecture. The reason for this quandary is a set of inherent difficulties beginning with the lack of *contemporary* sources regarding Pythagoras's work and his teachings. He appears not to have produced any writings, or if he did, they have not survived.

All this is explained in part by the fact that one of the fundamental rules of the Pythagorian brotherhood was *silence*. "What he said to his associates, nobody can say for certain, for silence with them was of no ordinary kind".[14A8a(19)] Early Pythagorean teachings remained *unspoken* and *confidential* For this reason in our examination of the basic features of the Pythagorean teachings, setting aside the foredoomed attempt to distinguish the founder from the followers, we shall consider the *Pythagoreans* as a whole.[1] On the other side, there existed the tendency – a common trait of all religious brotherhoods – reverently to attribute all theories and discoveries to the founder, even those of other members of the school: 'He sayeth', was the formula. Contributing, of course, to this unclear picture of the man, was the legend which surrounded the person of Pythagoras from the outset, presenting him as a superman or a demigod.

Who was Pythagoras, then? Was he a man "among them. . .of surpassing knowledge who possessed a vast wealth of understanding, master of all manner of

[1] The most important Pythagoreans are considered to be Petron, Hippasus of Metapontum, Alcmeon of Croton, Philolaus, Architas of Tarentum, and Ecphantus.

C.J. Vamvacas, *The Founders of Western Thought – The Presocratics*, Boston Studies in the Philosophy of Science 257, DOI 10.1007/978-1-4020-9791-1_7,

skills",[31B129] as he would be described with unbounded admiration by Empedocles a few years after his death? Or was he "prince of cheats"[22B81] who "having made a selection of these writings contrived a learning of his own, a polymathy, a worthless artifice" [22B129] as Heraclitus would contemptuously declare? Was he the firs philosopher in today's sense of the term, "the firs to use the term philosophy and call himself a philosopher"?[2] Or, was he "not a philosopher at all, but only a religious reformer",[3] or a "great Shaman"?[4] Was "his influenc on the ideas, and thereby on the destiny, of the human race probably greater than that of any single man before or after him",[5] as Koestler believes? Or, "in spite of his unsurpassed recognition for brilliance, was his wisdom not science and investigation, but revelation and illumination",[6] as K. Reinhardt claims? Does Russell's opinion stand -that "Mathematics, in the sense of demonstrative deductive argument, begins with him"?[7] Or does that of another great mathematician, B.L. van der Waerden, who would maintain: "In a word, while we know something about the musical teachings of Pythagoras [himself], of his arithmetic we know virtually nothing, of his astronomy even less and of his geometry -if one examines it well- absolutely nothing. A pitiable result"?[8]

So, who was the 'real' Pythagoras? A wise shaman?[9] A religious prophet? A social reformer? A visionary civic leader? A teacher of ethics? The firs true philosopher? Or a mathematical genius? Let us not be drawn into a one-sided answer. While each of the answers is, in any case, unprovable, it is most likely that they are all, more or less, correct. "Everything derives", W. Burkert -one of the most penetrating investigators of Pythagoras's work- would write, "from a 'wisdom' equally and undividedly committed to the *sacred* and the *worldly*, the *rational* and the *religious* -a wisdom of one whose 'knowledge transcends' that of the common man".[10]

---

[2] F. Wehrli, *Die Schule des Aristoteles. Herakleides Pontikos*, Fr.87.

[3] E. Rohde; in W.K.C. Guthrie, *A History of Greek Philosophy. The Earlier Presocratics and the Pythagoreans*, I, 168.

[4] E.R. Dodds, *The Greeks and the Irrational*, 143.

[5] A. Koestler, *The Sleepwalkers*, 25.

[6] K. Reinhardt, *Parmenides*, 232–233.

[7] B. Russell, *History of Western Philosophy*, 49.

[8] B.L. van der Waerden, *Erwachende Wissenschaft: Ägyptische, babylonische und griechische Mathematik*, I, 168.

[9] Shamanism is not, in the literal meaning of the term, a religion, but a collection of methods of ecstasy and therapy whose aim is contact with the parallel but invisible world of the spirits and the granting of their aim in the handling of human affairs (M. Eliade and I.P.Couliano, *Dictionaire des Religions*).

[10] W. Burkert, *Weisheit und Wissenschaft. Studien zu Pythagoras, Philolaos und Platon*, 173.

Life

Pythagoras was born around 570 B.C. at Samos, 'the firs of all cities' as it is called by Herodotus, the island which would later give rise to Melissus, Epicurus, the mathematician Aristarchus. Pythagoras must have visited the famed city of Miletus on the opposite shore of Asia Minor and have come to know the teachings of the three Milesian *physicoi* (natural philosophers). Later, hardly reliable legend would have the sage visiting a series of other countries like Egypt (most likely), Babylonia, and even India. Upon his return to Samos, he would found his firs school, the 'Semicircle'. In 532 B.C.., in his maturity, fleein the oppression of Polycrates's tyranny he would leave his homeland and move by way of Delphi to Croton in lower Italy. There he would fin a populace downhearted from their recent defeat by the Locrians, indifferent toward social values, surrendering to hedonism and sloth. Pythagoras's personal élan and prestige swiftly revived a frugal life, heralding the value of virtue.

Soon he would establish a political-religious community that engaged in integral philosophic and scientifi activities, maintaining strict secrecy. During the following twenty years, this Pythagorean brotherhood would gain great power, and under his influenc Croton would extend its rule to neighbouring cities. In 510, B.C., persecution incited by the opposition party of Cylon forced Pythagoras to take refuge in Metapontum, where he died several years later. Even during his lifetime he was a figur swiftly wrapped in legend.

Pythagoras was unquestionably a supreme genius and a figur of enormous personal dynamism and influence It is said that after one of his speeches six hundred Crotonians rushed forward to enrol in the Pythagorean community without even going home to bid good-bye to their families. Whatever may be historically confirme concerning these legends, lore clearly suggests that he must have possessed extraordinary spiritual powers. He preached a new theory and a new *way of life*, with religious, moral, social and ritualistic foundations.

The Way of Life

The teaching of the Pythagoreans comprised a special *mode of living*.[11] In attempting an initial overall approach to the Pythagorean stance, we fin that with Pythagoras we depart from the Ionian tradition in which powerful personalities sought the truth in a *fi st arche* of all things, and we venture toward a new spiritual horizon in which religious belief prevails and the central occupation of the closed Phythagorean community is the purifictio and education of the soul. The Pythagorean *way* is characterized by two basic axes of thought and action which may on firs sight appear unrelated to one another but are indissolubly linked, as will be seen below. The one is *mystical salvation* and the other *scientifi research*.

[11] Plato, *The Republic*, 600b3.

The firs moral-religious direction, based on the Orphic tradition of metempsychosis, aspires to the liberation of the soul from the cycle of reincarnations in which it is caught, so that it may return to its initial purity and bliss. Spiritual '*purificatio* ' thus constitutes the core and the goal of the Pythagorean life, and from it springs a set of austere practical rules, which envision the cultivation and development of man's spiritual and moral capacities to the level of perfection and harmony with the divine.

One may well wonder, however -how is it possible to link the *salvation of the soul* to *philosophic-scientifi* pursuits? On the one hand, the above way of life can be summarized in this statement about the Pythagoreans: "Every distinction they lay down as to what should be done, or should not be done, aims at communion with the divine. This is their starting-point; their whole life is ordered with a view to following and becoming one with God, and it is the governing principle of their philosophy".[58D2] It appears dissonant, then, when on the other hand Pythagoras states that he is a 'philosopher', adding: "In life to which we come from another life and nature, some enter the service of fame and others of money, but the best choice is the one of those few who spend their time in the contemplation of nature, as lovers of wisdom, that is, philosophers".[12]

Is it possible that *philosophy and science* can lead to the soul's 'participation' (*methexis*) in the *divine*? The Pythagoreans believe that this is the only path, difficul but feasible: "Avoid the broad, much-treaded roads"[58C4(83)] of the many and ignorant, and "take the footpath"[68C6(21δ)] ... of the few and the educated".[58C6[42]] To understand this fundamental Pythagorean conviction, we must recall a number of basic conceptions that were initially already deeply rooted in their spirit: The universe is 'harmony'; it is characterized by perfection; hence, its name *cosmos*, connoting order and beauty. The *cosmos* contains the divine, and an inborn 'kinship' and 'sympathy' exists between all the beings and things in nature: "heaven and earth and gods and men are held together by communion and friendship, by orderliness, temperance, and justice; and that is the reason...why they call the whole of this world by the name of *cosmos* [i.e., order], not of disorder or dissoluteness".[13] Within the framework of this cosmological 'affinity'-i contrast with the relationship of man and the Olympian gods in Homer- is established the possibility of the soul's union with the divine. Like recognises like, and the more the relationship deepens and multiplies ties, the greater the assimilation. Thus, through his relationship with that which is divine and decorous (*cosmio*), the *philosopher* tends toward assimilation into the divine, "to become like God, so far as this is possible".[14]

In contrast with the ancient Greek mysteries in which momentary union with the divine is attempted through ecstasy in orgiastic dionysiac rites or Orphic mystery rituals, the Pythagoreans blaze a new 'pathway' that is characterized by decency, moderation, and harmony. The means of *methexis* (participation) in God, "is no

---

[12] F. Wehrli, *Die Schule des Aristoteles. Herakleides Pontikos*, Fr. 88.

[13] Plato, *Gorgias*, 507e.

[14] Plato, *Theaetetus*, 176b.

longer ecstasy or sacrament", F.M. Cornford observes, "but *theoria*, intellectual contemplation of the universal order, whereby the microcosm comes to reproduce (*mimeisthai*) that order more perfectly and becomes *cosmios* (decorous), attuned to the celestial harmony".[15] Thus, *mental activity* contributes to ultimate *salvation* and *purification* There is a cleansing of the spirit that goes hand in hand with pure knowledge. Purificatio as a religious goal takes on a broader meaning so as to embrace ethical and intellectual purity through science and music, as well as by physical bodily soundness through gymnastics and hygiene. From this affiliatio between *theory* and *praxis* the uniqueness and the glory of Pythagorean thought derive.

'Acusmatici' – 'Mathematici'

In the second half of the 6th century, B.C. , Pythagoras himself inaugurated the movement mainly as political-religious organization -that is, a school. It possesses the features of a closed community (a 'society', an 'association', a working fellowship), the members of which live and act according to common set rules in a secretive, dignified ascetic, meritocratic way. Also, they function *collectively* -a practice quite new for the Greek spirit, which has always been characterized by extreme individuality. Soon the community will become a 'society', internally structured according to merit and ability, the main aim of which was the organization of the political society and the cultivation of science and philosophy which leads to redemption. The inherent *dual* character of the Pythagorean teachings, as discussed above, would soon also be imprinted on the internal structure of the Pythagorean community: "His teaching took two forms, and thus some of his disciples were called *mathematici*, and some *acusmatiki*".[18.2(37)] The *mathematici* (advanced students; mathematicians) would be those who developed the *mathemata* (sciences), such as arithmetic, geometry, astronomy, music. The *acusmatiki* (probationers; eager to hear, from *acuo:* hear) would meditate and cultivate the mystical redemptive Pythagorean practices, the *acusmata* (oral instructions and devotional sayings), which relate to faith, not to logical proof.

'Acusmata' – 'Symbola'

For the Pythagoreans, *acusmata* (oral instructions) or *symbola* (secret codes) are sayings, rules of action, allegorical aphorisms, moral precepts, "philosophic sayings unproved and without logical connection".[58C4] Most of the sayings are very old utterances in which the rational and the irrational, the divine, demonic and the human, the body and the soul, life and death, the heroes and the dead as yet constitute a dim, sombre world that surrounds human existence. In Aristotle's view,

[15] F.M. Cornford, *Mysticism and Science in the Pythagorean Tradition*, 141–142.

the *acusmata* refer to three basic questions: 'What is it'? 'What is it, to the highest degree'? and 'What must one do'? In turn, the *symbola* are similar -allegorical, symbolical exhortations accompanied by possible later explanations.

The Pythagoreans' lives are regulated by a set of austere practical rules that are accompanied by total '*discretion*'.[16] Dressed in white, they take daily walks, pray, share ideas, exercise, observe a spare diet -all this always in agreement with the ascetic dictates of the *acusmata*- and at night they return to their homes to sleep in beds of white. Without a priestly hierarchy, they live a serene, disciplined, religious life of self-restraint and contemplation, yet not -significantly "for its own sake," as W. Nestle notes, "but only as preparation for an active life, full of self-confidence" [17]

Immortality of the Soul

In his fundamental teaching concerning the *immortality of the soul*, Pythagoras brings about a decisive *point of departure* in the history of western thought. For the Homeric man, the soul after death constitutes a mere shadow, a hazy likeness to human existence, which has disappeared along with the mortal body. The Ionian 'natural philosophers' do not as yet distinguish between soul and body. Pythagoras would be the firs in Europe to preach that it is the *spirit* and not the *fles* that is the principle carrier of human existence. Man can be considered a 'microcosm' – a likeness and epitome of the universe, the 'macrocosm'. He is composed of the material *body*, which is subject to continual change, development and decay, and of the immaterial, ever-moving *soul*, a mental essence, which has its own existence. The human soul "is *immortal* because it is a portion of the immortal divine essence from which it was detached".[58A1a(28)]

For the firs time in Greek thought, the soul is viewed as a self-existent entity over against the mortal body, which is considered its ephemeral prison. Soon the meaning of the soul will be identifie with the *self-awareness* of the individual. This signifi cant point marks the *origin* of the attempt to conceive the spirit as an indestructible entity, which is in diametrical opposition to matter. Worthy of note is the fact that this process of the division between *spirit* and *matter*, which will have a catalytic influenc on the subsequent development of European thought, did not arise out of scientifi interest or practical need, but from a deep *religious* predisposition toward spiritual purificatio and the soul's assimilation into the divine.

The soul's existence precedes the body's, and is interred in the body in order to expiate for some undefine sin. The aim of human life is for the soul to return and identify with the *eucosmia* (decorum) and *harmonia* (human and cosmological concord) of the divine universe, from which it has fallen. This evolutional path of the soul toward the divine is achieved through ascetic life and purificatio within a series of reincarnations. According to legend, Pythagoras himself 'recollected'

---

16 For the firs f ve years, Pythagoras imposed total silence on his students, probably because, 'silence', as St Symeon, the New Theologian, will later affirm 'is the swiftest road to excellence'.

17 W. Nestle, *Vom Mythos zum Logos*, 108.

that in the past he lived in different human bodies.[31B129] The notion of reincarnation is not something new. The new element is the *ethical* dimension introduced by the Pythagorean teaching. The soul's undergoing a necessary *judgement* with each bodily death, determining its subsequent reincarnation -human, animal, or even vegetable- decisively projects for the firs time the feeling of *individual responsibility*, of purificatio and salvation from sin, of kinship with the beings of the universe, of justice, harmony, and fina identificatio with the divine. The ultimate and all-encompassing aim of the Pythagorean life demands the gradual abandonment of our personal self for the sake of our union with the divine.[18]

Homogeneity

The transmigration of souls from body to body presupposes a natural world which is *homogenous*, that is of the same kind and *akin to man*. The deathless soul of every man, ephemerally incarcerated in the body, is nothing short of an 'offshoot'[19] of the eternal, divine spirit, "for there is in fact one *pneuma* (breath; spirit) pervading the whole *cosmos* like the soul, and uniting us with them".[31B136] For this reason, "one must regard all animate beings as *homogeni* (of the same genus)",[14.8a(19)] since "heaven and earth and gods and men are held together by communion and friendship, by orderliness, temperance, and justice; and that is the reason ... why the Pythgoreans call the whole of this world by the name of *cosmos* [that is,] order and not disorder".[20]

Purificatio

Man, the *microcosm*, tends through imitation (*mimesis*) or participation (*methexis*)[21] to assimilate with the *macrocosm*, thus becoming himself *cosmios* (decorous). For the firs time, man is enjoined to achieve greater purity whereby on his subsequent reincarnation he will reach a higher level of life. 'Purification' then, constitutes one of the central concepts of the Pythagorean teachings as a whole. 'Purification is not limited to cleansing bodily influence on the soul by means of *gymnastics* and *hygiene*, but it also includes cleansing of the soul through *science* and *music*.

Harmony of Soul

Purificatio comes about within a harmonious world. The soul, attuned to cosmic harmony, 'possesses harmony, or is itself harmony'.[58B41] Harmony of soul can be viewed from various perspectives. In *outward* relationship, it is the harmony of the

---

[18] Hierocles, *The Pythagorean Golden Verses*, 70–71.

[19] Plato, *Phaedo*, 113b.

[20] Plato, *Gorgias*, 508a.

[21] Aristotle, *The Metaphysics*, 9877b9, 12.

microcosm in tending to 'imitate' the macrocosm. "All of heaven is harmony and number",[58B4] and the soul 'participates', inclining toward identificatio with this harmony. Considered as *inward* harmony, the soul achieves concord in its structural elements and form: "Our soul . . . is composed of a tetrad, for it is intellect, knowledge, imagination, sensation".[58B15]

Ethics

The harmony of the soul and the concept of metempsychosis and purificatio presuppose a fundamental *ethical* attitude of the individual toward himself and toward the surrounding world – a moral approach which guides and expresses the *personal*, *social*, and *political* life of the Pythagoreans. As the center of gravity of human existence now shifts for the firs time from the mortal Homeric body to the immortal soul, the latter undertakes the vital, responsible, guiding role in the individual's everyday conduct and quest for spiritual perfection. "Let not your eyes receive the sweetness of sleep", goes one of the Pythagorean sayings, "till you examine each of your acts of the day thrice: Where did I transgress? To whom did I do good? What should I have done which I did not do"?[22] *Respect* constitutes the loftiest ethical imperative of the Pythagoreans.[23] Wisdom and temperance in private life, respect towards one's fellow man in general, and honour toward nature and life as a whole lead to an experience attuned to the universe, and thereby decorous and orderly. The *social* behaviour of the Pythagoreans also springs precisely from this respect and kinship which a person feels toward all beings - the equality of the sexes, humane treatment of slaves, protection of animals and plants, deep friendship among members of the School, which may be raised to the highest level of initiation and lead even to self-sacrific (e.g. Damon and Phintias[58D7]).

Society

Harmony -as we theorize the universe, the 'macrocosm', and seek to experience it in ourselves as 'microcosm'- must be the ultimate goal in the shaping of the *society* herself,[24] by means of conscientious political thought and praxis. Pythagoras's "motive in acquiring power", notes W.K.C. Guthrie, "was not personal ambition but a zeal for reforming society according to his own moral ideas".[25] In most of the

---

22 Hierocles, *The Pythagorean Golden Verses*, 40–43.

23 Hierocles, *The Pythagorean Golden Verses*, 1–12.

24 "The word *polis* -etymologically allied to *polos* (pole)- implies elements in opposition. The art of balancing (equibalance) constitutes politics". E. Mikrogiannakis, *Pathologia ton politeumaton stin archaiotita(The Pathology of Political Systems in Antiquity)*, 18.

25 W.K.C. Guthrie, *A History of Greek Philosophy. The earlier Presocratics and the Pythagoreans*, I, 175.

cities of southern Italy, the early Pythagorean communities unquestionably played an active and leading role up to the second half of the 5th century, B.C.

The Pythagoreans consider mathematical and political thought to be indissolubly linked with one another. Arithmetical and geometrical relations can secure the harmonious symbiosis of individuals in a progressive society. The Pythagoreans' high ethical principles govern both political thought and action: ethos, responsibility, wisdom, knowledge, frankness, justice, self-criticism, friendship, solidarity, equality, meritocracy, brotherhood, initiative, discipline, respect for the law and one's fellow citizens, whereby the idea may be transmuted into action, not by force but as the result of a free choice through theoretical discussion and, above all, by the paradigmatic life of the leaders.

Education

The 'Pythagorean way of life' is based on proper guidance and education. Intellectual awareness and participation must exist in all periods of life: children do exercises in all the scientifi fields youth apply the customs and laws of the city, men observe behaviour and regulate civic affairs, while the widely experienced elders co-operate in studies, rulings, and consultations. In this manner neither will the children be babyish, nor the youth childish; neither will the men be youngsters, nor the aged once more turn toddlers. A continual energizing of the mind and exercising of the memory must take place, allied with a healthy, trained physique, a spirit of moderation and friendship, mutual aid, self-examination and self-control, humbleness, and ethical elevation. For the firs time, education commands a dominant place in society. Also, for the firs time there is division of educational field representing distinct rational systems of knowledge: arithmetic, geometry, astronomy, music. One can say that the Pythagorean 'societies' form the firs universities of the western world, which will soon supply Italy and Greece at large with philosophers, poets, and lawgivers. Plato was right in characterizing Pythagoras as the 'leader of education'.[26]

Theology

The Pythagoreans' attitudes and way of life are profoundly *religious*. The goal of life is accord with the divine. This is the essence of their thought; in other words they believe that man is foolish to seek the good in other sources, not from the gods: "Their whole life is ordered with a view to following God, and this is the governing principle of their philosophy".[58D2] Pythagoras would pronounce a theology based on faith in a just, eternal divinity, ruler of all things, whose form and thought transcend human dimensions. It is a theology that seeks moral perfection and ultimate union

---

[26] Plato, *The Republic*, 600a9.

with the divine, envisioning impartial judgement -punishment or reward. Thus, thus for the firs time there is a system of thought that offers not only ethical content but a sense of *hope* and *purpose* in human life. Pythagoras would bring down the hitherto insurmountable barrier between mortals and the divine, as man now receives the hopeful message that through purificatio it is possible for him to become like the divine, to become -himself- godly: "Man can become divine, insofar as this is possible".[27]

Philosophy

If A. Compte-Sponville's definitio holds -'To philosophize means to think my life and live my thought'- then Phythagoras, like Socrates, is *the supreme philosopher*. The core of Pythagorean philosophy unquestionably coincides with the Orphic myth of cyclic reincarnations, purification and discrimination between the soul and the body. Pythagoras would elevate the popular Orphic mystic worship to a conspicuous social-political, religious-philosophical system. At the same time, however, he would advance to a *rational* conception of reality. Beneath the gaze of the Pythian Apollo, not of Dionysus-Zagreus, he would reform his teachings in a cosmic theory centered upon the mathematical element of numerical relationships, of Apollonian harmony, measure, and order. The Pythagorean philosophical-epistemological system would not remain static. In dynamic dialectic interplay with the Eleatic philosophy there would be developments and renewals throughout the 5th century, B.C. At the end of the century the Pythagorean teachings embodied an integral system, indissolubly uniting *religion*, *philosophy*, and *science* – an amalgam appearing for the firs time in European thought.

How did this union between religion and science -which may at firs sight appear incompatible- come about?

Many hold that the principle of *catharsis* (purification leads directly from religion to science: "Purificatio consists. . .in separating, so far as possible, the soul from the body".[28] According to the Pythagoreans, it is precisely probing more deeply into arithmetic, geometry, astronomy, music that weakens the bond between the soul and the material body, and orients the soul away from the world of the senses toward the divine cosmos of universal spirit.

Others maintain that *mimisis* (imitation) or *methexis* (participation) constitutes the connecting link between Pythagorean theology and science. Examining and revealing the order that prevails in the *cosmos*, man himself becomes *cosmios* (decorous) like the *cosmos*, and this cosmic quality brings about his oneness with the divine: "Then the lover of wisdom associating with the divine order will himself become orderly and divine, insofar as this is possible".[29]

---

[27] Plato, *TheRepublic*, 500e.

[28] Plato, *Phaedo*, 67c.

[29] Plato, *The Republic*, 500e.

Still others seek in *anamnesis* (reminiscence) the contact point between religion and science. All true knowledge, as knowledge of the forms (*ideas*) in the Platonic sense, is a memory of the soul prior to being subjugated to the earthly body. Socrates will vividly prove this through the use of geometry as an example.[30] This theory is indeed indissolubly linked to the teachings on metempsychosis, philosophic knowledge, and scientifi inquiry.

All assumptions of this sort have a common feature: they present the current of Pythagorean thought moving *from religion to science.* Might we not, however, hazard the hypothesis that from the outset in Pythagorean thought there was also a counter-current *from early science to religion?* Pythagoras certainly knew the work of the three Milesian 'natural philosophers' and very likely he knew Thales, Anaximander, and Anaximenes themselves. From his early years the order, symmetry and harmony of the world must already have influence him, which were revealed for the firs time so lucidly and tellingly in Milesian thought. And he may well have experienced something felt by so many scientists from that time to our day: a profound, intense, transcendent attraction toward the divine. Einstein was to call it 'cosmic religiousness'. His words are worth listening to, for they might well be those that Pythagoras himself would have used, had he been able to break his *silence* and speak to us today: "The individual feels the insignificanc of human desires and ambitions and the grandeur and wondrous order which is revealed in nature and the world of the spirit. He perceives his individual existence as a sort of prison and wishes to experience the wholeness which is something unifie and comprehensive.... Thus we end in a perception of the relationship between science and religion.... The serious scholars, in today's generally materialistically inspired epoch, are the unique, deeply religious, individuals".[31]

Mathematics

"I believe", Bertrand Russell was to state, "that mathematics is . . . the chief source of the belief in eternal and exact truth, as well as in a super-sensible, intelligible world".[32] Precisely this 'eternal truth' and 'super-sensible, intelligible world' would firs be sought by the Pythagoreans through mathematics two and a half thousand years ago: "The so-called Pythagoreans applied themselves to mathematics, and were the firs to develop this science; and through studying it they came to believe that its principles are the principles of everything".[33]

It is universally acknowledged that the development and elevation of mathematics to a liberal science is the exclusive achievement of the Greeks. Euclidean mathematics -formulated in the 3rd century, B.C., and remaining virtually *unchanged* till

---

30 Plato, *Meno*, 82b.

31 A. Einstein, *Mein Weltbild*, 16, 17, 18.

32 B. Russell, *History of Western Philosophy*, 55.

33 Aristotle, *The Metaphysics*, 985b24.

the 19th century A.D.- certainly contains Babylonian and Egyptian elements. Nevertheless, "the contents of Babylonian mathematics remained profoundly elementary; ... in other words, Babylonian mathematics never transgressed and crossed the threshold of pre-scientifi thought",[34] notes O. Neugebauer, probably the top authority in our time on the early history of mathematics. Greek mathematics introduce the method of strict logical proof and evolve on an axiomatic-deductive basis.

The systematic development of mathematics by the Pythagoreans, involving the turn from *practical* application to pure *theory*, unfolds in four directions, which will also constitute the firs scientifi fields *arithmetic* (theory of numbers), *geometry*, *harmonics* (music instruction), and *astronomy*. To these four *mathemata* (mathematical disciplines), we will turn immediately below.

Arithmetic

"Pythagoras ... from the beginning cultivated mathematics and numbers".[14.7] But what portion of the Pythagoreans' early work on 'theory of numbers' involved philosophical theorizing, and what portion dealt with strict deductive provable methodology? The answer depends upon the degree to which one trusts the later commentators. Scholars today attempt a careful reconstruction of the work of early Pythagorean mathematicians.[35] By 'numbers' the ancient Greeks mean exclusively *natural numbers* -that is, integer, positive numbers.[36] The Pythagoreans represent numbers *graphically*, not symbolically, using dots [.] -initially *psiphi* (pebbles)- which facilitate the *visual* comprehension of mathematics, as well as the *geometrical* exploration of *numerical* relationships. Thus, they present triangular, square, rectangular, pentagonal numbers, which are accompanied by simple arithmetical sequences. Considering that the ancient Greeks used letters to represent numbers, the advantage of this Pythagorean *geometrical* representation of numbers is evident in calculating and relating numbers. Thus, for example, if odd numbers are arranged in a series of right triangles (*gnomons*), next to the unit (one dot), squares are formed where the sum of the odd numbers equals the squaring of the numbers' sequence.

Indeed, the Pythagoreans worked exhaustively with numbers and their relationships. They would defin as 'perfect' numbers those that equal the sum of all their divisors (for example, $28 = 1 + 2 + 4 + 7 + 14$), and as 'friendly' two numbers where the sum of the divisors of the one (except for itself) equals the other number and reversely; for example, the numbers 220 and 284, where the sum of the divisors of 220: $1 + 2 + 4 + 5 + 10 + 11 + 20 + 22 + 44 + 55 + 110 = 284$, and the

[34] O. Neugebauer, *The Exact Sciences in Antiquity*, 48.

[35] Among others, T. Heath, O. Becker, B.L. van der Waerden.

[36] The zero does not exist, and the one (1), the unit, is not a number, since it does not comprise *multipleness*, which is a characteristic sign of number.

sum of the divisions of 284: $1 + 2 + 4 + 71 + 142 = 220$.[37] The theory on *even* and *odd* numbers occupies a central place in Pythagorean arithmetic, a distinction that appears directly and visually in the representation of triangular numbers where even and odd series successively alternate: 2, 4, 6, ... and 3, 5, 7, ... The set of rules that arise from this theory is contained in Euclid's Book IX (21–34). Plato would distinguish in mathematics between 'method of arithmetic' and the 'method of calculation'.[38] *Theoretical arithmetic* involves the study of 'even and odd numbers', with practical application in counting, while *theoretical calculation* contains the study of arithmetic *ratios*, with *practical* extension to calculation of fractions. Along with the study of even and odd numbers, the Pythagoreans were involved in the systematic investigation of arithmetic *ratios*, which constituted the basis of their theory on *music*, as we will see below.

Geometry

"Pythagoras transformed the science of geometry into a form of liberal studies, gaining an overview of its principles and exploring theorems abstractly and mentally".[14.6a] The Pythagoreans consider that there is a close relationship between numbers and geometrical forms. The point is regarded to have magnitude and to be identifie with the unit: "*the point* $a(=1)$, the *line* $\beta(=2)$, the *triangle* $\gamma(=3)$, the *pyramid* $\delta(=4)$".[14A13] The early Pythagoreans prove certain simple geometrical theorems, like 'the sum of the angles of a triangle equals two right angles', and the familiar 'Pythagorean' theorem according to which, 'the square of the hypotenuse of a right triangle is equal to the sum of the squares of the two perpendicular sides'. The latter rule, with its arithmetic form, was already familiarto the Babylonians a millennium before, so that today its discovery by Pythagoras is questioned. Pythagorean is considered the rule that only 6 equilateral triangles, or 4 squares, or 3 equal-sided hexagons enclose the area around a point.

The Pythagoreans also contrived three of the f ve *regular* polyhedrons: the tetrahedron, the cube, and the dodecahedron (the two others -the octahedron and the icosahedron- were discovered by Theaetetus). The sides of a regular dodecahedron are regular pentagons. The diagonals of a regular pentagon form a f ve-pointed star, which constitutes the symbol of health as well as the sign of recognition among Pythagoreans. Each diagonal of the *pentagrammon* divides the two others at the 'golden mean', that is, the ratio of the small $(\alpha - x)$ to large sections (x) of this line equals the *ratio* of the large section (x) of the line to the whole diagonal (α). This ratio $(\alpha - x) : x = x : \alpha$ leads to the quadratic equation, $x^2 = \alpha(\alpha - x)$, which the Pythagoreans knew how to solve. The inscription of regular polygons in a circle of Euclid's Book IV is considered Pythagorean. Euclid's Book

---

[37] No mathematical formula has been discovered for the calculation of 'friendly numbers'. It is believed to be above 600.

[38] Plato, *Gorgias*, 451a–c.

II of *Elements* also contains a number of *algebraic* rules -of Babylonian origin- which the Pythagoreans set down and restated *geometrically*: *algebraic* rule II.1: $\alpha(\beta + \gamma + \delta + \ldots) = \alpha\beta + \alpha\gamma + \alpha\delta \ldots$ as well as the known *algebraic* relation II.4: $(\alpha + \beta)^2 = a^2 + b^2 + 2ab$ are proved *geometrically* through the areas of rectangles. In analogy, the similar relations II.7: $(\alpha - \beta)^2 = \alpha^2 + \beta^2 - 2\alpha\beta$, and II.5: $\alpha^2 - \beta^2 = (\alpha - \beta)(\alpha + \beta)$ can be proved. On the basis of the latter, problem II.11 can be solved -the findin of the 'golden mean' of a straight line. The Pythagoreans study conical sections and establish a set of theorems on 'elliptic' and 'hyperbolic' areas that are solved by the geometrical method of 'areas' comparison.[58B20] By the same geometrical method they solve linear equations of the form $\alpha x = \beta$, quadratic equations, as well as third degree equations of the form $x^3 = \alpha$. The latter would lead to the special case of the famous problem of the 'doubling of the cube', which would be solved by the Pythagorean, Archytas.

Incommensurability

What motivated the Pythagoreans to abandon the symbolic of algebra and to turn to *geometric* figures Is it their inborn tendency to grasp meanings and relationships through *vision* or is there also another basic reason? Indeed there is. Their turn to *geometric algebra* would be an ingenious recourse of the Pythagoreans for the successful confrontation with the problem that suddenly arose with the discovery of *asymmetria* (incommensurability). The development of the theory of *incommensurability* is considered one of the greatest achievements of the Greek mathematicians. Irrational, or 'unutterable', or *a-logos* (absurd) numbers are those that *cannot* be expressed as *logos* (ratio) of two whole numbers (e.g., $\sqrt{2} = 1,41421356\ldots$ or $\pi = 3,14159265\ldots$ characterized by unlimited, non-periodical decimal places), or from the *geometric* perspective, numbers whose length is calculated to be *incommensurable* with that of the number chosen as the unit (e.g., the ratio of the circumference to the diameter of a circle [π], or the ratio of the diagonal to the side of a square [$\sqrt{2}$]). These relationships were already known for centuries to the Babylonians who, however, provided practical solutions *by approximation*, thereby missing the unique properties of incommensurable relations. The Pythagorean mind -and here precisely lies its epistemological essence- was not concerned with *practical* applications, but with the *pure conceptualization* of numbers themselves and their relationships.[58B2] Thus, within an amazingly brief period from the time that Egyptian and Babylonian mathematics were introduced in Greece (6th century, B.C.), the Pythagoreans would pinpoint this problem. The discovery of incommensurability is attributed by tradition to Pythagoras's student Hippasus of Metapontum[39] in the firs half of the 5th century, B.C.

---

[39] According to tradition, Hippasus was drowned in the sea for an act of betrayal that was never clarified either because he discovered the 'irrational', 'unutterable' numbers or because he broke the Pythagorean vow of silence and made public their teaching.

We can only conjecture concerning the manner in which this discovery came about. It could have been *arithmetic*, on the basis of the Pythagorean theory of even and odd numbers.[40] More likely, however, the discovery of incommensurability was purely *geometric*. The relationship $d=\sqrt{2}$ cannot be solved exactly *arithmetically*. It can be, however, *geometrically*. The solution is precisely, *the diagonal d* itself of a square with side of unit 1, where according to the Pythagorean theorem $1^2+1^2 = d^2 \rightarrow d^2 = 2 \rightarrow d = \sqrt{2}$. Thus, geometry -as opposed to arithmetic- *can* express *incommensurable* magnitudes without ceasing to be an *exact* science.

The geometric solution of the problem of incommensurability is conjectured to be based on the method of *antanairesis* (corresponding diminution).[41] This method is applied to an *incommensurable* relation. Thus, attempting to apply the method geometrically to the relation 'diagonal to side of square', or 'diagonal to side of regular pentagon', they would fin that the subtraction continues indefinitel without ever concluding in the 'largest common measure', thus demonstrating a 'incommensurable' relation.

Geometrical Algebra

For the Pythagoreans the discovery of *incommensurability* would constitute an astonishing, shattering experience. The Pythagorean mind, which "likened all things to numbers"[58B2] and which discerned in simple arithmetic ratios (*logos*) the deepest meaning (*Logos*) of the harmony in all things, suddenly found itself confronted with magnitutes that were *a-logos*, absurd in their relations, and thus, 'unutterable', that is, 'inexpressible'. This logical 'scandal' would nevertheless be transformed into a 'challenge' to the Greek spirit which – *in contrast* to that of the Babylonians or Egyptians – is prepared to formulate deductively and to explore inductively bold general theories. Thus the immediate result of the discovery of *incommensurability* would be the extension of the theory of analogies to irrational numbers and the development of a scientifi theory of *approximations* with increased precision that would be based on deductive rules and would lead to the theory of limits from which would arise today's differential calculus. "This method of approximations", S. Sambursky observes, "opened to science the road to a deeper understanding of mathematical and scientifi reality by demonstrating that one can only gradually approach this reality through an endless series of approximations".[42] Zeno's

---

[40] The 'reductio ad absurdum' proof referred to by Aristotle (*Organon: Analytika Protera*) is described in Euclid's Book X. It is based on the Pythagorean theorem and the rule of the early Pythagoreans, 'multiplication of two odd numbers produces an odd number'.

[41] This method known to the ancient Greeks, leads to the discovery of the 'greatest common divisor' of two numbers by the continual subtraction of the smaller from the larger, until one reaches two identical number, which constitute the largest common measure. For example, between the numbers 72 and 40, the 'greatest common divisor' is 8:72–40=32, 40–32=8, 24–8=16, 16–8=8, 8=8.

[42] S. Sambursky, *Das physikalische Weltbild der Antike*, 54.

'paradoxes' -referring to the continuity and endlessness of time and space, not as to size, but in regard to divisibility- is; in turn, a result of the discovery of incommensurability. The latter would lead to profound mathematical changes. The Pythagorean arithmetical-geometrical expression with 'dots' would be replaced by linear geometrical figures Babylonian *algebra* and Pythagorean *arithmetic* would take the form of Greek *geometrical algebra*; and the need to establish a system of concrete and permanent basic rules would lead for the firs time to a strict *axiomatic* procedure of solving mathematical problems. Consequently, the view is correct that it was not so much the discovery of incommensurability in itself, but the development and impetus to mathematical science resulting from that discovery, which constitute one of the top creations of Greek mathematics.

Music

Has anyone observed that the more one becomes a musician, so much the more does one become a philosopher? asks F. Nietzsche; and B. Russell would write: "The pure mathematician, like the musician, is a free creator of his world of ordered beauty".[43] Pythagoras would be the firs one in the history of thought to explore that 'ordered beauty' of music *mathematically* and *experimentally*. *Objectively* measuring a physical quantity (the length of a string), and through a *mental* quantitative mathematical relationship (simple arithmetic ratios), he would establish a tie to a purely *subjective* psychological-aesthetic feeling (the enjoyment of musical harmony). Musical harmony means the combining of at least two tones that evoke pleasurable feeling. Pythagoras and his students experiment (it is significan that this is the firs time that *systematic experimentation* is introduced to establish scientifi truth) with strings of varying length and tension, with wind instruments of different sizes, with brass discs of the same diameter but different thicknesses, and with identical vases fille with amounts of water at different levels, in order to determine simple *quantitative ratios* between, for example *length* of string or pipe and harmonic musical frequencies (or tons), as well as *quantitative* relations between the *pitch* of musical notes and *frequency* of string vibration. More specificall , it is believed that Pythagoras himself discovered that the most harmonious musical intervals are created by the simple numerical ratios of the firs four integer numbers which derive respectively from the relations of string length: the eighth (1/2), the fift (2/3), the fourth (3/4).[44] In these

[43] B. Russell, *History of Western Philosophy*, 53.

[44] It is said that Pythagoras conducted his experiments using a *monochord*, a musical instrument of one string stretched above a sound board with a movable bridge to divide the string into desired lengths, thus marking simple relationships between the varying length of the string and the tones produced. Dividing the string in the middle produces a tone that is an octave higher, shortening it by 2/3 sounds the interval of a fifth and by 3/4 the interval of a fourth. These simple arithmetic relationships correspond to the frequency ratio of the respective tones. Today we know that on the diatonic scale the treble note A (to which the orchestra tunes) has a frequency of 440 Hertz, C has 524, D 588, E 660, F 699, G 785, high A 880, high C 1048 Hertz, etc. The Pythagorean ratio 1/2,

simple ratios, the Pythagoreans would discover the deepest mathematical relationships: The numbers 1, 2, 3 constitute an arithmetic analogy (the difference between the firs and the second number is the same as the difference between the third and the second: 2-1 = 3-2), which is termed the *arithmetic* '*mean*' of three numbers, α, β, γ: $\beta = 1/2(\alpha + \gamma)$. The numbers 1, 2, 4 constitute a *geometric* analogy (the quotient between the firs and the second number is equal to the quotient between the second and the third number: 1:2 = 2:4), which is termed the *geometric* '*mean*' of three numbers, α, β, γ: $\beta = 1/2\sqrt{\alpha . \gamma}$. Finally, the numbers 3, 4, 6 that represent the fourth (3:4), the fift (4:6), and the eighth (3:6) constitute the *harmonic* – or *apenantia* – analogy (the quotient of the difference between the second and firs number), divided by the firs is equal to the quotient of the difference between the third and the second, divided by the third: (4–3):3 = (6–4):6, which is termed the *harmonic* '*mean*' of the three numbers, $\alpha, \beta, \gamma : 1/\beta = 1/2(1/\alpha + 1/\gamma)$.[45] The sum of the four firs numbers 1+2+3+4 is 10, which would be considered the 'perfect' number by the Pythagoreans - the number that contains in itself 'the whole essential nature of numbers',[46] the source and root of eternal nature, and would be graphically represented by a triangular figure the *tetractys*, which became for them a sacred numerical symbol.

The above musical-arithmetical relations constitute the *fi st* historical example of the *mathematical* formulation of a *natural law*. "This discovery", Heisenberg observes, "is among the most powerful advances of human science".[47] This law enables measurement of sound in space. It is significan that this firs mathematical law of physics does not refer, as is usually the case, merely to the behavior of matter, but expresses what even today is inexplicable about music – that is, a strictly *mathematical* analysis proves capable of penetrating into the purely human, *sentimental* world, a fact that would play a decisive role in the development of the Pythagorean philosophy. The deeply paradoxical *mathematical* relationship between the *material* and *psychic* world would be a profoundly shaking discovery for Pythagoras and his students, in which they would perceive the fundamental essence of reality. Beethoven once said that 'music is the sole immaterial gate to a higher world of *knowledge* that embraces the whole mankind'. This immaterial portal would lead the Pythagoreans to the unshakable belief that the essence of all things is 'number', and that 'harmony' sustains the universe.

---

an octave, represents, for example the frequency relationship middle A:high A – 440:880 (= 1/2), or middle C to high C – 524:1048 (= 1/2), etc. The Pythagorean ratio 2/3, the fifth represents, for example, the relationship A:E – 440–660 (= 2/3), or middle C:G – 524:785 (= 2/3). The ratio 3/4, the fourth, represents, for example, the relationship A:D – 440–580 (= 3/4), or C:F – 524:699 (= 3/4).

45 It is significan that the Pythagoreans would discover the *harmonic* '*mean*' geometrically in the cube: "they speak of the geometric harmony of the cube . . . for everything about a cube reflect the mean".[14A24] The cube has 12 edges, 8 corners, and 6 sides. The number 8 is the harmonic mean between number 6 and number 12, since $1/8 = 1/2(1/6 + 1/12)$.

46 Aristotle, *The Metaphysics*, 986a9.

47 W. Heisenberg, *Wandlungen in den Grunlagen der Naturwissenschaft*, 81.

Harmony

Harmony signifie the "unificatio of a multifarious composition and the agreement of unlike spirits".[14B10] The Pythagorean sense of harmony has many meanings, with mathematical, medical, psychological, aesthetic, metaphysical, and cosmological implications. We have noted the harmonious analogy of numbers in Pythagorean thought, and we will proceed at once to see that the basic property of numbers is expressed in the harmonious interplay of opposite pairs – even and odd, limited and unlimited, matter and form. As in the *macrocosm*, so in the *microcosm - harmony* assures the balance of opposing forces: "What preserves health is the equilibrium of the powers wet, dry, cold, hot, bitter, sweet, and so forth, whereas the unchecked rule of any one of them engenders disease".[24B4] Ancient Greek medicine as a whole derives from this principle, which is also extended to the psychic world: "The soul is the blending and attunement of these [opposites] in the proper and due proportions".[48] Human illnesses are treated with special melodies and rhythms; somatic and psychological sicknesses recede, and "the harmony of the psychic forces return to its initial state".[49]

The aesthetic sense of harmony is not limited to music, but is evident in ancient Greek architecture. "The Greek temple is music in stone", J. Brun would write; "the *temporal*-intervals of music and the *spacial*-intervals of architecture blend in their rhythm, evoking the wholly harmonious union of extension and duration".[50] Beyond the strictly harmonic proportions of ancient Greek temples and theaters, which constitute the peak of architectural creativity, some scholars would discern the same simple ratios in the *angle* of the axes that determine the composition of a painting. It appears that the human being possesses an optical, aesthetic responsiveness to particular angles of the straight lines of an image when the tangent of the inclination angle is equal to the simple ratios occurring in the musical tones 1/2, 2/3, 3/4, etc. Harmony sustains and rules nature itself and makes universal order possible. "Nature in the cosmos was constructed out of unlimited and limiting parts, in the order of both the whole cosmos and all things in it".[44B1]

The convergence and metaphysical transcendence of the limiting and unlimited is achieved through harmony, which imposes order and beauty in the cosmos, while it also permits a logical approach to the truth through mathematical knowledge. This Pythagorean conception concerning cosmic harmony would form the basis of Kepler's 'Harmonice mundi', Leibniz's 'preestablished Harmony', and today's physical sciences. Many scientists -among them Einstein- believe that within the framework of the above 'preestablished Harmony', the *productive* unison between the *spiritual* and the *material* world is possible. Others favor the *evolutionary interplay* between *ideas* and *environment*. Both views, however, agree on the existence

---

[48] Plato, *Phaedo*, 86b.

[49] Iamblichus, *On the Pythagorean Way of Life*, 15, 64

[50] J. Brun, *Les Présocratiques*, 32.

of a harmonic relationship, which prevails between the *natural world* and *human thought*.

The opposites

According to the Pythagoreans, the two basic opposite principles are the undifferentiated *apeiron* (the unlimited, the infinite) and the *peras* (the boundary; the limit). From the harmonic union of these is derived the *cosmos* – the well-ordered, decorous universe. On the one hand, there is the Pythagorean *apeiron* (boundless) – in direct contrast to the Anaximanderean *divine apeiron* – is identifie with the irrational, the incomplete, the false, *matter itself*, which remains unshaped and indefinite On the other hand, there is the *peras* (limit), which shapes the *unlimited*, lending *form* to matter. *Extension* in space represents the *apeiron*, while geometrical *forms* made up of points, lines, and planes establish the *limits*. The Pythagoreans' statement that "sensible bodies are actually composed of numbers" supports precisely their conception that the geometrical *form* of things that makes up 'limit' over against 'unlimited' is measurable.

This central idea, which for the firs time makes nature as a whole subject to measurement, would form the starting point of scientific mathematical inquiry. In the fiel of music, the Pythagoreans would discern and verify this principle. Mere sound are an unlimited fiel of tones, undefine and cacophonous. The moment that 'limits' are introduced into this 'boundless' in the form of simple arithmetic ratios, a harmonic concord arises, a rhythm, as within the amorphous 'boundless' 'limit' imposes measure, introducing form and beauty. This explains why "the most beautiful and the most good do not exist from the beginning".[51]

As a consequence of this archetypal pair of opposites 'limited-unlimited', the Pythagoreans would discern a series of nine other antithetical constituents: 'odd-even', one-plurality', 'right-left', 'male-female', 'at rest-moving', 'straight-crooked', 'light-darkness', 'good-bad', 'square-oblong'. It is obvious that the 'unlimited' is identifie morally with the 'bad'; mathematically with the 'even' and 'plurality'; schematically with the 'crooked' and the 'oblong', etc; while respectively analogous to the 'limited' are the 'good', 'one', 'odd', 'straight', 'square', etc. Reality springs from these dynamic, dialectical opposites.

Number

"Pythagoras ... [who was] the firs to call philosophy by that name, named as firs principles numbers and the symmetries which inhere in them, calling them also harmonies".[58B15] The Pythagoreans would see "the properties and ratios of harmonies in numbers, and since all nature appeared portrayed in numbers, and

---

[51] Aristotle, *The Metaphysics*, 1072b32

numbers to rule throughout nature, they believed that the elements of numbers are the elements of all beings, and that the universe as a whole is harmony and number".[52] Thus, they would identify numbers with being itself: "number is the firs principle of beings",[53] in all their ramifications

- as regards essence, "number is the essence of all things";[54]
- as regards the material state of existence, number is "the material of beings";[55]
- as regards properties and conditions, number "constitutes the properties and states";[56]
- and as regards causes, "numbers are the cause of being in everything else".[57]

The Pythagoreans would relate numbers to pure intangible concepts, such as: the *one* related to intellect and being; the *two* to thought; the *four* to justice ($2 \times 2 = 4$; 'equally even'); but also as the *tetractys* (the triangular figur based on the number 4), related to the whole of nature; the *fiv* to marriage; the *six* to embodiment of the soul; the *seven* to weather, light, health; the *eight* to friendship and love; while the 'perfect' *ten* "comprises in itself the whole nature of number".[58] The dominant position they would give to the number *three*: "the Pythagoreans say the whole world and all things in it are summed up in the number three; for end, middle and beginning give the number of the whole, and their number is the triad".[59] This number has, in turn, an ethical dimension: "All is three and nothing more or less than that number three; the goodness of each person three -prudence, drive, and good fortune".[36B1] Similar correspondences were expressed, furthermore, between geometrical figure or their angles and various deities. Of course, this metaphysical numerology cannot be considered the exclusive creation of the Pythagoreans. They simply incorporated in their mathematical structure primordial archetypal arithmetic symbolisms which to this day we encounter in all cultures.

Cosmogony

"The whole universe is harmony and number".[60] Pythagorean cosmogony and cosmology are based on the principle of *number.* The Pythagoreans view the universe,

52 Aristotle, *The Metaphysics*, 985b32.

53 Aristotle, *The Metaphysics*, 986a16.

54 Aristotle, *The Metaphysics*, 987a19.

55 Aristotle, *The Metaphysics*, 986a16.

56 Aristotle, *The Metaphysics*, 986a17.

57 Aristotle, *The Metaphysics*, 987b24.

58 Aristotle, *The Metaphysics*, 986a8.

59 Aristotle, *On the Heavens*, 286a11.

60 Aristotle, *The Metaphysics*, 986a2.

"heaven ... [and] nature, as composed of numbers".[58B38] This was the firs attempt to understand the *cosmos* through the help of *mathematics* – in contrast to the Ionian thinkers, who sought its firs principle (*arche*) in a *material* state of existence.

The 'cosmogonic'*arche* is the monad: "the monad [is] ... the firs principle and beginning ... of everything".[44B8] The 'arithmogonic' *arche* also derives from the monad: "the monad [is] ... the firs principle of number".[58B26] Thus we have in Pythagorean thought an *identificatio* of the genesis of the *cosmos* with that of *numbers*. "When the One had been constituted... immediately the nearest part of the Infinit began to be drawn in and limited by the Limit".[61] The attraction of the unlimited by limit – which on the biological level is expressed in the union of the male, formative principle (the *one*; *limit*) with the female (*unlimited*) – constitutes at the same time "the introduction from the infinite *time, breath*, and *void* which separates forever the places of individual things".[58B30]

The void not only separates the *places* of physical beings *spatially* but contributes to the split of the monad and the generation of the other numbers: "From the monad [is created] the indefinit dyad, ... from the monad and the indefinit dyad the numbers, and from the numbers the points, and from them the lines, from which [are made] the plane figures and from the plane figure the solid figures and from them the tangible bodies, of which the elements are four, fire water, earth, air. . .and from them becomes a besouled mental world of spherical form".[58B1a(25)] The key point here is the unbreakable bond and identificatio between the genesis of *numbers* and *world*, which also attributes to numbers a *spatial* property.[58B30] The differentiation and identificatio of material elements with geometrical regular polyhedrons would become more specific "Earth is made from the cube, fir from the pyramid, air from the octahedron, and water from the icosahedron, and from the dodecahedron is made the sphere of the whole".[14A15] A clear parallel also seems to exist between the created world and a new-born creature. Both come into being from the union of the two opposites (limit-male with unlimited-female), both originate from a source of warmth ('the hearth of the universe' – the womb), both breathe, and – if we accept the conjecture of W.K.C. Guthrie – both are cooled by inhaling.[62]

Thus, according to the Pythagorean cosmogony, the universe was created from the *One*, and as it attracts and encloses the unlimited, it *expands*, *cools*, simultaneously introducing the concept of *time* and of *space*/void. Impressive indeed is the correspondence of this image with our contemporary scientifi concept of the creation of the universe.

---

61 Aristotle, *The Metaphysics*, 1091a15.

62 "Both seed and womb are hot, and so therefore is the whole body of the new-born creature. Hence, 'immediately after birth the animal draws in breath from the outside, which is cold'. This is done in order that the heat of the body may be cooled.... In cosmogony too, therefore, one purpose of the breathing of the nascent cosmos may have been to cool this fir in order to generate the other elements; but of this the sources say nothing". (W.K.C. Guthrie, *A History of Greek Philosophy. The earlier Presocratics and the Pythagoreans*, I, 278–279).

Cosmology

The Pythagorean cosmology would introduce a series of decisive innovations. We have seen that Anaximander was the firs to 'dare' to theorize that the cylindrical earth is not supported but suspended in space. The Pythagoreans would take the next bold steps. They would displace her from the center of the cosmos and give her a place among the other planets. This would unquestionably constitute an enormous contribution toward the liberation of man from the primordial presumption that he is the center of the universe. They would for the firs time attribute *motion* to the earth. And they would be the firs to see her as *spherical* – a view that was not accepted by their immediate successors, not even Democritus. They must have based their view of the sphericity of the earth as much on *observation* (earth's curved shadow on the moon during an eclipse) as upon the *theoretical* notion of the sphere as the perfect geometrical form, derived from the regular dodecahedron.[63] The whole universe, in turn, must be spherical; having in its center no longer the earth but a fier hearth of fire which we cannot see because it is continually on the opposite side of the Earth from the one on which we live. In circular orbits around this common 'hearth of all things'[44A6] revolve in order the earth, the moon, the sun, the other planets, and finall the spherical canopy of heaven with the fi ed stars, behind which is fier *aether*, like that in the center. These two fier sources illuminate the stars and the universe. The sun is not self-luminous but serves as a reflect ve lens. Life must exist on the moon – indeed much more highly developed life, for the duration of its illumination is greater. Here the direct relationship between *light* and the *creation of life* is evident, as well as the bold idea of the existence of life on other planets.

The known orbiting visible celestial bodies[64] were nine.[58B4] In the Pythagorean philosophic system, however, 'ten' is the 'perfect' number. Consequently, a tenth planet, invisible from the earth, *had to exist*, the *anti-chton* (anti-earth). Unscientifi fantasizing? On the contrary, the Pythagoreans thus inaugurated an extremely fertile method of developing scientifi knowledge, which to this day bears rich fruit: Using a mathematical theory as a model, they predict the existence of a body, the 'anti-earth', which with the means at their disposal they cannot locate, although they believe that they already have indications of it from the eclipse of the moon. Later it will become evident that such a body does not exist.[65] Yet that does not matter; what does matter is the fact that within the framework of a mathematical theory they develop, they confidentl follow its logical consequences so as to arrive at

---

[63] The dodecahedron, more than the four other regular polyhedrons, approaches the sphere. It is composed of twelve regular pentagons in which, as we saw above, appears the 'golden mean'.

[64] That is the f ve known planets, the sun and the moon, the fi ed stars, and the earth itself, as a planet revolving about the same centre as that about which the sun and the other planets revolve.

[65] Just recently, however, on July 29, 2005, Caltech Astronomer Michael E. Brown and his team announced the discovery of the 10th planet [2003 UB313], provisionally nicknamed Xenia. The IAU, however, did not agree that the object is indeed a planet.

predictions that are subject to later scientifi verificatio and experimental inquiry. This precisely is the scientifi 'path' which would lead then and now to significan discoveries. To mention only a single example: just as the Pythagoreans would predict the existence of an 'anti-earth' on the basis of their mathematical theory, so - two and a half thousand years later, in 1928 – P. A. M. Dirac would develop the firs relativistic theory of quantum mechanics, on the basis of which he would predict the existence of 'anti-matter', a concept so absurd at that time that it would provoke another great physicist, W. Pauli, to remark ironically that 'The attempt to support this theory in its present form appears futile in the light of its consequences'. Yet Dirac would prove more fortunate than the Pythagoreans. Only four years later, in 1932, 'anti-electrons' would be discovered in cosmic radiation, thus verifying what was initially a purely theoretical mathematical prediction.

The Pythagoreans introduced for the firs time simple mathematical ratios not only *statically* between the planets' distances, but also *dynamically*, between orbiting periods, which correspond to the harmonic tones of the musical scale. This imaginary 'harmony of the spheres' would later constitute for Kepler the ideal foundation upon which he was to construct modern astronomy.

Even in its encounter with the physical world, Pythagorean scientifi thought remains deeply spiritual. The immoral soul is closely joined with the star constellations and the heavens; it originates from them and returns to them once more. The fier center of the cosmos, the 'hearth of the universe', also forms the 'dwelling-place of Zeus' and of "the mother of the gods".[14A16] Although many of these conceptions have primeval roots, the indissoluble interweaving of religious and scientifi theorizing of the cosmos and of life through a harmonious holistic vision comprises the unique achievement of the Pythagoreans.

Overview

In Zen Buddhism there is a saying: 'The moment you speak of something, you miss your mark'. Completing here our outline of the Pythagorean teachings, we are deeply aware of the relevance of this thought. Analyzing Pythagorean thought according to today's academic disciplines – theological, ethical, social, psychological, mathematical, scientific aesthetic, etc. – we merely arbitrarily fragment the unifie Pythagorean spirit into diverse uncoordinated and disjoined ideas. We miss the mark and also commit an intellectual sacrilege. For when Pythagoras characterized his teachings as 'unutterable', he did not merely mean that they were 'secret', but that they were 'inexpressible', beyond words, something that cannot be verbally articulated but is *experienced* and manifested. "No doubt there is the unutterable. This *manifests* itself, it is the mystical",[66] Wittgenstein would say. Only with this attuned, *holistic* spirit can we approach Pythagorean thought.

---

[66] L. Wittgenstein, *Tractus Logico-Philosophicus*, 6.522.

Plato's theory of 'forms' (*idea*) and the Aristotelian philosophy were decisive in the further development of Pythagorean thinking. When Bernard Russell assures us that 'he did not know any other thinker who has been as influentia as Pythagoras in the sphere of thought', he hastens to explain that Platonism, in its enormous impact on subsequent intellectual evolution,[67] is at bottom Pythagorism. This identificatio constitutes, of course, an oversimplificatio and does not lack a certain hyperbole. Certainly there was a great influenc of Pythagorean thought on the formation of Platonic physical cosmology, psychology, ethics and political philosophy, as well as on the mathematical theory of 'forms'. On the other side, however, it was precisely the Platonic dialectic and ontology that exiled the Pythagorean *scientifi* spirit for the following two thousand years.[68] "The great influenc of Platonic philosophy", Sambursky notes, "which suppressed for so long the combination of experimental method and mathematics, can in good part be explained by the intellectual mentality of the Greeks, which overvalued the importance of the deductive method to the degree that the inductive method and empirical verificatio were finall considered superfluous" [69]

Though Aristotle -as opposed to Plato- would on the one hand accept the ontological reality of the world of the senses (*enylon eidos* – form involved in matter), on the other hand he would reject mathematics as a tool for investigation and understanding of the world. According to Aristotle, number constitutes simply a quantitative determinant and lacks any ontological value, either in itself as the Pythagoreans believed, or in relation to ideas, as in Plato. Aristotle is not interested in 'how' something is dependent on other factors, but in 'why' something comes about. In turn, influence by biology, within the framework of his logical studies he would mainly emphasize systematic classification leaving 'measure' utterly neglected.

These two great philosophic systems -the *Platonic* and the *Aristotelian*- would alternately prevail in European thought for the next two millennia (Neoplatonism up to the 12th century, A.D., and Aristotelianism from the 12th to the 16th centuries A.D.), suspending scientifi inquiry and experimental measurement. The harmonious Pythagorean synthesis of mysticism and science disappears. Neoplatonism overemphasizes the mystical, ridiculing scientifi research. Aristotelianism separates mathematics from physics, placing the latter in the realm of theology. The Earth is restored to the center of the universe and is denied any motion whatsoever. The cosmos ceases to be viewed as a besouled organism, and God is located outside it. Medieval scholasticism falls prey to the additional error of lending to

---

67 Alfred North Whitehead stated that the most accurate view of the overall development of the European philosophic tradition is that it is in essence a series of footnotes to Plato.

68 Indeed, with the exception of mathematics and a few brilliant flashes as in the work of Euclid, Archimedes, Aristarchus of Samos, a total scientifi darkness prevailed between the Pythagoreans (5th century, B.C.) and the firs Renaissance thinkers, Copernicus, Tycho de Brache, Kepler, Galilei (16th century, A.D.).

69 S. Sambursky, *Das physkalische Weltbild der Antike*, 67.

Aristotle's teachings the dogmatic character of the irrefutable truth, which is not subject to any subsequent examination or improvement. Thus, it is not exaggeration when Whitehead states: "Except in mathematics, the men of the Renaissance practically started from the position which Archimedes had reached"[70] eighteen centuries before.

Physical Sciences

With the decline of Aristotelianism in the 17th century, the Pythagorean spirit revives. Mathematics regains its importance and decisively influence philosophy as much as science. Through mathematics, Kepler, Galilei, Descartes, Huygens, and Newton postulate physical laws that reflec the inherent order of the universe. The mathematization of physical reality would reach its apex in the 20th century. In the words of Heisenberg, one of the pioneers in contemporary physics, "The successes of this mode of observing nature, which led in part to a true dominion over natural forces and thus contributes decisively to the development of humanity, in an unforeseen manner vindicated the Pythagorean faith".[71] This vindication, as we will see below, is versatile and functions on many levels:

"The universe. . .[and] nature consist of numbers."[58B32]

Twenty-one centuries later, Galilei would announce in this Pythagorean spirit that "the great book of nature is written in the language of mathematics". "Measure whatever is measurable, and whatever is not measurable, render it measurable", he would state, following the path which Pythagoras firs laid down toward the mathematical investigation of the natural world. The belief of the Pythagoreans that "sensible bodies are actually composed of numbers", and that all properties and mutations, as well as causes, can be expressed in numbers would constitute the foundation of the mathematization of nature on which the natural sciences have been based to this day.

"All things that can be known have a number."[14B4]

Presocratic philosophy bequeathed to contemporary physics two central ideas: that of *Democritus* regarding the composition of matter by indivisible particles, 'atoms'; and that of *Pythagoras* regarding the conception of reality by means of arithmetical relations. In mid-20th century, physics would bring a 'new rise of Pythagorism' and would "finall lead one to a universal perception," observes physicist B. D'Espagnat, "in which the material essence of things appears to dissolve into equations; into a view whereby materialism is increasingly obliged to develop toward mathematization, and, if one may put it thus, Democritus seeks asylum in Pythagoras".[72] Heisenberg agrees: "The contemporary abstract concept of the atom and of the mathematical forms used to represent the complexity of phenomena in today's atomic theory leads . . . to the fundamental thought of the *meaningful power*

---

[70] A.N.Whitehead, *Science and the Modern World*, 8.

[71] W. Heisenberg, *Wandlungen in den Grundlagen der Naturwissenschft*, 83–84.

[72] B. D'Espagnat, *Auf der Suche nach dem Wirklichen*, 11.

of mathematical structures. This belief of ours clearly firs appeared in the teachings of the Pythagoreans".[73]

"The whole of the universe is harmony and number."[58B4]

The Pythagoreans extended musical harmony expressed in mathematical relations to a *universal* law of order and beauty, expressed by simple mathematical laws. According to contemporary quantum theory, for every particle there is a corresponding harmonic movement described by a wave function. "The Pythagoreans' harmony", notes Heisenberg, "which Kepler believed he found even the planetary orbits, Physics seeks from the time of Newton in the mathematical structure of the *dynamic* law, in the equation expressing that law. This turning point signifie a consequent realisation of the Pythagoreans' program, so that thereby the infinit complexity of natural events find its exact mathematical expression in the multiple solutions of an equation. ... This discovery of the mathematical dependence of harmony is one of the most powerful impetuses ever given to human science".[74]

"Pythagoras. . .[considers] numbers fundamental and their inherent symmetries, which he also calls harmonies."[58B15]

The concept of 'symmetry' in nature is of purely *Greek* origin.[75] Plato would raise this Pythagorean principle to the most fundamental structural law of the universe.[76] As with Pythagoras and Plato, so in contemporary Physics, 'symmetry' constitutes the fundamental basis for understanding the universe. Its ultimate goal of containment and unificatio of all elementary particles and natural powers in *one* mathematical formula depends precisely on the discovery of the proper 'symmetry'[77] which would determine the final universal mathematical equation. In the firs infinitesima instant of the world's creation (10^{-43} sec) the highest form of symmetry was maintained. In the sequel, with the gradual fall of temperature and successive dynamic breaking of the symmetry, the plurality and the shaped structure of today's natural world would be created.

In his book, *Symmetry, the Architectural Design of Nature*, theoretical physicist H. Genz introduces a series of examples showing that symmetry constitutes a fundamental principle: "Natural laws obey the rules of *symmetry*. *Symmetry* determines which particles can exist and which interactions can take place between

---

[73] W. Heisenberg, *Wandlungen in den Grundlagen der Naturwissenschaft*, 81.

[74] W. Heisenberg, *Wandlungen in den Grundlagen der Naturwissenschaft*, 82–83, 81.

[75] F. Capra states in this regard: "The attitude of Eastern philosophy with regard to symmetry is in striking contrast to that of the ancient Greeks. . .The concept of symmetry does not seem to play any major role in their philosophy. Like geometry, it is thought to be a construct of the mind, rather than a property of nature, and thus of no fundamental importance". (F. Capra, *The Tao of Physics*, 245).

[76] Plato, *Timaeus*, 69b.

[77] The concept of 'symmetry' from the purely mathematical and geometrical point of view has been expanded in contemporary physics to involve various physical 'inner' symmetries, as well as 'gauge' symmetries, that do not act on the points of ordinary space. Basically, when the equations which describe a physical system remain unaltered in a group of transformations, we say that this group comprises a 'symmetry' of the system.

them. Through *symmetry* we can foresee what natural laws should be like, without knowing the laws themselves. The *symmetry* of natural laws determines that magnitudes of like energy and electron charge are not changed in the course of time, or that, for example, the [elementary particles], quarks and gluons are forever bonded together".[78] The impressive effort in recent years to discover a 'Grand Unifie Theory' (GUT) is based on the fundamental principle that *symmetry* exists in the universe. Physicist E. Wigner's points out that in the future physical laws will not determine the symmetries, but the symmetries the physical laws.

Pythagoras was the firs to teach the fundamental importance of reflectio on abstract ideas, introducing *number* as the determining factor in the periodicity of musical tones. "The importance of the abstract idea of periodicity was thus present at the very beginning both of mathematics and of European philosophy", Whitehead observes. "Truly, Pythagoras in founding European philosophy and European mathematics, endowed them with the luckiest of lucky guesses – or, was it a flas of divine genius, penetrating to the inmost nature of things"?[79]

"From the infinit is introduced time and breath and void."[58B30]

According to the Pythagoreans, time and space do not pre-exist independent of matter, but are created upon the genesis of the universe, as "the parts of the infinit nearest [to the unit] at once began to be drawn in and limited by the limit".[58B30] Today we encounter this position -that space and time spread, 'stretch out' simultaneously with the expanding universe- in Einstein's General Theory of Relativity, according to which space-time geometry and matter defin each other. "When Einstein and his followers proclaim that physical facts, such as gravitation, are to be construed as exhibitions of local peculiarities of spatio-temporal properties", Whitehead notes, "they are following the pure Pythagorean tradition".[80]

"The void divides the nature [of the numbers]."[58B30]

The Pythagoreans consider the 'void' to be the means by which numbers are separated. In contemporary mathematics the void, in the form of irrational 'gaps', enters among the rational numbers of the Dedekind 'cuts'.

"Pythagoras transformed the science of geometry into a form of liberal education."[14.6a]

"Everything is geometry' proclaim today's supporters of the General Theory of Relativity".[81] Einstein hoped that he would be able to reduce to a common denominator all dynamic field through the interrelation of matter and spacio-temporal geometry.

"It is evident that the universe and everything in it were shaped from the elements of limit (finite and the unlimited (infinite)"[14B]

---

[78] H. Genz, *Symmetrie, Bauplan der Natur*, Introduction.

[79] A.N. Whitehead, *Science and the Modern World*, 48.

[80] A.N. Whitehead, *Science and the Modern World*, 36.

[81] B. D'Espagnat, *Auf der Suche nach dem Wirklichen*, 12.

We have seen how the Pythagorearn 'limit' *shapes* the 'unlimited' -rendering it measurable, beautiful and harmonic. The Pythagoreans would also verify this spatial principle in music. In a parallel fashion, the quantum theory of physics would introduce a 'limit' into the up to then unformed *energy*, definin a measure -Planck's constant- for the understanding of the microcosm. A form of motion marks each elementary particle, with a characteristic frequency and a definit amount of energy.

"The opposites are the origin of beings."[24A3]

As with the Pythagoreans, we fin opposite pairs in contemporary physics, like matter-antimatter, positive-negative electrical charge, etc.

> *"The experiment was then extended to various instruments, that is, to the vibration of vases, to tubes, to reeds, to mono-chords, to triangles, and the like. And in all these was found a consistent harmony in ratio with numbers."*[82]

The deficiencie most often attributed by certain scholars to ancient Greek philosophy are two. The firs is that the ancient Greeks did not use *experiment* to verify their theories. This, however, does not hold for the Pythagoreans. In spite of the rudimentary technical means at their disposal, they arrived at the laws of musical harmony through a series of systematic experiments, observations, and measurements.

> *"They agreed with others that the ratios lay in movements, a swift motion being high-pitched since it produces a continuous succession of blows and stabs the air more sharply, and a slow motion low-pitched because more sluggish."*[47A19a]

The second inadequacy attributed to ancient Greek thought is that it was limited to a *static*, spatial view of nature, ignoring her dynamic properties in space and *time*. In this, too, the Pythagoreans were the exception: They attempted -again with the scanty means at hand- to introduce the element of *time* into their experiments -to establish, that is, a *dynamic* interpretation in their theory of harmony.

> *"Pythagoras was the fi st to apply the name cosmos [well ordered; decorous] to the world in recognition of the order which is displayed."*[14.21]

The Pythagoreans discovered the harmony and order of the universe through the simple harmonic mathematical relations in music. The order and beauty of the universe also constitute the fundamental belief of contemporary physicists. It is significan that two of the greatest physicists of our age end their books with similar phrases: "I think [that] nature has a simplicity and therefore great beauty",[83] writes Feynman. And Heisenberg adds: "Through the understanding of the atomic structure of matter by means of simple mathematical properties of symmetry . . . one arrives at structures of utterly rare simplicity, wholeness and beauty – structures which seem especially meaningful, for they no longer have to do with one specialized area of physics, but with the entire universe".[84]

---

82 Iamblichus, *On the Pythagorean Way of Life*, 26, 119.

83 R. Feynman, *The Character of Physical Law*, 173.

84 W. Heisenberg, *Wandlungen in den Grundlangen der Naturwissenschaft*, 183

Philosophy – Holism

'Divine genius', above all epithets, may best epitomize the personality of Pythagoras. Pythagorism played a crucial role in the advancement of European thought. It would draw the decisive and irreversible *dividing line* between western and eastern thinking, for the firs time introducing measurement, "measure",[85] as a means of investigating the world. This would be done, however, on the basis of three presuppositions:

(a) 'Measure' is not something absolute, independent of conditions. It remains at bottom, as the ancient Greek sages put it, the *metron* aris *ton* (the excellence lies in the golden mean) which expresses *inner* relationship and proper analogy: *harmonia, logos – harmonia* as spiritual, aesthetic balance; *logos*, not as a mathematical relation alone, but as an expression of the ordered reason and symmetry of the universe.
(b) 'Measure' does not necessarily lead to the 'truth'. The Pythagoreans have such a profound understanding of the *limits* of measurable knowledge, that for them it merely constitutes a means of *approaching* absolute truth, which is unutterable, inexpressible. "About invisible things as well as the earthly, the gods alone are certain; we humans are allowed only to conjecture, by means of clues",[24B1] they would say.
(c) Universe and man comprise a unifie and unbreakable whole. Religious, philosophic, scholarly, scientific and aesthetic concerns are *one and indivisible*, indissolubly linked to the spiritual, ethical, political experience of the individual, within the framework of universal order and harmony, as determined by 'measure'.

These three basic canons of the Pythagorean teaching

(a) 'measure' as inner and universal relationship,
(b) the incapacity of rational thought to discover the absolute truth, and
(c) the holistic view of the *cosmos*

would soon be misinterpreted. During the medieval period and up to the time of Kant, inspired spirits like Augustine, Aquinas, Descartes, Spinosa, Pascal, and Leibniz would connect rational thought to religious feeling and moral order. However, from the enlightenment onward, although the Pythagorean mathematical spirit revived, the three basic rules that accompany it were driven into complete oblivion, as

---

[85] D. Bohm states categorically, "It is clear that the different ways the two societies have developed fi in with their different attitudes to measure. Thus, in the West, society has mainly emphasized the development of science and technology (dependent on *measure*) while in the East, the main emphasis has gone to religion and philosophy (which are directed ultimately toward the *immeasurable*". (Bohm, D. *Wholeness and the Implicate Order*, 23).

scientifi research aimed mainly at man's dominion over nature in accordance with Bacon's slogan, 'knowledge is power':

(a) Measurement often becomes thoughtless and ontologically *a-logos* (ir-rational) in that it violates the crucial *inner relation and meaning* (*logos*) of things. Our time, with its ubiquitous mathematization of everything, seems to have forgotten the other dimension of Pythagorism -the ethical, social harmonious 'measure' taught by Pythagoras. And this is the cause of contemporary tragedy: "Our *measurements* become ever more *without measure*", notes J. Brun. "Facilely concluding that 'to calculate is to know', we arrogantly proclaim ourselves 'lord and master of nature' ".
(b) Thinking in numbers leads to the notion that mathematical symbols comprise the ultimate objective reality. As a consequence, mathematical concepts and relations are no longer viewed as suggestive forms but are considered absolute expressions of reality.
(c) Indifference toward the search for inner measure and relation, along with the real need for thorough investigation of scientifi questions, fragments scholarly and scientifi pursuits into innumerable specializations. Theology entrenches itself in its own dogmatic world, alienated from human and social problems, viewing the progress of physical science as a quasi threat. Mysticism appears to lack a steady line of development, since the mystics can agree with one another in regard neither to their ethical position, nor to their stance toward science and politics.[86] Philosophy, which once embraced all the sciences, goes off on her own to the extent that she is scorned by and alienated from the other disciplines.

At the beginning of the third millennium A.D., contemporary man is beginning to realize the consequences of the excessive quantificatio and unwise fragmentation of knowledge. Physicists, biologists, sociologists, philosophers, are beginning to ask whether there may be a way of arriving at a fuller synthesis of truth, perhaps of the Whole, which contains the observations and experiences of man rather than simply the sum of fragmented knowledge. Could it be that the fragmented character of scientifi investigation does not constitute the most rewarding means of comprehending the whole of the world? We are beginning to hear from all quarters an intense demand for a renewed, unifie theorizing about the world.

(a) As regards the mathematization of nature, it is true that today all the knowledge we have about the world is expressed in mathematical forms and mathematical symbols. Nevertheless, it has been realized that mathematical relations cannot relate to the essence, the 'whatness' of an entity, but merely to the 'how' of its behavior, and that in a symbolic language. When Newton was originally

---

[86] A. Schopenhauer, F. Nietzsche, M. Weber, E. Bloch. (See P. Giatzakis, *I Kritiki tis mystikis empeirias sti neoteri skepsi (Critique of Mystical Experience in Modern Thought)*, 69–76).

asked about his Law of Gravitation and his famous equation $F=G.M.m./r^2$, he said, "I have told you *how* it moves, not *why*".[87] Physicist D. Bohm notes: "Our theories are not 'descriptions of reality as it is' but, rather, ever-changing forms of insight, which can point to or indicate a reality that is implicit and not describable or specifiabl in its totality".[88]

(b) Bohm continues: "Reality, however, is the whole. Fragmentation does not represent reality itself; . . . it only constitutes a manner of thinking about things."[89] Along with Dürr, Schrödinger was to write: "Plurality is various reflection of the One".[90]

(c) It is evident that science, religion and philosophy are moving once more on *converging* tracks, while during the entire course of the 19th century and the beginning of the 20th century they appeared to be diverging to ever-greater distances. The theory of relativity and quantum physics instigated the firs inpulse in this new direction. "The structures of mathematical physics", writes B. d'Espagnat, "constitute a knot in which man and Being meet. And in this way 'theories' arise for the former regarding the latter – uncertain and mysterious, to be sure, but not illusory".[91] One can say -paraphrasing Popper's appeal[92]- that contemporary thought is calling 'Back to the Pythagoreans'! Of course, 'back to the Pythagoreans' does not mean the literal adoption of their teachings, but openness to their way of thinking and retrieval of the lost '*inner measure*', *logos* (rational relation), and *'harmony'* that sustain the world.

The loftiest conception of the harmonic symbiosis of the natural and the transcendental inner spirituality and scientifi rationality was achieved for the firs time in the history of European thought by Pythagoras. What was said of Heisenberg surely also holds for Pythagoras: "The quest for the deepest well-spring of all understanding was for him the common starting-point of religion and science".[93] Pythagoras would also agree wholeheartedly with Einstein: "The religiousness [of the researcher] derives from his ecstatic wonder before the *harmony* of natural laws which reveal such a superior *Mind* that, in contrast, the entire sum of human wisdom and scientifi classificatio constitutes a trivial glimmer".[94] The words, however, which perhaps reflec the immortal spirit of Pythagoras in the most eloquent way, were those of Heisenberg in a manuscript written in the fall of 1942 and published after his death: "The most significan areas of pure science are those in which there

---

[87] Cited in: R.P.Feynman, *The Character of Physical Law*, 37.

[88] D. Bohm, *Wholeness and the Implicate Order*, 17.

[89] H.P. Dürr, *Vorwort, Physik und Transzendenz*, 14.

[90] H.P. Dürr, *Vorwort, Physik und Transzendenz*, 14.

[91] B. D'Espagnat, *Auf der Suche nach dem Wirklichen*, 165.

[92] K. Popper, *Back to the Presocratics*.

[93] W. Heisenberg, *Wandlungen in den Grundlagen der Naturwissenschaft*, XXX.

[94] A, Einstein, *Mein Weltbild*, 18.

is no longer any concern for applications, in which pure intellect, chiefl , traces the hidden harmonies of the universe. This most inner region, on which science and art can scarcely be distinguished from one another any longer, is perhaps for mankind today the sole place that reveals the truth with utter clarity . . . through which unfolds a glimpse of portions of the universe about which one can speak only in parables".[95]

[95] W. Heisenberg, *Letter, Fall, 1942*. Cited in: H.-P Dürr, *Physik und Transzendenz*, 18–19.

Xenophanes of Colophon (ca. 570–470, B.C.)

Tragic man is the opposite of the man
who is engaged in the harmony of the universe and at one with the necessity of nature.
This means that a different sort of freedom
has already arisen within him...
Thus, we no longer have man simply in his biological dimension,
but the person in his spiritual form.

Christos Malevitsis

Life

Xenophanes was a poet. A friendly sojourner of scant means, "tossing about . . . from city to city", he was well received in aristocratic circles of western Greece to whom he recited his poems. His temperament was Ionian; restless, curious, many-sided, critical as well as biting, he would rightly be considered the head of the ancient enlightenment – the man who would trace new paths in crucial theological, philosophical, and gnoseological areas.

He was born in 570, B.C. , at Colophon, Ionia. It was said that he was Anaximander's student and the teacher of Parmenides, whose works, however, would later influenc his own thought. He left his homeland at the age of twenty-f ve, fleein the 'abominable' of the Persians. Wandering thereafter through cities of Sicily and lower Italy reciting poems, it is likely that at the Syracuse court of the tyrant, Hiero, who was friendly to the muses, he meet the poets Pindar, Simonides, Epicharmus and Aeschylus. He would receive – without fully embracing – the influenc of Pythagorism, which he would strive to combine with Ionian cosmology. He died extremely advanced in years, around the age of one hundred.

Personality

Xenophanes was both a poet and a profound thinker who through his elegies, epics and the *Silloi* communicated his philosophical, theological, ethical, and cosmological teachings. Xenophanes "philosophized through his poetry".[21A18] In contrast to the three Ionians, who wrote in prose, as well as Pythagoras, who communicated through the spoken word alone, Xenophanes would write in meter, facilitating memory, and increasing the range of his teachings. As for his ideas *Peri physeos* (On Nature) it appears that they were mainly based on the rich tradition of the Ionian

C.J. Vamvacas, *The Founders of Western Thought – The Presocratics*, Boston Studies in the Philosophy of Science 257, DOI 10.1007/978-1-4020-9791-1_8,

physicists as well as his own observations. Xenophanes was above all a *reformer*. Decisively and optimistically he would be the firs to challenge man to take his future and fortune in his own hands, seeing that personal and social progress are *strictly determined by man himself*, setting, at the same time, the proper limits, according to 'due measure'.

Cosmology

Xenophanes's natural philosophy appears at firs glance to derive from a conscious *reaction* to the deductive, Ionian thinking about the world, which appears fragmentary and often arbitrary. He aims firs at *demythologizing* and secondly at *explaining* natural phenomena. This would lead Reinhardt to note, not without exaggeration, that Xenophanes's thinking about nature is 'crudely empirical' and "should, compared to the cosmological thinking of Anaximander, appear inconceivably retrogressive, if one does not do injustice to both in hazarding such a comparison".[1] Xenophanes must have been a very capable observer. He would rightly attribute the saltiness of the sea to the presence of 'mixtures' (salts we would call them today) that have accumulated in the sea after being washed off through the ground: "He says that the sea is salty because of the many mixtures fl wing in it. ... He thinks that a mingling of earth with sea takes place, and that in the course of time it is dissolved by the wet element".[21A33(4)] Here we have a surprising early recognition physical chemical processes such as *mixing, washing, extraction, filt ation, dissolving, evaporation, condensation*.

For Xenophanes there is neither a cosmogonic temporal origin, nor a creator. Therefore he is also not concerned with specifying any material 'firs cause', which we saw in the Milesian natural philosophers. There are 'infinit universes', but -contrary to Anaximander's cosmology- they are all the same, 'preciselysimilar'.

Biology

The same powers of observation that Xenophanes exercises to draw scientifi conclusions are also found in his examination of living things. He would consider light to be an indispensable condition for the existence of earthly life, indirectly recognizing that the moon is not self-illuminated: "The sun is useful in generating the cosmos and living creatures thereof, but the moon is redundant".[21A42] It is significan that the thinker speaks not only of the generation but also of the *organization* of living beings, believing that they are contained in an *organized ecological* system. From the examination of a number of specifi sea organisms remaining petrifie on land,[2]

---

[1] K. Reinhardt, *Parmenides*, 145.

[2] "He claims as proofs that shells are found in the midst of the land and on mountains; and in quarries at Syracuse, he says, the impressions of a fis and of seaweed have been found, on Paros

he would conclude that where land now exists, once there was water, and given the familiar Ionian tendency to generalization, he would surmise that land and sea must be involved in a perpetual competitive cycle, the land sometimes dominating, draught leading to the temporary disappearance of life, and sometimes the water, whereby life would be regenerated out of the mud (the 'clay'). "For we are all born out of earth and water",[21B33] he would say. "All that is born and grows is earth and water".[21B29] And he would conclude: "From earth come all things, and all things end in earth",[21B27] recalling the familiar Christian saying, "Dust thou art, and to dust shalt return".[3]

Physics

We can ask ourselves, nevertheless: was Xenophanes's scientifi thinking really so simplistic and naïve? We have already seen that many of his conclusions are based on solid *observation*. A closer examination, however, leads to the new understanding that Xenophanes wished to develop a *unifie* cosmology based on the concept that all of the heavenly bodies and meteorological phenomena are in fact *clouds* of varying density and heat: "the sun is composed of ignited clouds;[21A40] the moon is a condensed cloud;[21A43] the stars are made of ignited clouds;[21A38] all these things (comets, meteors and falling stars) are systems or mov+ements of ignited clouds;[21A44] and what men call *Iris* [a rainbow] is by nature a cloud;[21B32] lightning occurs when clouds are made bright by movement;[21A45] those phenomena that appear on ships like stars [St. Elmo's fire which some call *Dioscuri*, are little clouds".[21A39]

This *unifie* cosmological perception of Xenophanes appears to set him apart from the ideas of the Ionian 'physical philosophers' in the sense that while they referred to a material firs cause (e.g., water, air) as the origin of everything, Xenophanes would view the various heavenly bodies and atmospheric phenomena mainly as *a common sort of manifestation of clouds* -as events, occurrences, or phenomena, rather than things. Xenophanes was the firs to advance to a bold generalization, relating a series of separate observations to a single, all-encompassing theory of 'clouds'. This 'inter-theoretic identity'[4] (A is in reality B), as formulated for the firs time in the above fragments by Xenophanes, in which 'sun', 'moon', 'stars', etc. (A) *are* in reality 'clouds' (B), constitutes today a fundamental mode of contemporary scientifi expression, like, for example, 'light', 'X-rays', 'radio waves' (A) *are*, in reality, 'electromagnetic waves' (B).

---

the impression of a bay leaf in the depth of the stone, and at Malta flattene shapes of all sorts of sea creatures".[21A33(5)] This reference is considered to be the *fi st* scientifi geological observation on the continent of Europe.

[3] Old Testament, *Genesis*, 3, 19.

[4] A. Mourelatos, *"X is realy Y": Ionian Origins of a Thought Pattern*, 288.

Theology

Xenophanes's was a deeply religious spirit. At the same time, however, he possessed a rational, innovative mind, which introduced the firs enlightenment in Greece. The result of this combination of attitudes would be the rejection of what had been till then the conventional religion and the establishment of a belief in a single divinity ('God is one') who is "in no way like mortals either in body or in mind".[21B23] Xenophanes's merciless attack upon the Olympian gods was conducted in a critical, probative way, and therefore is amazingly convincing. His rational argument would be based on two essential points, which concerned *appearance* and *ethos*.

The anthropomorphic character of the gods could not be authentic, but is clearly a human invention. "Men suppose that gods are brought to birth, and have clothes and voice and shape like their own".[21B14] That this conception is purely subjective and relative is proved by two unshakable arguments: Among men, "the Ethiopians imagine their gods as black and snub-nosed, Thracians as blue-eyed and red-haired";[21B16] and, "if oxen and horses or lions had hands, horses would draw the gods shaped like horses and lions like lions, making the bodies of the gods resemble their own forms".[21B15] Nor could the ephemeral life of man apply also to god. "The same sacrilege is committed by those who say that the gods are born as by those who say that they die; in both cases it is as if they do not exist for a period of time".[21A12] The mythological conception that the gods can be born yet are immortal at the same time is absurd. Whatever is born will inevitably die.

Nor "does it befi god to go here and there at different times",[21B26] as the Olympian gods are wont to do. For god "without toil . . . makes all things shiver by the impulse of his mind".[21B25] In turn unsuitable to god are human weaknesses: "Both Homer and Hesiod have ascribed to the gods all deeds that among men are a reproach and disgrace: thieving, adultery, and mutual deception".[21B11] Finally, on the basis of logical argumentation he would also reject the polytheism of traditional religion. Polytheism presupposes a hierarchy and consequently degrees of submission, which are not in accord with the omnipotence of god. For if there were two or more gods, they would either be *unequal,* whereby one would be inferior to the other -a condition unsuitable to a god- or they would be *equal*, possessing like powers, which again is not suitable to the nature of a god, who must be 'almighty'.

Xenophanes's theology is fundamentally an *apophatic* (negative) one. As we have seen, he rejects the association of polytheism, movement, immorality, temporariness, and anthropormorphism with the divine. What, then, is the nature of god? First of all, Xenophanes would realize that the answer to this question rests not upon *knowledge*, but upon *belief.* "Certain truth has no man seen, nor will there ever be a man who knows about the gods and about everything of which I speak; . . . but in all things there is opinion",[21B34] he would state. Thus, Xenophanes believes in *one* god who is the sole divinity: "God is one.";[21B23] everlasting: unborn and immortal; "unchanging and rational";[21A35] of one substance, without specialized anthropomorphic organs,[21B24] and thus, according to this conception, like a sphere;[21A28(7)] "omniscient";[21A31(9)] "omnipotent".[21A31(3)] God is the all-powerful. For the firs time we have the conception of god ruling the universe exclusively with his 'mind': "Effortlessly he sets everything in motion through the power of his thoughts";[21B25] "most excellent in all things";[21A31] ethical. Xenophanes's God is perfect and above all spiritual, yet -being transcendent- is also inconceivable and incomprehensible.

Xenophanes was the thinker who "would conceive a unifie cosmos",[21A30] as Aristotle puts it. The allusion in some of his lines to 'gods' (plural), as for example, "It is good always to hold the gods in high regard",[21A30] does not express polytheism, but suggests that Xenophanes's 'one God' -derived from his philosophic vision- is not a god to be worshipped, nor does he replace the many gods of traditional religion, who -along with human beings- are of an inferior category: "God is one, greatest among gods and men".[21B23] The phrase 'among gods' is used to distinguish the 'one god' who stands out in sharp contrast to lesser beings. While Xenophanes's *monotheism* is generally accepted, opinion is divided as to whether the 'one god' is to be identifie with the universe.

Anthropocentric Theory

Xenophanes's god is totally *uninvolved* in human actions. This constitutes a fundamental new view of man's role in both his personal actions and in his contribution to the development of civilization. Up to this time, man believed himself to be subordinate to the powers of the divine. In the Homeric epics there are numerous instances of the intervention of the gods in the lives and actions of individuals. Hesiod heard the word of truth from the muses, and the goddess would take Parmenides by the hand to grant revelations. People prayed in expectation of the 'blessings' of happiness and wealth, the 'reward' of goods, honor, wisdom and glory from the gods, while they also feared 'Justice' – divine judgment and punishment. Man reached the point of negotiating with the divine so as to gains good fortune. As Epicharmus would write, "In payment for our efforts, the gods grant us all good things".[23B36]

Xenophanes would be the firs European thinker to reject the involvement of god as supervisor, or instructor, or instigator, or benefactor, or collaborator in human affairs, thus rendering man exclusively responsible for his actions. So, he would point in a new direction, leading for the firs time to the revelation of the Ego and the lifting of man to self-awareness and self-confidenc in his strength and role as a personality. No longer viewed as merely a biological entity, man is established as a *person*, as the free and responsible creator of his historical destiny. This bold, innovative position of Xenophanes would prove decisive in the entire subsequent shaping of European vision.

Progress

Xenophanes's liberal spirit was not limited, however, to the denunciation of the traditional deities as the determining factor in man's life. He would attempt yet another revolutionary stride, pioneering in European thought the idea of *progress in civilization* as the exclusive result of *human striving*. "The gods do not reveal everything to men in the beginning, but in the course of time, by searching, they fin out better".[23B18] This bold conception of man as the exclusive creator of all spiritual and material goods of civilization and continual progress contributes both a decisive and optimistic quality to the history of cosmic becoming.

Denying divinity's intervention in the development of civilization, Xenophanes limits its influenc strictly to the beginning, and then not as direct and concrete involvement but as *indication* of the divine, which implies hidden, indirect hints and signs. On the basis of this initial implied divine 'indication', man progresses thenceforth in time, seeking and inventing the 'better'. This striving is no longer considered irreverence. Only two hundred years before, Hesiod had described the horrendous divine retribution upon 'many-desired' Prometheus, who dared to steal fir from the gods and teach mankind the acquisition of goods. In contrast, Xenophanes considers it a duty for man continuously to look to progressive improvements: "It is an obligation", he would say,[5] "for men always to be concerned with progress".[21B1,24]

Continuous striving and searching are the only means of human beings achieving 'better' goals, and, indeed, this is a *common* effort, since Xenophanes uses the plural, 'men'. Human seeking and discovery do not bear fruit immediately but progressively through the course of time. Xenophanes stresses that people achieve progress within the dual framework of *seeking* and *time*. Seeking is an inner drive which may be the eternal echo of the initial divine 'indication' while time is the outward condition which permits the dynamic unfolding of the 'better'. Time ceases to be the mastering force of inescapable 'fate' in life, the controlling power "according to the assessment of time" in Anaximander, or Thales's "wisest time which discovers everything", as an autonomous entity, independent of man. Innovatively, time is now viewed as a valuable collaborator in the human striving toward sustained progress, the 'helpful coadjutor'[6] as Aristotle would put it. For the firs time, as A. Kelessidou observes, Xenophanes "opens up the prospect of freedom in human existence. History appears as a progressive becoming in which the acting force is man".[7]

Xenophanes's is an optimistic spirit. Of course others also recognized the progress of civilization. They have generally seen it, however, like Hesiod with his f ve epochs in the pessimistic light of increasing decline. Xenophanes, conversely, maintains that in his sustained quest within time, man always discovers and enhances the 'better'. This is not 'perfection' -the superlative *ariston*- but that which is a *relative* improvement, the *ameinon*. Thus, his optimism would not tend to carry man away into pride and arrogance, *hybris* (insolence) – the notion that he could attain the

---

[5] Here we adopt H. Fränkel's suggestion for the reading of the last line of Fragment 1, which reads the firs word as *chreon* (obligation) rather than *theon* (of the gods). *Dichtung und Philosophie des frühen Griechentums* , 422, note 3; see also, A. Markos, *Paratiriseis pano ston stiho, B1, 24 tou Xenophani, (Remarks on Xenopahnes's verse B1, 24)*, 283–294.

[6] Aristotle, *Nicomachean Ethics*, 1098a28.

[7] A. Kelessidou, *Le Temps et l'escpace chez Xénophane*, 88.

ultimate. On the contrary, the goal of betterment enjoins a moderate path of perpetual quest and discovery which never leads to self-congratulation or rest, but to continual self-refutation and liberation from complacency.

Ethics

The sense of moderation, 'due measure', which we have encountered in the Milesian thinkers and Pythagoras pervades the work of Xenophanes. He would state, for example, "It is not wrong to drink as much as allows any but an aged man/to reach home without a servant's aid".[21B1,17–18] In turn, he would reject the traditional accolades to the mythical Titans, Giants and Centaurs as "fiction of old" which are of "no use".[21B1,23]

For Xenophanes the 'righteous man' is the model of the ethical individual, who combines two fundamental traits: happiness and delight in life, on the one hand; and, on the other, wisdom and prudence. Like Democritus later, Xenophanes does not cut the prudent and thoughtful man off from the joys of life. On the contrary, he wishes him felicity, at the same time that he bears certain ethical responsibilities. The firs is to praise God with gratitude in words fille with reverence and moral purity: "First righteous men must hymn the god/with reverent words and pure speech".[21B1,13–14] Good, he adds, is "always to hold the gods in high regard".[21B1,24] If we adopt here Fränkel's suggestion of *chreon* (it is necessary) in place of *theon* (of gods), then the line carries the meaning that 'one is continuously obligated to provide good and useful things'.

Xenophanes's 'righteous man' prays to, and beseeches God. What is significant however, in Xenophanes's thought, is the *object of man's beseeching*. The Psalmist prays: "Teach me the way I should go".[8] In contrast, Xenophanes already *knows* his path: it is the path of obligation to perform 'just acts'. What he asks of God is not, therefore, "Teach me to do thy will",[9] but to help him to be "strength to act aright".[21B1,15] With amazing confidence Xenophanes believes that man is able on his own to fin the road he should take. How does one gain this knowledge? There are two sources that pronounce moral law: the one is the reason, wisdom and prudence, which characterize the 'righteous man'. The other is the "noble deeds that are brought to light by memory and a striving for virtue".[21B1,19–20] It is not by chance that Xenophanes uses here words like 'bring to light', 'memory', 'striving', which one can maintain refer directly to those few things the divine "reveals to man in the beginning"[21B18] –to those 'unwritten and inviolable laws of the gods' of Sophocles's Antigone: "For these do not only exist now or yesterday, but eternally and forever and no one knows when they were brought to light".[10] Xenophanes seems to be inspired by human knowledge of 'just acts' derived on the one hand

---

[8] Old Testament, *Psalm*, 143,8.

[9] Old Testament, *Psalm*, 143,10,

[10] Sophocles, *Antigone*, 454.

from *pure reason* and, on the other, from the *unwritten moral law* that was indicated once and for all, "from the beginning", by God. For this reason, man prays for *strength* to *act* justly and correctly and not for *divine illumination* as to what is just and proper. This masterful conception of Xenophanes's for the firs time establishes man as the exclusively responsible and conscious guide and critic of his own actions. At the same time, it implies that he is tragically alone on his path through life.

Community

Xenophanes's ethics are not a theoretical matter. 'Righteous men', engaged in community life, act as responsible citizens; 'usefulness', 'justice', 'kindness', 'virtue', and 'due measure' are the ethical principles on which the individual actively participates in public affairs. As we have seen, Xenophanes would choose as the basis of action -contrary to the Homeric standard of bodily strength- prudence and knowledge, as he states in his second Elegy: "it is right not to prefer strength to this good expertise [in moral wisdom]".[21B2.13–14] Setting up this sharp contrast between bodily and mental power, Xenophanes supports the latter as the exclusive source of social progress. Having worked to that day exclusively for *personal* happiness and fame, man is now urged in another direction -the concern for, and pursuit of, *social* progress.

Knowledge

Xenophanes's saying, "The gods do not reveal everything to men in the beginning, but through seeking men uncover what is better in the course of time",[21B18] applies above all to the achievement of *knowledge*. "A new idea is put forth here"; Snell would note, "that man gains knowledge through personal inquiry. . .. In Xenophanes we fin a new insight into the significanc of intellectual activity – into individual questioning and the effort to bridge the chasm between the human and the divine sphere".[11] Setting out with a deep doubt "about everything",[21A35] the poet would seek through steady pondering and examination of ideas to clarify and extend his knowledge. His approach would be fundamentally empirical – observations based on the senses (we have seen above Xenophanes's power of observation and explanations of natural phenomena). Thus, many would call him an *empiricist.* But this is a one-sided conception. Considering "the senses illusory," he judges that "we must reject the senses and fantasy and trust in *logos* (reason) alone".[21A32] Thus, he is also a *rationalist* who always submits empirical data to logical examination and evaluation.

Yet Xenophanes has doubts "concerning reason (*logos*) itself".[21A39] Having irrevocably denounced the mythical, and having chosen ongoing inquiry and invention

[11] B. Snell, *Die Entdeckung des Geistes*, 129.

for the conquest of knowledge, he also clearly realizes the limits of knowledge: "No man has seen, nor ever will know, the absolute truth about the gods and about everything of which I speak. For even if one happened to utter perfect truth, he himself would not know it. *Dokos* (guesses; opinions) accompany everything".[21B34] For Popper "there is nothing in the whole literature of philosophy that is so critical, so self-critical, so correct and so true, as that saying".[12]

Xenophanes is the firs to develop the theory of *objective knowledge*, as he distinguishes between *objective* truth and *subjective* certainty. *Hypothetical* knowledge alone can exist, within a perpetual *approach* to absolute truth. This "firs explicit confrontation, as two separate things, of knowledge and seeming", Guthrie notes, "was of tremendous consequence for the development of Greek thought".[13] In contrast to god, who "knows the truth",[21A24] who "hears and sees the whole and not a part,"[21A32] man by nature is capable of perceiving *only* "so much as is revealed to observation and understanding".[21B36] Man will always possess only a *fragmentary* knowledge of objective truth, of that which is 'clear'. Searching, he will tread steadily within time toward 'clarity', toward ultimate truth and perfect knowledge, which objectively exists (for god) but will remain forever beyond the mere mortal. Thus, man remains limited each moment to the 'guess' (*dokos*) – to transient knowledge, seemingly valid and probable, to the degree that it "resembles the truth".[21B35]

However, Xenophanes does not emphasize only the phenomenological view that knowledge cannot penetrate to absolute truth, being limited strictly to that which *reveals itself* to us. He also recognizes *relativity* in the quality of sensorial perception generally: "If God had not created amber honey, men would believe fig to be much sweeter".[21B38] Xenophanes conceives the progress of civilization and knowledge to be an exclusively human achievement, positing *apriori*, nevertheless, human limits. In sum, mankind is on an extending road, ever transcending 'better' achievements and 'guesses' toward an eternally remote and unattainable *ariston,* the ultimate 'best' and 'clear'.

The Tragic Element

Does Xenophanes's philosophy -in spite of its ebullient spirit- contain the element of tragedy? Could the optimistic Xenophanes, at bottom, be fundamentally tragic in attitude? Though this question has been given little consideration by contemporary commentators, Xenophanes can be perceived as a profoundly tragic personality. We meet the tragic feeling on many levels in his teachings: in his placing on man full responsibility for his actions, in his irrevocable choice of reason over myth, in his unending quest through the course of time, in his setting a limit on human possibilities.

---

[12] K. Popper, *The World of Parmenides*, 46.

[13] W.K.C. Guthrie, *A History of Greek Philosophy. The earlier Presocratics and the Pythagoreans*, I, 399.

As we have seen, Xenophanes is the firs to view man as voluntarily and consciously taking the responsibility for his personal and social life into his own hands, making him *exclusively responsible* for his actions. The word *harmony*, which we have so often encountered in the Ionian thinkers and Pythagoras, does not appear one single time in the extant work of Xenophanes. For him, man has ceased to fl w along passively attuned to the *cosmos*; he has become, rather, the dynamic creator of his civilization. The assumption of this personal responsibility *individualizes* man for the firs time, constitutes him as a *person*, and offers him a freedom that entails not only tragedy but grandeur.

This presupposition of man's personal responsibility precludes divine intervention within cosmic becoming. Xenophanes places his *one God* beyond any human actions. "This, however, . . . implies an enormous, unrelenting loneliness. The cosmic loneliness of the *person*, his hovering over the abyss of nothingness, the sensation of a boundless wilderness within the desert of time and space, constitute the essence of the tragic, which remains man's bitter reward for his personal involvement in the *cosmos*".[14]

Besides the tragedy of personal responsibility and existential loneliness, there is also the awareness of the *limits* of human striving and understanding. Xenophanes's man will never achieve 'clarity' and the 'best', and he knows this very well. Tragic feeling is precisely "the awareness of man's crushing at his limits".[15] In spite of all this, the person does not submit or resign herself, but continually strives and creates anew, realizing deeply that each moment acquired knowledge bears within itself the seed of transcendence and self-correction -a fact that constitutes the tragic heroism of the human spirit.[16]

Xenophanes can justly be characterized as "the tragic physical philosopher-poet[17]"21A9. "The century of *Tragedy* is also the century of *Philosophy*," Malevitsis states. "And these two expressions of the human spirit, ontologically of equal weight, function on the same level, which is the awakening of consciousness and its liberation from the horrible residue of the powers of the mass unconscious. When we state that in Greece the passing from *myth* to *reason* comes about, we are referring to the passing from the unconscious to the conscious, . . . an event that bears witness to the birth of autonomous consciousness. This is the historic glory [of the Greeks]. This event means that *Logos* (reason) declared war against 'mere opinion' (*doxa*)

---

[14] C. Malevitsis, *Peri tou tragikou (On the Tragic)*, 29.

[15] C. Malevitsis, *Peri tou tragikou (On the Tragic)*, 28.

[16] In the light of this reality, the conquerors of Mount Everest, for example, at the culminating moment following years of strenuous effort, reaching the world's highest peak, are not as one would expect, elated and triumphantly moved; instead they experience "the feeling of the lonely beauty of the evening . . . a hint of fear of a vast unknown . . . a fleetin feeling of disappointment. . .and the suspicion that maybe there was something more, something beyond the three-dimensional form of the moment. If only it could be perceived"! (P. Davis and P. Park, *No Way. The Nature of the Impossible*, 9).

[17] It is not certain whether, the poet referred to here is Xenophanes or Thespis.

on the rational level, and against 'complexes' on the existential level. The former battle was mounted by Philosophy; the latter by Tragedy".[18]

Overview

"It was one of Xenophanes' greatest achievements", writes Popper, "that he anticipated and strongly represented all the main ideas of the European Enlightenment. Among these were the ideas of fightin for the truth against obscurity; of talking and writing lucidly and modestly; of practicing irony and especially self-irony; of avoiding the pose of a deep thinker; of looking critically at society; and of looking upon the world with wonder, and with an infectious curiosity".[19]

Athens University professor John Theodorakopoulos once said that two features above all characterize Greek civilization: intellectual *vision* and *heroism*. Xenophanes -the 'revolutionary of the intellect', as he has been styled- can be viewed as one of the major representatives of these two tendencies. As regards intellectual vision, it is he who resolutely chooses the path of intellect and logic which will be taken by subsequent Greek thinkers, the firs key figure being Parmenides and Heraclitus. As for heroism, he is the one who will invite man to take his fate courageously in his own hands so as to responsibly forge his personal life and social progress, without divine aid, alone and tragically free.

This event marks, moreover, the fina and irreversible separation from Myth -a separation which consolidates the unbridgeable gap between Eastern and Western thought. "Yet without Myth", Nietzsche warns, "any culture loses its healthy, creative natural power".[20] What were the consequences of this crucial Xenophanean choice of rationalism as the guide for human progress in the evolution European civilization, and how responsible can Xenophanes be held for the later turn it would take? "It is a great achievement to have the life of the individual and the community regulated according to the demands of reason", Malevitsis observes. On the other hand, "when man attempts to fil his loneliness with knowledge, he will end in committing crimes. And knowledge has deeply sunk our century in crime".[21] What has gone wrong during the two thousand years between the Xenophanes's optimistic message, "In the course of time, by searching, [men] may fin out better [things]"[21B18] and today's tragic deadlocks? During the evolution of Western European civilization, Xenophanes's position has been violated in at least three basic respects:

---

[18] C. Malevitsis, *Peri tou tragikou (On the tragic)*, 121.

[19] K. Popper, *The World of Parmenides*, 35

[20] F. Nietzsche, *Die Geburt der Tragödie*, 179.

[21] C. Malevitsis, *Peri tou tragikou (On the tragic)*, 25, 17.

(a) Knowledge has ceased to exclusively serve man's deepest ontological need for the search for truth and has gradually begun to submit to the increasing utilitarian exploitation of the natural world.

The uniqueness of the Greek attitude toward the world derives from the *point of departure,* which the Greeks took toward thought and the *goal* they posited for it. The point of departure was 'wonder',[22] Xenophanes's '*wonder* toward all things'.[21A35] "It is through *wonder* that men now and from the beginning began to philosophize".[23] And the goal of philosophic inquiry was comprised from the beginning by the active pursuit of Xenophanes's 'clarity', apart from any exploitative interest. "It is obvious clear that they pursued knowledge for its own sake and not for any practical purpose".[24]

From the Western middle ages onward, this Greek tendency was overturned. The chief purpose of knowledge ceases to be the selfles search for truth and takes on a technocratic, utilitarian, worldly character. Francis Bacon, perhaps the chief advocate of this new tendency, would proclaim: "Let mankind regain their rights over nature, assigned to them by the gift of God, and obtain that power".[25] The ancient Greeks wished to *understand* nature, not to *change* it. According to Bacon, this delayed the development of humanity for more than two millennia.[26] The reason was that the Greeks lacked a goal: "There is a powerful and great cause of the little advancement of the sciences, which is this: it is impossible to advance properly in the course when the goal is not properly fi ed. But the real and legitimate goal of the sciences, is the endowment of human life with new inventions and riches".[27] In any case, it is generally believed that today the main thrust of scientifi research has shifted from the search for basic natural realities to the systematic support of technology.

(b) The conscious Xenophanean humility toward the impossibility of 'absolute knowledge' and 'clarity' has been replaced by the arrogant as well as simplistic conviction that science can lead to the absolute and objective truth.

Xenophanes's saying, "Certain truth has no man seen, nor will there ever be a man who knows [absolute truth]"[21B34] reflect the sage humility of the Greeks, who always possessed an understanding of human limitations. Through the course of time, these Xenophanean *limits* of rationalism would be cast aside, resulting in

---

[22] Twenty-f ve centuries later, the now elderly Goethe would write: "The greatest height man can reach is wonder (*das Staunen*)". (Letter to his friend, Eckermann, 2-18-1829).

[23] Aristotle, *The Metaphysics*, 982b12.

[24] Aristotle, *The Metaphysics*, 982b20.

[25] F. Bacon, *Novum Organum*, 366.

[26] F. Bacon, *Novum Organum*, 334.

[27] F. Bacon, *Novum Organum*, 339.

Descartes' pronouncement that "all science is certain, evident knowledge".[28] Upon this arrogant assumption of science's potential to achieve absolute truth scientists would build their work for the following three hundred years, and this confidenc would fin its ultimate philosophical expression in the neo-positivism of Wiener Kreis (M. Schlick, R. Carnap, H. Reichenbach, et al.), according to which *empirical proof* (the verificatio principle) constitutes the unquestionable criterion of truth.[29] Consequently, the metaphysical, primary and crucial question concerning the 'being'of beings would become 'meaningless' (*sinnlos*).[30]

Extreme optimism and self-confidenc would even lead some outstanding scientists to maintain that in certain areas, man has already covered the entire fiel of knowledge! The physicist Lord Kelvin is said to have claimed at the end of the nineteenth century that the sole thing left for physics to do was to fil in the next decimal place.[31] When the young graduate Max Planck sought advice about the prospects of the profession from von Jolly, Professor of Physics at the University of Munich, the dissuasive response he received was that in physics virtually everything has been done and what remains is only a few gaps to be fille in and some constants to be determined with greater accuracy.[32] Even Hawking, with reserved optimism, repeated in 1980 the question, "Is the *end* in sight for theoretical physics"?[33]

(c) The realization of the Greek conception of social experience, which consists in the participation of the individual in the decision making process of civil life, has been distorted by the reduction of the citizen to an im-personal social unit.

The concern of the ancient Greek is directed toward the *polis* (the city-state). Xenopohanes would designate 'wisdom' as the means toward harmonious communal action and civic progress. In contrast to theocratic societies, where the law is imposed by force, in the Greek *polis* laws are enacted by free, thinking citizens whose main concern, according to Xenophanes, is "to act aright",[21B1,15–16] which means a continual selfles care for one's fellow man, the 'thou'. On the contrary, today's Western man has been entrapped by the dramatic deadlocks of the 'I'. In turn, in the place of acting *persons* who justly direct society, we fin a will-less, empty shell, prey to ideology, politics, technology, consumerism.

Today the innovative insights of Xenophanes are timelier than ever. World wars, racial discrimination, genocide, ecological catastrophes, which are the direct consequences of the above distortion of Presocratic conceptions, challenge every

---

[28] F. Capra, *The turning point*, 57.

[29] W. Stegmüller, *Hauptströmungen der Gegenwartsphilosophie*, I, 381–382.

[30] W. Stegmüller, *Hauptströmengen der Gegenwartsphilosophie*, I, 383.

[31] L. Badash, *The Completeness of Nineteenth-century Science*, 48–58.

[32] A. Hermann, *Max Planck*.

[33] S. Hawking, *Is the End in Sight for Theoretical Physics*?, 1, note 1. Nevertheless, two decades later in a lecture at the University of Crete in Herakleion during the summer of 1998, Hawking corrected himself: "Could it be that the answer to this question always lies *a bit beyond the horizon*"?

thoughtful person to self-examination and to a re-orientation of values. The effects of a revival have begun to appear on all levels of human endeavor -practical, social, philosophical.

Sober judgments of outstanding contemporary technological leaders once more recall Xenophanes's measured words: "You should not be proud of something that you have accomplished. You must always go forward and strive to achieve something better". These words, which might have been Xenophanes's own, are from the lips of D. Packard, one of the two founders of one of the leading computer manufacturers -Hewlett Packard.

The contemporary physical sciences return us to Xenophanean *relativity* and the *impossibility* of discovering absolute objective truth. In *mathematics* the two theorems of O. Gödel establish that the mathematical expression of physical theories of physics does not necessarily lead to a fina distinction between truth and falsehood.[34] In *physics* quantum mechanics introduce ontological probabilism and 'non-local' correlations (EPR paradox), which suggest the unity of the entire universe, excluding any possibility of absolute knowledge and prediction. Similarly, contemporary physics of chaos (the physics of self-organization) shows that the process of creating an ordered structure (a cell, an animal, a galaxy) remains to a certain extent unpredictable. Thus, the Xenophanean insight that, "Certain truth has no man seen, nor will there ever be a man who knows. . .but in all things there are mere opinions"[21B34] again takes on a dramatic timeliness.

Just as contemporary physics transcends the rational limits of Newtonian physics, certain sociologists and ecologists today reveal the limitations and consequences of the Cartesian cosmo-theory, proposing a new *holistic* conception of the universe, once more uniting man and nature.

Paralleling recent developments in the fiel of physics, contemporary *critical epistemology* (K. Popper, S. Kuhn, I. Lakatos, P. Feyerabend, et al.) directs decisive criticism against the simplistic concepts of the Neo-positivists. To the principle of 'Verificationism' Popper opposes the principle of 'Refutationism', according to which scientifi theories cannot be proved by verificatio but can only be confronted by refutation.[35] This concept returns to the Xenophanean perception that men "through the course of time by searching fin out better",[21B12] never that which is *perfect.*

"Based on Xenophanes' theory of truth and connected with his moralism and intellectual honesty",[36] Popper develops twelve professional *ethical* principles, which do not apply to scholars alone. "There is one element of rationality in our attempts to know the world", he observes: "It is the *critical* examination of our

---

[34] E. Nagel and J.R. Newman, *Gödel's Proof.*

[35] This critical position of K. Popper would later lead R. Carnap to support the replacement of these two principles of research with the broader concepts of 'capacity of confirmation (Bestätigungsfähigkeit) and 'capacity of proof' (Prüfbarkeit). (W. Stegmüller, *Hauptströmungen der Gegenwartsphilosophie*, I, 404).

[36] K. Popper, *The World of Parmenides*, 62–65.

theories . . . Xenophanes knew that our knowledge is guesswork, opinion – *doxa* [opinion] rather than *episteme* [scientifi knowledge] -as shown by his verses:

"The gods did not reveal, from the beginning, / All things to us; but in the course of time, /Through seeking we may learn, and know things better."[21B18]

" But as for certain truth, no man has known it, / Nor will he know it; neither of the gods, / Nor yet of all the things of which I speak. / And if perchance he were to utter / The perfect truth, he would himself not know it: / For all is but a woven web of guesses."[21B34]

Popper continues: "The proper answer to my question 'How can we hope to detect and eliminate error?' is, I believe, 'By *criticizing* the theories or guesses of others and our own'. This answer sums up a position, which I propose to call 'critical rationalism'. It is a view, an attitude, and a tradition, which we owe to the Greeks".[37]

[37] K. Popper, *Conjectures and Refutations*, 152, 26.

Heraclitus of Ephesus (ca. 540–480 B.C.)

The new form of insight can perhaps best be called
'Undivided wholeness in flowin movement'.

David Bohm

Introduction

Heraclitus and Parmenides mark the decisive turning point in the course of Presocratic thought. Although they live on opposite sides of the Greek world, they spring from the same Ionian spiritual heritage in their quest for the deeper meanings of *becoming* and *being*. Their thought, however, breaks free and leaps the bounds of this heritage, marking the origin of philosophy.

The Ionian 'natural philosophers' envisioned *change* as a transition from one opposite to the other -the Anaximanderian 'coming-to-be'- 'perishing', for example. However, in the enthusiasm of the neophyte, they never questioned, or doubted, the very existence of beings, nor wondered whether the senses convey true reality. Heraclitus -and then Parmenides, shortly thereafter- would be two great thinkers who would firs pinpoint the problem of 'being', of 'becoming', and of 'knowledge'. They would provide -each from his own perspective- a response to those three fundamental issues, which thenceforth would challenge both philosophy and the natural sciences.

Life

Heraclitus would live in extremely troubled times – a period of revolutions, changes in forms of government, wars, natural catastrophes. These clashes of cultures, great powers, ethnic groups, cities and systems of government, would profoundly influ ence the thought of Heraclitus. The perpetual strife in which the fall of the one marks the rise of the other -and vice-versa- recalls his enigmatic words: "Immortals are mortal, mortals are immortal, living the death of the others and dying their life".[22B62] Each day of his life brings home to him his saying: "War is the father of all and king of all...".[22B53]

Scion of an aristocratic family and ecclesiastical offic holder, he would soon perceive in the corrupt political establishment the futility of his involvement in the community. He would hand over all his inherited privileges to his brother, reject an appointment by King Darius, and thenceforth withdraw into a solitary, self-styled,

C.J. Vamvacas, *The Founders of Western Thought – The Presocratics*, Boston Studies in the Philosophy of Science 257, DOI 10.1007/978-1-4020-9791-1_9,

deeply thoughtful life. His disgust for civil affairs -even toward democracy, which had in the mean time been established- would be transformed into complete scorn and outrage toward the mob when the city exiled a capable legislator from his offic as oligarch![22B121] On the one hand, he breaks his bond with the masses, an event that would later lead many irresponsibly to label him arrogant, condescending, misanthropic. Withdrawing, he prefers to play dice with children outside the famed temple of Artemis. To the jibes of his fellow citizens, he snaps: "Surprised are you, you good-for-nothings? Isn't this better than playing politics with you"?[22A1(3)] On the other hand, he desires to aid them, knowing nevertheless that words are futile. Wittgenstein was to say, "that which *can* be demonstrated *cannot* be put into words".[1] Accordingly, Heraclitus preferred silence. When asked why this was so, he responded sardonically: "So you can do the babbling".[22A1(12)] Thus, when asked his opinion as to how the city can live in harmony, he did not resort to words but to a 'dumb show' in one act -an image: "he went to the rostrum, took a pitcher of cold water, poured ground grain into it, stirred it with a branch, and having drunk it, left. Thus he showed them that letting what we have suffic and having no need for luxuries, preserves the city through peace and concord".[22A3b]

Heraclitus not only taught the unity of opposites but was also himself an amalgam of contraries. "While as a youth he said he knew nothing, when he reached maturity he claimed he knew everything".[22A1(5)] Nietzsche observed that "Heraclitus was proud. And where the pride of a philosopher is involved, that is a very great pride indeed".[2] But Heraclitus was not a philosopher *only*. He possessed a profoundly mystical, prophetic spirit. And these combined traits led to extreme humility: "Listen not to me but to the *Logos*",[22B50] he exhorts. Though he condemns and scorns the 'many' for their irresponsibility and corruption, he also wishes to awaken them and enlighten them with his aphoristic message. Throughout his life among them, the attitudes of Ephesians toward him ranged from reserved to hostile, but after his death they would acknowledge his achievement, even stamping coins '*Heraclitus of Ephesus*'.

Writings

The extant work of Heraclitus comprises 139 brief fragments with no evident order among themselves. This leads many to doubt that Heraclitus ever wrote a unifie study and to believe that the book attributed to him, *On nature*, is nothing more than a posthumous collection of oral sayings.

Concerning Heraclitus, Nietzsche remarks, "perhaps no man ever wrote more eloquently and luminously".[3] Conversely, others would fin him 'dark'[22A3a] and 'enigmatic'.[22A1(6)] This conflic springs from attitudes toward his teachings. If one

---

[1] L. Wittgenstein, *Tractatus Logico Philosophicus*, 4.1212.

[2] F. Nietzsche, *Die Philosophie im tragischen Zeitalter der Griechen*, 295.

[3] F. Nietzsche, *Die Philosophie im tragischen Zeitalter der Griechen*, 294.

demands reasoned discourse -of the sort we encounter for the firs time a few years later in Parmenides- then one would be obliged, along with Aristotle, to protest that he thinks absurdly, violating the logical principle of 'identity'. Yet, Heraclitus does not violate – he transcends the rules of logic. He would be the *fi st* to ponder the problem of *change*. His concept of the 'oneness' of opposites, which so disturbed Aristotle, is not an absurd identificatio but -in spite of change- a continuous preservation of its identity of being, which can express many things. The *union of opposites* is at the core of his thought. This is a conception, as much poetic as philosophic, in which primordial *unity* remains yet unshaken and unbroken between universe and consciousness, the general and the particular, the objective and the subjective, the sensorial and the theoretical.

His language is personal, concise, immediate, suggestive, intensely significant polysemic. In his time, 'purports' had not yet been petrifie in language. For this reason the meaning of each fragment is not exhausted or even approximated in one interpretation. The text invites one to further probing and creative contemplation. Heraclitus attempts to render in each statement a full, dynamic image as he has conceived it intuitively – its center the union of opposites, the absolute unity of all things: "all things one".[22B50] It is that very inapproachable nature of Heraclitean meditation, which renders his *Logos* ambiguous. "The very question simultaneously suggests answers", Jaspers states. "The questions are not developed but pronounce responses".[4] His statements present many-faceted *images* referring *simultaneously* to a host of 'signs', such as: *en* (one, in a numerical or axiological sense, or in the sense of unity, or as an ontological or cosmological allusion); *idios* (one's own; private; personal; distinct; strange; unusual; distinguishing feature); *apanta* (quite all; the whole; everything; altogether); *Logos* (computation; account; measure; relation; correspondence; proportion; ratio, analogy; explanation; pretext; plea; statement; argument; proposition; rule; principle; law; thesis; hypothesis; reason; ground; formula; language; narrative; fable; legend; tale; speech; utterance, expression; word; phrase; report; rumor; description; mention; discussion; debate; dialogue; saying; oracle; proverb; assertion; subject-matter); *tauton* (identical; oneness), *heteron* (different, another), *eris* (strife; discord; quarrel; contention; rivalry); *sophon* (wise; judicious; prudent; worthy); *nomos* (usage; custom; law; ordinance); *xynon* (common; public; usual, indifferent; impartial); *dike* (usage; custom; order; right; judgment; justice; law); *metron* (measure; due measure; rule; proportion); *cosmos* (order, well-ordered world; universe; harmonious arrangement; ordered structure; ordered society; ornament; adornment); *pyr* (*f re*, fir as cosmic principle), etc. Accordingly, an attempted analytical paraphrase and interpretation of the *entirety* of the image leads to an arbitrary breaking of its unity and ceases to render the original Heraclitean thought. Examining Heraclitus, perhaps we must follow the advice of Wittgenstein: "A statement says something only as long as it remains an image… Regarding something about which one cannot speak, one must remain silent".[5]

---

[4] K. Jaspers, *Aus dem Ursprung denkende Metaphysiker*, 30.

[5] L. Wittgenstein, *Tractatus Logico Pilosolphicus*, 4.03, 7.

He consciously undermines the *single meaning* of a word or of a sentence. Thus *one* word can lead to a *pair* of opposites, as in *mortal-immortal, visible-invisible, brought together-brought apart, in tune-out of tune*; or a *pair* of opposites leads to identification as in "The road up and the road down are one and the same",[22B60] or "On a circle beginning and end are common".[22B103] A single word is developed etymologically as carrier of a profounder antithesis: "the name of the bow (*bios*) is life (*bios*), but its working is death".[22B48]

Consequently, his language is not analytical -deductive or inductive- but still deeply archaic, solid, concise, unifie and polysemic, whereby experience and thought, acting and apprehending, coexist. It is a language based more on intuition and less on rational observation, as we have come to expect from the Milesian 'natural philosophers' – a Logos surprisingly condensed and elliptical, symbolic, which does not elaborate but projects images. "The meaning", E. Ioannidou observes, "is not developed step-by-step by the statement but is captured at once through an image, in an unexpected vision – or it is not captured at all".[6] Heraclitus does not intend to demonstrate but to manifest; he does not discuss but prophesies; he does not instruct but initiate; he does not strive to simplify, but to weave an image, complex and many-sided; he does not explain but signifies practicing what he would say of the god Apollo: "the Lord who owns the oracle at Delphi neither speaks nor hides his meaning but indicates it with a sign".[22B93] Thus, as we proceed in our approach to Heraclitus, we should not as much aspire to be tutored as to be touched by his thought.

Heraclitus's thought cannot be fathomed passively, but only in an active, inquisitive, exuberant engagement. The poignant confession which he pronounced, "I searched myself"[22B101] -in response to the Delphic injunction, 'Know thyself' - is also recommended to his fellows: "all people possess a share of self-knowledge and wisdom",[22B116] he would state. For, "those who seek gold dig much earth and fin a little".[22B22] Be of good cheer! he counsels: "If you do not hope, you will not fin the unexpected, for it is unexplored and inaccessible",[22B18] Here is a message fille with passion -a call to a meeting of visions, to an immersion in the supreme truth which is deeply hidden, which cannot be 'said' but is 'manifested'. Thus, Socrates would later state bemusedly that for the understanding of Heraclitean thought, "it takes a masterly diver"[22A4] to capture the valuable prize from the dark depths. Heraclitus is a inspired revisionist, a spiritual luminary, a singular initiate.

Union of Opposites

Pairs of opposites are not an invention of Heraclitus's; they constitute a basic feature of almost all the Presocratics as well as the Greek physicians of his age. It appears, as H. Fränkel notes, that there was a "mode of thought during the archaic period in

[6] E.I. Ioannidou, *Herakleitos, Glossa kai Skepsi. (Heraclitus, Language and Thought)*, 14.

Greece after the age of Homer which was the prevailing one – that is, a bi-polar form of thinking: qualities could not be conceived except together with their opposite".[7]

What is the innovation, then, which Heraclitus introduces? He rejects the Anaximanderian conception of the eternal battle between 'coming-to-be' and 'perishing', involving an injustice that demands recompense. He ridicules the Pythagorean teachings concerning a conciliatory, unresisting reunion of the opposites in a *peaceful* combination as well as the attribution to them of certain *ethical* values, identifying, for example, 'at rest' with 'good', and 'moving' with 'bad'. He would deny the ontological doctrine of Parmenides -in the event that he knew it- that the opposites belong to the apparent, non-existent world of 'opinion', while true 'Being' is indivisible, unchanging and unmovable.

Heraclitus's decisive fundamental message, the nucleus of his thought, is the *unity of all* – "all things are one".[22B50] This conception arises from the *union of opposites* between which there is an eternal struggle and dynamic balance. Thus there is a *unity of all things in tension.* Humans are troubled because "They do not grasp how something being at odds with itself is in agreement with itself".[22B51] If there were no opposites, with tension between the two, the universe could not exist. A lack of tension would mean stagnation and the end of the world, "for musical harmony would not exist without high and low tones, nor living creatures without male and female which are opposites. Everything would disappear".[22A22] The joining of opposites constitutes a unity in which there is both an opposition and identificatio that is totally alien to formal logic, as it will later be developed as an academic discipline. Aristotle ignores this fact and becomes indignant, unjustly accusing Heraclitus of violating the fundamental logical axiom of 'identity' or 'contradiction'.[8] But judging at his time -two centuries after Heraclitus- Aristotle is biased, thinking within the framework of a logical system crystallized in the meantime.

Heraclitus is still on the arduous course from *poetic* to *philosophic* thought. Meanings are not yet crystallized but inhere in integral images and visions, grasped as an indissoluble whole in which the distinction between the visible and the conceivable has barely begun to emerge. His thought is at once both highly *abstract* in its generality and extremely *concrete* in its evocation of *opposites*, which he dares to carry all the way to their ultimate form of a *contradiction* – 'mortal-immortal', 'just-unjust', 'whole-not whole', 'we step-we do not step'. Heraclitus does not support his doctrine regarding the oneness of all things through reasoned, systematic arguments but through examples, images, symbolism and comparisons taken from all the domains of life: human experience, nature, the divine, geometry, ethics, society, language, knowledge. Inspired by a *dialectical* intuition -rather than formal conceptual logic (which, in any case, has not yet developed) – he grasps actuality whole as a *process* containing bipolar antitheses in a dynamic balance that in this sense constitutes their mutual 'identity'. He does not apprehend this *tauton* (identical) in a narrow, formal sense. The conception of the union of opposites points to a

---

[7] H. Fränkel, *Dichtung und Philosophie des frühen Griechentums*, 77.

[8] Aristotle, *The Metaphysics*, 1005b23-24.

deep bonding of the two because the one is determined by the other and co-exists with it. *Tauton* (identical; oneness) is manifested in various forms and is founded on:

(a) *Interdependency of opposites*: Something is valued only by virtue of being compared to its opposite. The existence of the one presupposes the existence of the other. Thus, "it is disease that makes health pleasant and good, hunger satiety, weariness rest";[22B111] "they would not have known the name of right if these [wrong] things did not exist";[22B23] "if there were no sun. . .it would be night".[22B99]
(b) *Reciprocal succession:* The transition of the one exclusively to its particular opposite -not to some other oppositional entity- in a reciprocal manner: "[day and night] . . . are one [thing]";[22B57] "cold things grow hot, hot things cold, moist dry, dry wet";[22B126] "the same thing [in us] is living and dead,[9] and the waking and the sleeping, and young and old; for these things when they have changed are those, and those when they have changed are these";[22B88] "the immortals are mortals, the mortals immortals, since the life of these is the death of those, and the life of those is the death of these".[22B62]
(c) *The opposed appearances of the same thing relative to different perceptions or points of view:* "sea water is the purest and most polluted water, drinkable and salutary for fish undrinkable and deadly for men";[22B61] "for one thing is the happiness [felt by] a horse and another by a dog and a man . . . Donkeys would prefer a stack of hay over gold";[22B9] "[humans] should not fin pleasure in mud. Pigs fin greater pleasure in mud than in pure water";[22B13] "the road up and the road down are one and the same";[22B60] "on a circle beginning and end are common";[22B103] "the most beautiful ape is ugly compared to man";[22B82] "compared with God, the wisest of men will appear an ape, in wisdom, beauty and all else".[22B83]
(d) *Total unity, in spite of the apparent plurality in all opposite pairs*: "things taken together are whole and not whole, something which is being brought together and brought apart, which is in tune and out of tune; out of all things there comes a unity, and out of a unity all things";[22B10] "to God all things are fair and good and just, but men have assumed some to be unjust and some just";[22B102] "god is day and night, winter and summer, war and peace, satiety and hunger . . . . This god [*nous*] is all the opposites, [or:] All the opposites – that is the meaning [*nous*]".[22B67]

Each pair of contraries can be placed within the whole, but it retains its dynamic, bipolar essential nature. Its *unity*, hidden in the depths, appears on the surface as plurality: "invisible harmony is more powerful than visible harmony".[22B54] All the diverse pairs of opposites lead to the lofty divine *arche*, which can also be identifie

[9] For Heraclitus death does not lead to complete disappearance but means a change from one condition to another.

with the *Logos* or *Wisdom*, gathering and bonding in wholeness. While humans perceive the apparent plurality, incapable of fathoming the 'invisible harmony' of the unity of all things, since "nature loves concealment",[22B123] nevertheless, for the *divine* (the *Logos, Wisdom*) seeming differences dissolve, giving place to an undivided, interconnected, all-encompassing wholeness: "it is wise to agree that all things are one".[22B50]

Perpetual Strife of Opposites

The union of opposites is not paired in a peaceful, stable co-existence. On the contrary: The opposites are engaged in ongoing dynamic tension, in contention, in conflic and war – in strife (*eris*), which, however, contains the element of necessity and justice: "We need to know that war is common (*xynon*) and right is strife and that all things happen by strife and necessity",[22B80] Heraclitus writes. Thus, conflic constitutes the supreme universal principle that governs the world: "War is father of all and king of all".[22B53] In the firs statement, *xynon* conveys the meaning not only of the 'common' but also of that, which is impartial, belonging to both of the opponents.[10] In this sense while strife is ubiquitous, it is not merely random and arbitrary, since as a common element of all opposing parties, it is linked to *justice* (the struggle for justice in a peaceful world would be superfluous and is ruled by *necessity.*

Becoming

"And all things come into being by way of strife".[22B8] 'Becoming' is the consequence of the ceaseless clash of opposites. 'All things' are involved in *endless* movement and change. Any pause or stillness would unavoidably result in disintegration and disappearance of the world. Heraclitus illustrates this doctrine through the example of *kykeon*, an ancient drink prepared by stirring barley and grated cheese into Pramneion wine: "The *kykeon* falls apart if it is not being stirred".[22B125] As long as movement and tension are sustained, beings continue to be what they are – the '*kykeon*' kykeon. Thus, both movement and change can be perceived directly (e.g., "cold becomes hot, hot cold, moist dry, dry wet"[22B126]), or they may not be apparent in daily manifestations (e.g., the maturing of a child). While many things seem stable and unchanging, in the fina analysis *all* are subject to unending motion and change: "It is not that some things are in motion and others not; *all things are in eternal motion* which, however, evades our notice".[11]

---

[10] Homer, *The Iliad,* 18, 309 "Aris [the god of war] is impartial (*xynos*) and kills the man who goes to kill".

[11] Aristotle, *The Physics*, 253b9.

This is Heraclitus's central message, vividly illustrated by the *river* image, expressing the uninterrupted continuity of endless movement and change: "Heraclitus says that all things move on and nothing remains still, and likening things to the current of a river he says that you cannot step into the same river twice",[12] for clearly, "those who step into the same river are continually wet by different and again different waters".[22B12] Indeed, Heraclitus's supporter, Cratylus, would outdo the Ephesian thinker, "reproaching him for saying that it is impossible for one to step twice into the same river because it is not even possible for one to step in the same river once,[13] since from the moment you touch the stream's surface till your foot reaches the bottom, the waters at that point have changed. The current is never the same at two succeeding temporal points; it is always changing, and the same is true of ourselves who enter the stream, though we may not realize it. Accordingly, Heraclitus will also state: "we step and do not step into the same rivers, we are and are not".[22B49A] Unceasing flu and variation of the river symbolize the perpetual change of all things: "It is impossible to step into the same river twice, just as it is also impossible to touch twice a substance in the same state, since because of the impetuosity and speed of change, it scatters and again (or rather, not again, nor later, but simultaneously) gathers, approaches and retreats".[22B91] Much later a commentator would record what was incorrectly attributed to Heraclitus himself yet epigrammatically renders his message: *'aei panta rei'*[14] (everything is forever in flux) In Heraclitus, 'Being' is contained in 'becoming'.[15]

Heraclitus is the firs to introduce the problem of *change, identity*, and by extension, *knowledge.* Does the river in which the water is ceaselessly fl wing remain *tauton* (identical) or does it each moment become *heteron* (different; another)? Keeping its name, a stable bed and fl w, is it 'that very' river, or do the endless changes of the water continuously make it 'another'? According to Heraclitus, all things are ever changing, nothing remains stable. Yet could it be that something that is changing could simultaneously stay the same, sustaining its identity? 'Becoming' is a passage from one state to another. *In the course* of this passage, 'something has ceased to be what it was' and 'is not yet what it will become'. The moment it 'is' that next thing, it ceases to 'become'. To have retained its identity would mean change did not take place, but again if it did, the thing is no longer that thing which has changed; it has ceased to be the same – *becoming* something else. On the other hand, however, the concept of *change* presupposes that the thing that is *changing* remains *identical with itself.* The person growing up and eventually aging does not

---

[12] Plato, *Cratylus*, 402a.

[13] Aristotle, *The Metaphysics*, 1010a14.

[14] Simplicius, *Aristotelis Physicorum Commentaria*, 1313.11.

[15] The deep, archetypal kinship between *einai* (being) and *gignesthai* (becoming) is reflecte in the Greek language itself. K. Axelos, in his book, *Vers la pensée planétaire*, 83–84, observes: *einai* (to be; being) is literally joined with *gignesthai* in that in two tenses of the verb *einai* (the past and the present perfect), Greek uses the verb *gignesthai* (*egenomin, gegona*).

cease to be that person *him-self*, does not become *other*, even though he *changes* each day and is completely different from when he was a child.

In spite of the poverty of verbal means at his disposal during that time,[16] Heraclitus was the firs to comprehensively express -in a moving, vital and penetrating archetypal voice- cosmic 'reality' as 'ceaseless becoming', thus planting the seed for all subsequent elaborations on this central problem, be they philosophical, metaphysical, logical, epistemological, theological, or existential. Through the river image he perceives the deeper 'selfsameness', the *tautotita* (maintenance of identity) which resides within 'otherness' (*heterotita*). He apprehends *self-identity* not as an inert condition but as a *dynamic* movement in time, a *process* which comes to be realized only by way of the *bi-polarity* of opposites and their deeper *self-unity*: "all things are one".[22B50] Things are not stable; they just seem to be so, as detached 'frozen sections' within 'temporal becoming', taking on significanc and real existence only through dialectical relation with things before and after. Reality consists of a continuous *process*. "The greatness of Heraclitus", Popper states, "lies in the fact that he discovered the central problem of the natural sciences and of cosmology: *the problem of change*".[17] Things that 'change' yet maintain their 'self-identity' do not exist; 'self-identity' lies only in 'processes'; they include *opposites* which are also *identical*, since they *can exist solely* as poles of a change; thus, change itself is self-identical; "Heraclitus," Popper notes, "was the great thinker who firs realized that men are flame and that things are *processes* ... [and for this reason, he was] perhaps the greatest and boldest thinker among the Presocratics".[18]

Fire

"This well-ordered world [cosmos], the same for all, none of the gods nor of men has made, but it was always and is and shall be: an ever-living fi e, which is being kindled in measures and extinguished in measures."[22B30]

One can scarcely imagine a more pertinent image than the one Heraclitus chose to epitomize the endless cosmic *process*. Process, like fire is characterized by a material state of existence, perpetual motion of vortexes, whirlwinds, rekindlings and recedings, an inborn clash but also balance, *due measure*. What other symbol in the world besides fir could better render the sum of these properties? Fire may *appear* stable and motionless, but in reality, as we know today, it is simply the phenomenon of combustion – that is precisely a *process* of chemical *reaction* in which molecules of oxygen *collide with* molecules of other compounds which *break up*, their atoms then *combining* in specifi *proportions* with oxygen, in new oxides.

---

[16] M. Heidegger and E. Fink, *Heraklit*, 16. "Before the period of Plato and of Aristotle, 'concepts' *did not as yet exist*".

[17] K. Popper, *The World of Parmenides*, 218–219.

[18] K. Popper, *Conjectures and Refutations*, 148.

Chemical process: 'collision'→'separation'→ 'combination' ['simple proportions']. 'Heraclitus' 'Becoming': 'strife'→ 'brought apart'→ 'brought together' ['due measure'].

Of course the science of chemistry was centuries off. Heraclitus simply 'saw' the flam giving off smoke, the log altering and disappearing, the oil in the lamp going down.... What spark of inspiration, what intuitive power, led this man to render 'eternal becoming', the perpetual process characterized by *due measure*, through the image of *f re* – a singular trope which is in accordance with today's scientifi view?

In Heraclitus's enigmatic language, 'fire is not clearly defined "is not clarifie at all".[22A1(8)] *Pyr* (fire should *not* be considered another 'primary substance' like the natural *arche* of the Milesian 'natural philosophers'. It can be regarded as a reference to the changing, colliding, dividing and combining of *material* elements, but it also may represent the substantive existence of an invisible determinism which governs timeless 'becoming'. "All things transact with fire and fir transacts with all things, as goods are in transaction with gold and gold with goods".[22B90] Transactions involving gold and goods always proceed, of course, on the basis of some *measure*, some *analogy* (*ana-Logos*: according to a proportion). The deeper symbolism of the visible *f re* is precisely this eternal cosmic *Logos*.

Logos

"Listening not to me but to the Logos it is wise to agree that all things are one."[22B50]

The *Logos*, which Heraclitus is the firs to introduce into philosophy, constitutes the keystone of his thought. His insight is polysemic, and no contemporary interpretation can render it in a single word. Before we attempt to defin it, let us describe it through the extant Heraclitus fragments themselves:

"Listening not to me but to the Logos it is wise to agree...."[22B50]

The *Logos* is different from and independent of Heraclitus's *Logos* (word, reasoning). It is the primordial *Logos* (behest language) to which one must hearken and assent.

"One must follow what is common; but although the Logos is common, most men live as if they had a private understanding of their own."[22B2]

The *Logos* with which we must assent is 'common', the same for everyone, as opposed to 'private' – the personal thought and opinion of the individual.

"And he proclaims Logos the criterion of truth, not of any (Logos) whatsoever, but of that which is common and divine."[22A16(127)]

This common *Logos* is the truth and divinity itself.

"Although this Logos exists for ever, men prove as unable to understand it once they have heard it as before they had heard it."[22B1]

Albeit the *Logos* is common and accessible to all, people cannot comprehend it. Ever-present though it be, they are ignorant of it; "they divorce themselves from that with which most of all they are in continuous contact. And that which they meet daily seems strange to them".[22B72]

"All things come to pass in accordance with this Logos. . . ."[22B1]

The *Logos* is cosmic law, the supreme principle which ordains the becoming of the universe – "the *Logos* governs all things".[22B72]

"One must speak with intelligence and trust in what is common to all."[22B114]

This *common* cosmic law -that is, the *Logos*- is also the law of human *thought*: Nevertheless this *Logos* remains deeply hidden in the recesses of the human spirit: "The limits of soul thou wouldst not discover, though thou shouldst travel every road; so deep a *Logos* it has".[22B45]

This initial tracing of the multiple significanc of the Heraclitean *Logos* raises the question of the special meaning of this word for Heraclitus. The word *Logos* conveyed and continues to convey many shades of meaning in the Greek tongue. *Logos* derives etymologically from *legein*. *Legein* initially meant 'collecting', 'collection', and, in the narrow sense, 'computation'. The sense of the word 'computation' develops in two directions. The one leads to 'recitation' > 'narrative' > 'fable' > 'tale' > 'story' > 'oration' > 'rumour' > 'speech'> 'talk' > 'proposition' > 'statement' > 'commandment' > 'behest language' > 'principle' > 'saying' > 'proverb' > 'maxim'. The other direction leads to 'account' (calculation) > 'estimation' > 'compensation' (balancing an account) > 'report' > 'syllogism' > 'clarification > 'discussion' > 'debate' > 'argumentation', as well as 'reckoning' > 'explanation' > 'cause' > 'reason' > 'pretext' > 'pleas' > 'relation' > 'analogy' > 'correspondence' > 'ration' > 'symmetry' > 'measure' > 'rule'. As for the Heraclitean word *Logos*, does it refer only to 'speech', 'relation', 'measure', or to something more subtle as well? Could *Logos* connote, perhaps, the deeper implications of thought and language and at the same time the significanc of the cosmic firs principle? Could it be that *Logos* is the 'cosmic order', which is likewise the 'order of language and thought'? Or, could *Logos*, perhaps, be the 'rational' and 'wise', which "men prove unable to understand"?[22B1]

In his ontological approach, Heidegger would lend a deeper dimension to the study of the Hericlitan *Logos*.[19] We have seen that *legein* initially signifie 'collecting'. *Col-lection* presupposes the *se-lection* of something, which remains concealed. The *Logos*, then, is the collected and presented (both as pronouncement and as revelation) unconcealment of beings, that is of the 'Being of beings'. Precisely this revealing of beings from concealment (*lethe*) is *a-lethia* (un-concealment: the truth). Truth comes forth out of the perpetual opposition and conflic of beings, where the gathering and presenting *Logos* unconceals them as a unity ('all [are] one'). "In that great era", Heidegger states, "the saying of the Being of beings contained within itself the [concealed] essence of Being of which it spoke. The secret of greatness consists of such historical necessity".[20] *The Heraclitean Logos is not about 'being' but is 'Being' itself. Logos presents 'Being', the 'Being of beings', and is identifie* with *Truth. Logos = Being = Truth (unconcealment).*

---

[19] M. Heidegger, *Vortäge und Aufsätze, Teil III, Logos (Heraklit, Fragment 50)*, 3–25.

[20] M. Heidegger, *Introduction to Metaphysics*, 103.

Truth

Truth (*a-lethia*) in ancient Greek thought means un-concealment, the letting-step-forth to appearance. "Only through the question, what does all this mean and how can it become", notes Heidegger, "did wonder *begin*".[21] And this is precisely the question that Heraclitus poses: "How could one escape the notice of that which never sets"?[22B16] 'Never-setting' is identical with that which endures, emerges, rises, grows – the 'ever growing' (*aei phyon*), that is *physis* (nature). The significanc of appearing, stepping into 'unconcealment', can only be felt through a deep interrelation with 'concealment'. *Physis* signifie on the one hand the 'never-setting' – that which 'ever stept into unconcealment', yet always with a silent, dynamic reference to 'concealment', *lethe.* Thus is explained Heraclitus's saying, 'that which stands ever in unconcealment loves to remain concealed': "*Physis* [i.e., the constitution of things] loves concealment".[22B123]

The *Logos*, the Truth, the 'never-setting', that is 'nature', the 'ever growing', is identical with the 'everliving'. In the Heraclitean phrase, 'everliving fire' 'everliving' does not mean simply 'eternal', but the fir that is endlessly coming into appearance, symbolizing precisely *Physis* and *Logos.* In this lighting up of truth, 'neither the gods nor men' can remain hidden. 'Truth' *has not yet* taken on the meaning of 'correctness', as it will later, nor is truth *subjective*. Referring to wholeness, the *Logos* is *common* to all, is *for* all, and discloses to those who are *circumspect*: "one must speak with intelligence and trust what is common to all".[22B114]

Logos: Common-Private

"The *Logos* is common".[22B2] The *Logos* as 'common' is the supreme principle that combines and sustains everything – the directive power, which pervades, governs, and arranges the 'single and common world'. The *Logos* is "the thunderbolt that steers all things".[22B64] "Wisdom is one thing, to know the thought by which all things are steered through all things[22B41]".[22] As it is "wise to speak the truth",[22B112] and the "judge of truth" is "common and divine",[22A16(127)] Heraclitus urges mankind to adhere to the 'common': "one must follow what is common".[22B2] Heraclitus uses the word *xynos* (common) intentionally, extending it by word-play to *xyn noo* (with intelligence), that is, an identificatio of the *cosmic* 'common' *Logos* with the 'intelligence, common to all'. Thus, he invites humans who speak rationally (logically – with *Logos*) to place their trust in that which is common, namely the *Logos*: "speak with intelligence and trust what is common to all".[22B114] Heraclitus admits, however, that, "although this *Logos* exists forever, men prove as unable to understand it once they have heard it as before they heard it. For, though all

[21] M. Heidegger, *Vortäge und Aufsätze, Teil III, Aletheia (Heraklit, Fragment 16)*, 55.

[22] According to the reading by G. Vlastos and others [Vlastos, G., *On Heraclitus*, American Journal of Philology, 76, 352–353, 1955].

things come to pass in accordance with the *Logos*, men seem as if ignorant".[22B1] The reason is that "although the *Logos* is common, most men live as if they had a private understanding of their own".[22B2] Against the 'common' *Logos*, the 'common intelligence', the vast majority of mankind set their own 'private understanding'. The arbitrary private *Logos* (private reason; language) -as opposed to the common cosmic *Logos* to which Heraclitus invites us to assent- prevents us from discerning unity within diversity, from opening ourselves to the un-concealment of truth, from apprehending the 'invisible harmony' that is "stronger than [the] visible".[22B54] What appears to the many to be random and untidy is in reality an integral portion of a deeper universal order and cosmic beauty: To the many "that most beautiful world [appears] like a heap of disorderly, scattered things".[22B124]

Man's tragedy is to live and act in endless proximity to that *Logos* which is common and accessible to all -the very *Truth of beings*- yet to respond with absolute ignorance: "The *Logos* with which they are ever in closest contact, they oppose. And those things which they encounter every day, are precisely that which seems to them foreign".[22B72] They believe that personal, subjective truth and the notions of individuals can supersede the one and only truth that pervades the universe. Thus, the many are in a position neither to detect, nor apprehend, nor utter this *common* truth: "when they hear [the *Logos* they] are like the deaf. The saying describes them: though *present* they are *absent*".[22B34] The fina words here etch Heraclitus's merciless critique of the tragic deportment of mankind. Though ostensibly *present*, human beings are *absent*: they know "neither how to hear attentively, nor how to speak".[22B19]

Logos: Law-Measure

The 'common' *Logos* sustaining all things discloses the deep harmony, the *homologia* (agreement), between human *reason* and the rational cosmic *order*. On the one hand, the *Logos* is identifie with human rationality: "thought is common to all".[22B113] The *Logos* is the law of the reasoned *thought* of man. On the other hand, the *Logos* controls *cosmic* becoming: "all things come to pass in accordance with this *Logos*".[22B1] Heraclitus calls upon man to apprehend this cosmic law: "Wisdom is one thing, to know the Thought by which all things are steered through all things[23]". [22B41] Such knowledge is possible to the degree that the individual inward rhythm (*Logos*) of our *syl-logism* (thought) is attuned and *homo-Logos* (in unison) with the immanent rhythm (*Logos*) of cosmic becoming. "Wisdom is the highest virtue, and wisdom is to profess the truth and act in accordance with the cosmic law, once we fathom it".[22B112]

Heraclitus sets himself apart from the basic concerns of the firs Ionian 'natural philosophers' in regard to the material and temporal *arche* of the world. What concerns him is the *interrelationship* of beings, the inherent regulating law – that

---

[23] According to the reading of G. Vlastos and others.

very *Logos* which pervades the universe. Positing the *Logos* as the guiding principle, he affirm an immanent deterministic *order* in the *cosmos* – eternal becoming, which precludes any *transcendent* intervention or *random* development. Here *Logos* means *relation*, the invisible harmonious *ana-logia* (proportion), the 'due measure', the 'rules', which prevail mutually in human thought, community life and cosmic order: "One must speak with intelligence and trust what is common to all, as a city in its law and much more firmly for all human laws are nourished by one, the divine, which extends its sway as far as it will and is sufficien for all and more than sufficient" [22B114]

Cosmic order is secured "according to irresistible destined necessity",[22A5] and is watched over by Justice on both the personal and communal levels -"justice will convict those who fabricate and testify to lies"[22B28]- as well as in the cosmic sphere: "the sun will not overstep his measures; otherwise the *Erinyes* (avenging deities and restorers of order), servants of Justice, will fin him out".[22B94] Excess leads to untruth, to arrogance and corruption, to *hybris* (insolence). "Insolence", Heraclitus warns, "must be quelled more promptly than a conflagration" [22B43]

A man can apprehend the *Logos* to the degree that the inward *Logos* of his mind rejects all traces of the 'personal', the 'private', and becomes *homo-Logos* (in unison) with the cosmic wisdom. Unconcealment will then come about, and he will apprehend the single Truth of beings, that *all things are one*: "Listening not to me but to the *Logos* it is wise to agree that all things are one".[22B50]

Logos: One-All

The *Logos* is finall epitomized in these two precious words: 'All One'.[22B50] The 'One' is at the same time 'Wisdom': "Wise is only the One, which knows everything, that is: how to steer all things through all things[24]".[22B41] The 'One', 'Wisdom', which co-ordinates and controls the 'All', is clearly detached from and independent of beings as a whole: "*Wisdom* is separate from *all* things".[(22B108)25]

The human approach to the interrelation "All [are] one", is different from the stance of the divine. Humans must approach the 'One' through reason, trusting in what "is common to all".[22B114] Man discerns the differences among the entire range of beings: "Even if all beings went up in smoke, the nose [i.e., the senses] can detect the difference [between them]".[22B7] Only man differentiates between justice and injustice. On the contrary, for the divine, the unifying power, the 'One', cancels any distinction: "For the divine everything is beautiful, good and just, while men are the ones who consider some things just, others unjust".[22B102] Thus, the 'One' is not a part of the 'All'. It is a totally separate factor, which achieves the *unity* of the 'All'. Consequently, the 'All' is not 'wholeness' (*Ganzheit*), but the 'entirety'

---

[24] According to the reading of K. Reinhardt.

[25] The term 'separate' constitutes 'the most difficul issue in Heraclitus', Heidegger observes. (M. Heidegger and M. Fink, *Heraklit*, 45).

(*Gesamtheit*[26]), as Heidegger puts it. From the entirety of beings arises the One, and from the One the variety within the entirety: "from all things one and from one all things".[22B10]

Wisdom

"*The one wise thing...*".[22B32] The 'One', the union of all things, is 'Wisdom' alone. Heraclitus appears to attribute to the word wisdom (*sophia*) a double meaning:

(a) On the one hand 'Wisdom' tends to be identifie with the cosmic *Logos*, which is independent of human thought. Thus, 'Wisdom' shares the traits of the *Logos*: unity ("the one wise thing"[22B32]), distinction from all other beings ("Wisdom is separate from all things"[22B108]), Truth itself ("It is wise to speak the truth"[22B112]), and the controlling power of the universe, "Wise is only the One, which knows everything, that is: how to steer all things through all things[27]". [22B41]
(b) This last saying of Heraclitus contains the transition to the second meaning of *wisdom*. For there are many who interpret this fragment as follows: "Wisdom is one thing, to be skilled in true judgement how all things are steered through all things[28]".[22B41] Here 'Wisdom' takes on its human dimension, signifying 'correct and wise thought, perfect and deep knowledge'. According to Heraclitus, human wisdom lies in the profoundly significan key term *homo-Logos* (in unison), which expresses the accord and tuning between *human* 'thought' and the *cosmic Logos*: "Listening not to me but to the *Logos*, it is wise to agree that all things are one...".[22B50] Only through this consent, acceptance, and *homo-logia* (agreement; concession), to the degree that man can expel 'private' thought and embrace 'common' Wisdom, can true virtue come about: "Wisdom...the highest virtue...to tell the truth and act in accordance with the cosmic law, having come to understand it".[22B112] To what degree, however, is such 'understanding' possible for mortal man?

Knowledge

Heraclitus firml believes that man possesses the *potential* to achieve knowledge. Reason is accessible to all men.[22B113] Reason applies both to self-knowledge and understanding of other beings: "all men have a portion of self-knowledge and

---

[26] M. Heidegger and E. Fink, *Heraklit*, 50–51.

[27] According to the reading of K. Reinhardt.

[28] According to the reading of G.S. Kirk, et al.

wisdom".[22B116] The achievement of that 'common wisdom', however, is extremely demanding, and success depends on the individual himself: "Those who seek gold dig much earth and fin a little".[22B22] Unfortunately, most "act and speak like men asleep".[22B73] Heraclitus would use sharp words about people like this. They are dull and brutish.[22B29] Only a few, the 'best' (the *aristoi*) set themselves apart from the crowd – those, who 'love wisdom', the *philo-sophoi.* "Men who are philosophers must be inquirers into many things indeed".[22B35] For such searching and insight, it is evident that one must work with the senses as well as the mind: "Things that can be seen, heard, learned -these are what I prefer".[22B55]

For Heraclitus there is not as yet a split between the mental, spiritual and sensuous world. Man understands and experiences with his whole being. There is no distinction between rational concepts and empirical knowledge. On the contrary, there is full co-operation, so that the proper processing of sense data is based on the sensitive capacity of the mind and soul to correctly assess them. Thus, Heraclitus would write: "eyes and ears are bad witnesses to men if they have souls that understand not the language".[22B107] However, even the knowledge of the *few* who are *philosophers* remains tragically problematic and relative: "unfortunately, for men appearances are deceptive";[22B56] and "the most trustworthy of men perceives and assimilates deceptive appearance".[22B28] Neither, of course, does extensive learning -however useful- lead to reason and perfected insight: "erudition does not teach sense",[22B40] Heraclitus observes.

How, then, can we arrive at absolute knowledge, at 'Wisdom' and *homo-logein* (agreement), given that "divine things for the most part escape recognition because of unbelief"?[22B86] Unlike Parmenides, Heraclitus does not suggest a systematic method; he simply points to the road: "[Listen] not to me but to the *Logos*, it is wise to agree that all things are one".[22B50] The utter realization of our total ignorance constitutes the sole springboard whereby we may approach and attune ourselves to the 'common and divine' *Logos,* which discloses and illumines the only Truth -the unity of all beings.

The Soul

The divine, cosmic *Logos* is the very soul of man. Unapproachable and unsearchable, the *Logos* resides at soul's bottomless depths: "The limits of the soul wouldst thou not discover though thou shouldst travel every road: so deep a *Logos* it has".[22B45] The road to the *homo-logia* (agreement; concession), to the concord with the cosmic *Logos*, inevitably leads through the deeps of the human soul. Thus, Heraclitus says, "I searched myself".[22B101] This response to the Delphic injunction, 'Know thyself' is very careful. The philosopher does not claim, 'I came to know myself' -a condition which he has realized is unachievable- but with deepest humility he states that he has *sought* and *inquired of* himself. This quest will lead away from any trace 'private thinking' toward the 'common' *Logos*, thus coming himself to be one with the order and rhythm of the Universe, in the acknowledgement and experience of 'All-One'.

Heraclitus would be the firs of the Presocratics to speak exhaustively concerning the soul. In him as yet resides the age-old idea of the affinit and union of the 'objective' outward and 'subjective' inward truth. But ceasing to be that Homeric 'smoke', 'shade', or 'dream', the soul has become for the firs time the in-formed, reasoned bearer of thought, the well-spring of man's inexhaustible intellectual development: "The soul possesses reason which grows of itself",[22B115] he would write. This polysemic Heraclitean saying Hegel would place at the opening of his firs great work, *Phänomenologie des Geistes.*

"The soul is a spark of the stellar substance".[22A15] Spiritual and material elements are thus engaged in common in the unending cycle of cosmic becoming. Later Aristotle and the Stoics would attribute to Heraclitus the concept of 'cosmic soul'. The soul of the *cosmos* and the soul of living things are *homo-genoi* (congeners).[22A15] The 'homogeneity' of the ever-moving and ever-living essence of the *cosmic* soul with that of *human* soul, enables the latter to be in tune with, and consent to, the former. "For the firs time", K. Reinhardt observes, "we encounter here the interrelation of microcosm and macrocosm as process and principle – as a conscious fulfilmen by means of an intellectual quest and pursuit of the heart's deepest need for the divine".[29]

Death

The 'homogeneity' of *microcosm* and *macrocosm* also extends through the dimension of time. While human activity and sleep are mirrored by the periodicity of day and night, the larger cycle of time corresponds to life and death. Thus, sleep is an intermediary state between life and death; and in both cases the soul leaves the body and the senses – in sleep, partly (breathing its sole contact with the outer world), totally in death. Clearly death plays a crucial role in Heraclitus's thought, thus leading to the definitio later to be articulated by Plato: "Philosophy is the study of death".[30] Yet, at bottom, death remains beyond our ken: "When men die there awaits them what they do not expect or think".[22B27]

Heraclitus's precise conception of the 'immortality' of the soul is difficul to reconstruct from the extant fragments. His total scorn toward the deceased body ("Corpses are more fi to be cast out than dung"[22B96]) suggests the conception of an imperishable immortal *soul*. However, any trace of *individuality* appears to vanish in death, when the soul as an impersonal, ethereal breath joins and assimilates to the cosmic Whole: "The soul is immortal; for when it departs, it returns to the soul of the all, to its likeness".[22B96] So long as the 'immortal' soul resides in the 'living' mortal body (*soma–sima*: body-grave), it tends toward a condition of 'death', while on the advent of 'natural death', it once more find its 'immortal' nature, reuniting with everliving' fire This appears to be the implication of Heraclitus's oracular

[29] K. Reinhardt, *Parmenides*, 193–194.

[30] Plato, *Phaedo*, 64a.

saying: "Immortal mortals, mortal immortals, living the death of others and dying their life".[22B62] On earth we experience the soul's death; its life comes through the body's death. So long as the mortal (body) lives, the immortal (soul) remains dead. "We live the death of their souls and they live our own death".[22B77] *Microcosm* and *macrocosm* are inseparable constituents of one living Whole ruled by a single, *Logos* -common for all- in the antithetical union of all things.

Cosmology-Physics

It is precisely the antithetical union of all things that is the core of Heraclitus's cosmology. What is Heraclitus's conception of the natural world? Is there validity in the disheartening opinion of some commentators that "Nothing [on this matter] is stated with clarity"?[22A1(8)] This view holds in the event that one is looking for an extension of the Ionian cosmological tradition entailing a specifi element as *arche* of the universe. Heraclitus, however, is no longer concerned with the Milesians' cosmo-*gonies*, i.e., the *origin* of the universe; instead, he focuses on the dialectic relation 'One-All', upon which he would systematically construct a new cosmo-*logy*, i.e., the *structure* of the universe. Heraclitus would perceive the universe as a vital organic whole, eternal, orderly, common to all, characterized by perpetual cycles of alternation in which one facet arises from the transformation of another. This continuous process is governed by strict determinism. Let us examine what is involved in the Heraclitean theory of the natural world.

"Eternal, orderly, common to all."

Heraclitus states, "this well-ordered world, the same for all, none of the gods nor of men has made, but it was always and is and shall be: an everliving fire which is being kindled in measures and extinguished in measures".[22B30] There is neither beginning nor end in the *order* of the *cosmos*. The universal *order* neither arose from -nor will it conducts to- a condition of *disorder* through human or divine intervention. Harmonious cosmic order existed before, exists now, and will always exist.

"Perpetual process of cyclical alternation, arising from the dialectical interrelation, life death."

Heraclitus is not so much concerned with the arrangement of matter in the universe, as with the *process,* which goes on within it. His ideas are not *static*, but *dynamic*. He is not concerned with *space*, but with *time*. To the Anaximanderean *spatial* arrangement of the universe, Heraclitus opposes the perpetual *temporal*, periodic alternation "day-night, winter-summer, war-peace, satiety-hunger".[22B67] For Heraclitus *cosmos* is not characterized by its seemingly crystallized *static material structure* but by a deeper *continuous process and alternation*. The alternation is temporally retrogressive, not linear as it may at firs appear. This unceasing re-cycling in which beginning and end coincide, derives -and this is one of the fundamental features of Heraclitus's cosmology- from the dialectical interrelation, *life-death*. The birth of one element springs from the death of another, "fir lives the death of earth and air lives the death of fire water lives the death of air, earth that of water".[22B76]

"The interrelation 'fi e-cosmos' as expression of the Heraclitean Logos 'One-All'."

The *opposed* interrelation is expressed by the 'turning of fire to the plurality of the world and vice-versa: "from the one all, and from all the one".[22B10] *Fire* does not constitute a part of the world's composition but is above and beyond the continuous physical cosmic processes, precisely as precious gold is beyond the other material things: "All things are an exchange for fir and fir for all things, as goods for gold and gold for goods".[22B90] On the other hand, however, the *cosmos* is *'fi e'*. Underlying the visible world of endless change and the clash of opposites lies the hidden inner *identificatio* of the cosmos with *ever-living fi e*, symbolizing precisely that which above all epitomises the universe: its unquenchable power of life and motion, continuous transformation, vital activity.

"'Fire' as symbol of the strict determinism governing the universe."

There is a determinism based on measure and balance: The world is "ever-living fir ... kindled in measures and extinguished in measures".[22B30] Heraclitus was, in turn, to choose another highly visual illustration to prove the rule of measure that pervades the world: "The sun will not overstep his measures; otherwise the *Erinyes* (avenging deities and restorers of order), servants of Justice, will fin him out".[22B74] Any violation of balance is detected and rectifie by *Dike* (justice). Elsewhere Heraclitus would refer to the preservation of order "in accordance with the inescapable fate of necessity".[22A5] This is a supreme law which, however, should not be conceived as an outside dominant power that intervenes in cosmic becoming, but as an intrinsic deterministic principle that inheres within the *cosmos*; as K. Axelos would put it, "the universe is its own destiny".[31] The terms *metron* (due measure), *Dike* (justice), *Erinyes* (avengers of unfilia conduct and upholders of the natural and moral order), *Eimarmeni* (destiny) show that the supreme rule that governs the *universe* is for Heraclitus the very same rule that controls the ethical and social behavior of *mankind.*

Ethics – Community – Law

One can summarize Heraclitus's ethical categorical imperative with his saying, "we should tell the truth and act in accordance with cosmic law [that is, the *Logos*], having come to understand it".[22B112] The philosopher warns his fellow men that "although this *Logos* exists forever, men prove unable to understand it[22B1] ... [and] although the *Logos* is common, most men behave as if they had a private understanding of their own".[22B2]

On the basis of this statement, Heraclitus would bring such a pitiless indictment against the behavior of the crowd that one can understand the distance his fellow citizens kept from him throughout his life. He condemns their softness and sensuality.[22B4] He condemns passivity and apathy.[22B29] He condemns inertia and blind conformity to tradition.[22B74] He censures wealth and also distinctions

---

[31] K. Axelos, *Héraclite et la Philosophie,* 106.

as causes of depravity.[2B125a; 22B132] He considers malice a direct expression of falsity and fabrication: "Bad people are averse to true men".[22B133] He attacks arrogance.[22B131] He recognizes, however, what an enormous effort it takes for one to reign in his passions and desires: "To figh with desire is hard: whatever it wishes it buys at the cost of soul".[22B85] In spite of his disdain for the careless throng, Heraclitus is concerned about his fellow man who has lost the way to virtue.[22B71] Accordingly, he would attempt with various imperatives -'one must not' and 'one must'- to nudge the mindless toward rumination and the truth: "*One must not* act and speak like men asleep";[22B73] instead, "*one must follow* what is common, the Logos,[22B72] ... for all things come to pass in accordance with this very *Logos*".[22B1]

Since Heraclitus considers man and the community indestructibly enrolled in cosmic *becoming*, the ethics of individual and social relations are harmonized and co-ordinated by the rational natural order, the 'common *Logos*'. To act and behave ethically, then, is living 'according to nature'. "One must trust in what is common to all"[22B114] – that is, the *Logos*. Drawing power from the *Logos* the individual breaks the bonds of "illusory private perceptions"[22A16(134)] and develops his attunement with the eternal *Logos*: "The best renounce all for one thing -eternal glory rather than mortal possessions".[22B29] Above all, we must harmonize with the *Logos* as the 'due measure', which determines the limits of the personal, social and natural world. All overbearing pride (*hybris*), must shrink: "insolence must be quelled more promptly than a conflagration" [22B43]

In Heraclitus, as in Xenophanes, the individual himself is held exclusively responsible for his actions. The archaic Greek myth viewed man's life as arbitrarily determined by *Moira,* the goddess of destiny – an external demonic power. Heraclitus locates the *daemon* (spiritual, guardian powers) within man: "A man's character is his *daemon*".[22B119] Inescapable destiny continues to exist for Heraclitus. "Without question there always are destined events",[22B137] he declares. However -and here lies the fundamental difference from the mythic tradition- this does not mean that man is not free. Personal *freedom* and inescapable *fate* are not incompatible but complementary: man is *free and responsible* to choose between *self-entrapment* within his illusory 'private thinking' and bestial life, on the one hand, and on the other, *self-opening up* toward the "common and divine *Logos*".[22A16(127)] The latter choice leads to *homo-logia* (to the agreement and oneness with *Logos*), *conscious* consent and submission of one's individual being to the common and divine, cosmic *Logos,* which comprises the *due measure* and the *supreme law* -natural and ethical- of all beings.

Just as in nature "the sun will not overstep his measures",[22B34] so in the city "a right-minded people will defend the law as they would their city's walls[32]".[22B44] In turn, as in the universe cosmic Law imposes natural order (if the sun oversteps measure, "the *Erinyes*, servants of Justice, will fin him out"[22B94]), so in the city society's *Dike* (Justice) will punish the offender: "It is certain that Justice will con-

---

[32] Many would perceive here a limit upon individual freedom, in accordance with Heraclitus's other saying, "It is also a law to obey the counsel of the One".[22B33]

vict those who fabricate and testify to lies".[22B28] *Natural* laws and *social* justice are identical, since perpetual strife and conflic between opposites comprise not only a *natural* but also *social* phenomena: "One must know that war is common, and justice strife, and that all things come about by way of strife and necessity".[22B80]

Heraclitus is very clear on this point. *Justice is strife.* Justice *is* finall inherent in the conflic itself; otherwise there would be no reason for the latter to exist. The endless change that comes from continuous strife is governed by Justice as a necessity imposed by Law in its broadest possible meaning: ethos, custom, civil law, and, of course, natural law. For the firs time ethical principles are expressed as a reflectio of cosmic, divine law: "All human laws are nourished by one, the divine".[22B114]

The Divine

The 'divine' could be approached and described from three perspectives: the divine in relation to *religious worship*; the divine in relation to the *individual*; and the divine in relation to *attributes* ascribed to it by Heraclitus.

Heraclitus does not hesitate to level harsh criticism against the religious rituals and popular cults of his time.[22B5; 22B14] He attempts to exploit logical argumentation in his attack against popular religiousness: He asks the Egyptians who are mourning in honor of Adonis, "if these are really gods, why are you mourning for them, and if you mourn them, they are not gods".[22B127] In turn, to Greeks bringing offerings to the gods, he says: "They shower honors on statues of gods that cannot hear them, as if they could hear them, upon statues of gods that do not respond in any way, just as they also make no demands".[22B128] He would be the firs to perceive the *symbolic* character of all these popular religious rituals, and only in this respect would he accept their meaning.[22B15] "In his discovery of the symbolic character of religion", Nestle observes, "Heraclitus brought to light a source of endless misunderstandings on religious matters, preserving nevertheless the justificatio of religion through the open recognition of its nature".[33] Also perceiving the *psychological* influence that religious rites may exercise upon the worshipper, Heraclitus refers to 'curing', "as tending to cure our troubles and the disasters attendant on generation".[22B68]

Mortal man cannot be compared to the grandeur of the divine: "Man is infantile in the eyes of a god, as a child in the eyes of a man[22B79] ... Compared with God, the wisest of men will appear an ape, in wisdom, beauty and all else".[22B83] In turn, "Human nature has no insight, but divine nature does".[22B78] This explains why "to God all things are fair and good and just, but men have assumed some to be unjust and some just".[22B102] Yet man does not live estranged from the divine. As we have seen, "all human laws are nourished by one, the divine".[22B114] Accordingly, Heidegger would gloss the Heraclitean saying, "a Man's character is his daemon,"[22B119] as "a man -to the measure that he is a man- lives close to the god".[34] For Heraclitus,

---

[33] W. Nestle, *Vom Mythos zum Logos*, 101.

[34] M. Heidegger, *Platons Lehre von der Wahrheit*, 106

man lives close to god to the degree that he is *homo-Logos* – in unison and in agreement with the *Logos* that 'the One is all things'. The crucial difference between god and man is that the god is 'the only wisdom',[22B32] while man at best remains a *philo-sophos* (a 'lover of wisdom'). Thus, we see that Heraclitus's thought builds a bridge between philosophy and religion, since man's love of wisdom leads him to "the only wisdom",[22B32] god.

Who then, is *god* (singular) for Heraclitus? A man can only answer this question indirectly and according to the depth of his faith, since "Divine things for the most part escape recognition because of unbelief".[22B86] As we have seen, god is "The One, the only Wisdom"[22B32] – the One who has insight".[22B78] God is the righteous judge, and thus if we accept the identificatio of god with fire fir brings down universal judgement: "For fire [Heraclitus says,] will come and judge and convict all things".[22B66] "God is the one who directs the whole".[22B41][35] God is the transcendent being, "the wisdom that is separate from all things".[22B108] In God becoming and the union of opposites are articulated.[22B67] Thus, while men see things as just and unjust, "to god all things are fair and good".[22B102]

Since the 'divine', '*Logos*', and 'fire are described by Heraclitus in similar terms, the question arises as to the interrelationship between the three. The answer is difficult for nowhere in the extant fragments is there an *explicit* formula, divine = *Logos* = fire[36] Here one must resort to Heraclitus's allegorical images that *signify* but do not *name*. "God . . . changes just as fire which when it is mingled with fragrances is named according to the scent of each".[22B67] 'God', '*Logos*', 'fire' are simple 'names', constituting expressions of a *unity*, without being identifie with one another.[37] Divinity is *transcendent*, "separate from all things",[22B108] yet it is also inherent to the *cosmos* as the 'common' *Logos* and fire "Heraclitus's thought", notes Axelos, "strives to surpass opposites through the conception of a unifying divinity. Opposites nevertheless remain in place for mortals for whom their being is an unceasing becoming. . .Each authentic conception contains a contradiction or an antithesis – that is, a tragic element".[38]

The Tragic Element

In the person of Heraclitus we encounter one of the rare moments of humanity, which blend ultimate human *tragedy* with the greatness of man's *dignity*. Every presence of life bears within it the roots of the tragic. Yet this is only understood

---

[35] According to the reading of K. Reinhardt.

[36] Only some later commentators clearly influence by stoic philosophy and Christian thought perceived such identification

[37] Some scholars, like K. Axelos (*Héraclite et la Philosophie* 124) and K. Reinhardt (*Parmenides*, 205–206), would describe divinity as the 'cosmic significance' and 'cosmic reason' (*Weltvernunft*) – but again, not as identifie with the *Logos*, but as union with the *Logos*.

[38] K. Axelos, *Héraclite et la Philosophie*, 133.

by those keen intellects which glimpse in each existence the seed of its inescapable demise. The consciousness of this pernicious fate evokes a profound awe. In the face of existential anguish, striving to master fear and recover his lost feeling of security, man reacts either by *forgetfulness* -as if one could indeed forget- by putting himself under the wing of the crowd, who live a life of illusion, acting and speaking "like men asleep",[22B73] "though present . . . absent";[22B34] or, by *transcendence* -as if one could transcend this tragic awareness- by questing for the *ultimate meaning* through a higher mental and psychic consciousness yet to be known, manifested through *art, meditation*, or *worship.* Heraclitus is the tragic figur who probes these three modes of realization with deep intensity. He is the *poetic thinker* musing on a horizon charged with the *divine* spirit.

If existence itself, 'being', is incomprehensible and inconceivable, then its vanishing is doubly inexplicable and uninterpretable. Heraclitus would feel compelled to capture transience of being in the most shockingly tragic way: Not only as a *fina* , necessary end ("unquestionably there are destined things"[22B137]), but as *continuous change*, a ceaseless, pitiless, relentless *becoming* in which, according to Aristotle's interpretive words, one cannot "lay hold twice on any mortal substance in one permanent state".[22B91] It must have been an exasperation that the philosopher felt in the tragic perception that nothing -nothing whatsoever- remains steady for a moment, that "one cannot . . . step twice into the same river".[22B91] Everything is shifting, transforming; all things are fragmentary, clashing, rending one another. Few enlightened men would be able to accommodate and live with such tragic awareness. Parmenides, the other supreme thinker who became conscious of the problem of 'becoming' at the same time as Heraclitus, would prefer through a set of cold syllogisms to shelve the issue by claiming that in essence there is *no such thing as* 'becoming'; 'change' and 'movement' are only an illusion, since 'Being' is one – eternal, immortal, indivisible, immovable, immutable.

Conversely, Heraclitus -showing an incomparable height of will, power and faith- would decide to fathom the nature of 'becoming'. He had the courage, in the face of this "*cosmos* that seemed like a heap of disordered scattered things",[22B124] to hold on to *hope*; for "if one does not hope, he will not discover that which is unexpected, since it would remain unsought and out of reach".[22B18] He also had the *will* to probe beyond superficia conventional harmony for the *unseen. Hope* and *will* are the key elements, which constitute man's personal *freedom*. From the outset, Heraclitus indeed oppose this personal *freedom* to *destiny* in an anguished search for the ultimate meaning, a constant point of reference within the "endlessly shifting currents"[11B12] of cosmic becoming. "I searched myself",[22B101] he would declare. He would turn to the innermost recesses of his soul, "so deep a *Logos*"[22B45] it has. He would labor greatly in the search, for "those who seek gold dig much earth and fin a little".[22B22] And through an intense, painful effort he would achieve transcendence in an ultimate reconciliation between human *liberty* and human *destiny:* "A man's character is his destiny".[22B119] This was his *homo-logia,* his concession and accord with divine law – a *conscious* choice of acceptance and submission to necessity. Thus, Heraclitus preserves and elevates man's dignity to a new tragic, ontological level, expressed through his freedom.

Indeed it was a titanic struggle to discern the hidden *unity* within cosmic transience, fragmentation and dissension, finall perceiving that "it is wise to agree that all things are one".[22B50] Through the fl wed means of the illusory sense organs ("eyes and ears are bad witnesses"[22B107]) and an elliptical language that forced him to *signify* rather than *name*; in an inconstant environment ruled by fragmentation, separateness, selfishness incoherence, instability and conflict he would discover and express with unparalleled eloquence the depths of a unity and 'most lovely harmony' concealed beneath plurality and flux If this realization of a deeper unity beneath surface multiplicity comprises a lofty intellectual achievement, the tragic insight remains that finall cosmic *conflic* is not transient, not an evil that will pass with time, but is the *fundamental and supreme law* of the universal 'becoming' in which we have lived, live now, and will live till the end of time. The elevation of strife -"everything comes into being by way of strife"[22B8]- to a supreme arbiter of the universe, since "war is father and the king of all",[22B53] is a decisive dramatic element in Heraclitus's thought.

Tragic time comprises the necessary dimension within which cosmic becoming is played out. Heraclitus fully recognizes the hegemony that time wields over events. It constitutes the foundation of reality like a calculated yet fated eternal game of change and movement played by an uninfluence and innocent child: "Time is a child playing with pebbles; the kingdom belongs to the child",[22B52] that is to the time. The game of time, a perpetual succession of generations, of life and death, is dramatically projected by Heraclitus: "When they are born they wish to live and to meet with their doom -or rather, to repose- and they leave children behind them to meet with their doom in turn"[22B20] in an inescapable circle that will forever have neither beginning nor end, since, "on a circle beginning and end are common".[22B103]

Tragic loneliness was the high price that Heraclitus would pay for his ardent seeking and discovery of the ultimate truth in that *common and divine Logos*, in which 'All things are One'. His stature is shown in that his personal loneliness did not make him a misanthrope. His passion and love for wisdom are "the philosophy of one who suffers", U. Hölscher was to write. "It is not amazing, then, that his gaze never again became pleasant; what is amazing is that it did not bear malice".[39]

Overview

Many would bend with admiration over the work of Heraclitus. But none would ever render unalloyed the meaning of the 'enigmatic' Ephesian's utterances. After all, he himself cautioned: "Let us not make random conjectures about the greatest matters".[22B47]

Aside from its juxtaposition to Parmenides's doctrine, Heraclitus's thought does not appear to have had a significan influenc on the subsequent Presocratics, since Empedocles, Anaxagoras, and the Atomists would focus their attention on the prob-

[39] U. Hölscher, *Anfängliches Fragen*, 172.

lem that would be raised by the anti-empirical 'logical' conclusions of Parmenides. The Sophists, by reason of certain positions of Heraclitus concerning perpetual flu and the relativity of tangible perception, would end up in extreme relativism and ontological nihilism. For Protagoras (ca. 480–411 B.C.) the rubric is, "Man is the measure of all things",[80B1] which denies the existence of objective truth. Gorgias (ca. 483–385 B.C.)would arrive at a total denial of any gnoseological value. Plato and Aristotle, influence by the absolute nihilism of Cratylus, would consider ceaseless 'becoming' a weak foundation for knowledge, since with "the whole sensible world ... always in a state of flu ... there is no scientifi knowledge of it".[65.3] Thus, they would limit 'becoming' exclusively to the world of the senses, creating the transcendental realm of *forms* and *divine Mind* as the constant point of reference for Platonic and Aristotelian philosophy.

The Stoics (3rd century, B.C, to 2nd century, A.D) revived Heraclitean thought. They would be the ones who would rescue and comment on many of his fragments, without being able to avoid certain extreme interpretations. Beyond the impact on ancient Greek philosophy, especially on Hellenistic thought, Heraclitus would not be uninfluentia on the western mystics of the Medieval period, Meister Eckhart and Nicolaus Cusanus, as well as on great thinkers of the Renaissance and 17th century like Giordano Bruno, Böhme and Spinosa.

At the beginning of the 19th century, philosopher-theologian F.D.E. Schleiermacher, together with A. Wolf, would collect and publish a collection of extant fragments of Heraclitus, dedicated to Goethe. This valuable edition would for the firs time make Heraclitus's thought directly accessible to the European intellectual world. Heraclitus's 'union of opposites' would have a decisive influenc in the shaping of the dialectical philosophy of Hegel, who views the Ephesian as the founder of dialectic: "Heraclitus conceives the absolute ... as the dialectic itself",[40] he writes in his study of the *History of Philosophy*, acknowledging that "there is not one sentence of Heraclitus that I have not included in my 'Logic' ".[41] To a great degree this position would determine the European philosophical thought of the 19th and 20th centuries. This development would be present in Hegel's followers in both the socialist and conservative camps. Marx's colleague F. Lassalle would publish the study, *The Obscure Heraclitus of Ephesus* in 1857. Proudhon, leader of the French anarchist movement, seems also to be indirectly influence by Heraclitean thought. Marx -and even more so- Lenin, consider Heraclitus to be the father of dialectical materialism. Kierkegaard views him as his mentor, and Nietzsche calls him the forerunner of his own philosophy, 'beyond good and evil', at the same time lending ethical and aesthetic implications to the Heraclitean relation, 'One-Everything'. With unbounded admiration he proclaims that his message "should pervade future millennia. For the world eternally needs truth; consequently, it eternally needs Heraclitus".[42]

---

[40] G. W.F. Hegel, 18 *Vortesungen über die Geschichte der Philosophie*, 319.

[41] G. W.F. Hegel, 18 *Vortesungen über die Geschichte der Philosophie*, 320.

[42] F. Nietzsche, *Die Philosophie im tragischen Zeitalter der Griechen*, 297.

During the 20th century distinguished philologists and philosophers such as K. Reinhardt, O. Gigon, K. Cherniss, W. Jaeger, Hölderlin, Novalis, T.S. Eliot, Char, Breton, Saint-John Perse and others return to the inexhaustible wellspring of Heraclitus. The man who through the thought of Heraclitus as well as that of the other Presocratics would aspire to return to the roots of the primary and genuine meaning of 'Being', 'becoming' and 'truth' would be one of the major proponents of contemporary existential ontology, the German thinker Heidegger. Heraclitean sayings and terms like *Logos, aletheia* (truth), *physis* (nature, the real constitution of things) and *en-panta* (One-All) would constitute objects of profoundest study in Heidegger's philosophy. He would view the Ephesian as "one of the greatest among the great Greeks". Concerning him who wrote, "The thunderbolt steers all things",[22B64] he stated: "Once, at the very beginning of Western thought, the essence of language flashe the light-burst of Being. One time, when Heraclitus pondered the *Logos* as word-guide to meditate by that word the Being of being. But the lightning abruptly died. No one grasped its gleam, nor that which lit up around it".[43]

Heraclitus was the thinker of relation (*Logos*). In the light of *dialectical relations*, we would do well to examine briefl the *natural sciences.*

Given the mystical element in Heraclitus, could it be that his thought is wholly alien to contemporary natural science? The opposite is true. Perhaps no other Presocratic thinker is as timely as Heraclitus. The Heraclitean attitude clearly pervades the entire structure of *contemporary physics and chemistry*. In his study *To Have or to Be*, Fromm observes: " 'What is Being?' has been one of the crucial questions of Western philosophy ... I shall mention only one crucial point: the concept of *process, activity, and movement as an element in being.* As George Simmel has pointed out, the idea that being implies change, i.e., that being is *becoming*, has its two greatest and most uncompromising representatives at the beginning and at the zenith of Western philosophy: in Heraclitus and in Hegel... Heraclitus' and Hegel's radical concept of life as a process and not as a substance I paralleled in the Eastern world by the philosophy of Buddha ... Contemporary scientifi thought has brought about a renaissance in the philosophical concepts of 'process thinking' by discovering and applying them to the natural sciences".[44]

Some writers have attempted to draw parallels between the concept of the world in contemporary physics and in Eastern philosophy.[45] The similarities, however, can only be superficial In contrast, both *historically* and *essentially*, Heraclitean thinking comprises the creative nucleus of contemporary scientifi theory. We will confin ourselves to noting only three basic points of difference between Eastern mysticism on the one hand, and on the other, Heraclitean thought and Western science.

---

43 M. Heidegger, *Vortäge und Aufsätze, Teil III, Logos (Heraklit, Fragment 50)*, 25.

44 E. Fromm, *To Have or to Be*? 25–26.

45 For example, F. Capra, *The Tao of Physics*.

- Eastern mysticism is concerned above all with *experiencing*, not with *describing* and *interpreting* reality.[46] Such 'experiences' are not referred to in Heraclitean writings or in Western thought.
- The Eastern systems contain an element of salvation, recommending and showing the paths of redemption (marg, raja yoga, karma yoga, bhakti, tri-ratna) toward ultimate deliverance (moksha, kaivalya, nirvana). The idea of salvation and deliverance is alien to Heraclitus and Western science.
- Eastern mysticism views plurality as an illusion and its various expressions as degrees toward the ultimate union with the divine. Heraclitus does not privilege unity, but centers his attention on the *inter-relation as such* between opposites, between plurality and unity. This is also the attitude of contemporary physical science. Pauli, one of the leaders in the shaping of physics in the 20th century, writes: "Mysticism seeks the unity of all outward things, and the unity of the inward world of man with those things, entailing the plurality of phenomena as illusory and unreal. . . . Thus the unity of man with the divine is created step by step – in China with the Tao, in India with Samadhi or Nirvana. Conversely, it is in accordance with *Western scientifi attitudes*, or in a certain sense, one can say, with *Greek* ideas, to pose the question, for example: Why is One reflecte in the Many? What is that which is reflected and what is the reflector Why did the One not remain alone? What is that causes what we call illusion"?[47] These are precisely the questions, raised firs by Heraclitus, that remain unanswered after two and a half thousand years.

The dialectical thought of Heraclitus turns upon two axes: (a) the unseen 'union' of all things, and (b) the endless 'becoming' of all things. Both *union-plurality* and *process* stand within a dynamic dialectical relation (*Logos*) of balance. On these two very axes, contemporary natural science also functions. Let us attempt a brief historical review: The crucial problem facing physics at the beginning of the 20th century derived from the fundamental question raised by Zeno concerning the 'continuous' and the 'discrete'. As to the nature of *matter* Leucippus and Democritus taught that the universe consists of invisible and indivisible particles called 'atoms', and this atomic theory held intact up to the beginning of the 20th century. As to the nature of *light*, developments were more complex. During the 17th century, Huygens developed the *wave* theory of light, while conversely Newton supported the *corpuscular* nature of light. A hundred years later, Young and Fresnel experimentally verifie its wave character, which was established by the electromagnetic equations of Maxwell. Consequently, at the end of the 19th century, we fin ourselves with two *independent* established approaches to physics – the physics of the 'continuous' (with properties like: wave, frequency, wave-length, diffraction, interference, etc.) and the physics of the 'discrete' (with properties like: particles, mass, energy, collision, momentum, electrical charge, etc.).

---

46 F. Capra, *The Tao of Physics*, 29, 281.

47 W. Pauli, *Die Wissenschaft und das abendländische Denken*, 103.

In the meantime great strides have been achieved toward the *unificatio* of seemingly totally different theories, following the Heraclitean dialectic method that leads to the union -not the rejection- of two different separate phenomena within an *integral* theory. The firs stride would be taken by Newton who in the theory of universal attraction united two hitherto separate phenomena, falling on earth caused by gravity and the motions of the planets in the sky. Thereafter would come the unificatio of mechanics and the theory of heat by Boltzmann, and the unificatio of electricity and magnetism in the electromagnetic theory of Maxwell.

No physicist, however, could have imagined prior to 1900, that a unificatio of the *corpuscular* theory of matter and the *wave* theory of light would have been possible. The reason is simple: *Particle* and *wave* are two concepts that are not only contrary, but are *contradictory,* since the one excludes the other, while there is no third solution (tertium non datur). Consequently, the whole neatly-built edific of 19th century physics would crumble when Planck confirme in 1900 that light is emitted in discrete quantities of energy (quanta) which lend to light, and to electromagnetic waves generally, corpuscular properties, since energy (E) consists of discrete multiple elemental quantities of hv (photons).[48] The years immediately following experimentally established both the 'discrete', corpuscular nature of light (the photoelectric effect explained by Einstein, the Compton effect, etc.) as well as its *wave* nature (interference, diffraction, etc.). On the other hand, it would be discovered that matter -the electron, for example- exhibits a mixed behavior: Though it is *a particle*, it does not follow an orbit but it is described by a wave function ψ, the square of which ψ^2 determines the probability of locating the electron in a certain position. The *wave-like* character of the electron would also soon be verifie experimentally (1927: Davisson & Germer; G. Thompson). Thus, contemporary physics discovers the dual nature *wave-particle* both of matter and of light (electromagnetic waves). The formerly contradictory concepts 'wave' and 'particle' are unifie into the *wave-particle* notion.

During the firs quarter of the 20th century, the new theory of quantum mechanics would establish the theoretical mathematical interpretation of the *synthesis* of these two *contradictory* concepts. At the same time it would lead to the *interrelation* of different phenomena and forces under *unifie theories*, such as the *unificatio* of the various forces in physics and chemistry, the *unificatio* of kinetic mechanics and electrodynamics through Einstein's special theory of relativity, the *unificatio* of quantum mechanics and the special theory of relativity through the quantum electrodynamics of Dirac, the *unificatio* of electromagnetic and weak interactions – through the breaking of the gauge symmetry between them. The attempt to unify the four fundamental forces (electromagnetic, weak, strong and gravitational) into a single 'Theory of Everything' (TOE) is the goal and dream of physicists today. In

---

[48] Planck's familiar formula E = hv for the firs time relates the physical values of two previously different natural domains, that of the 'discrete' (E = energy) and the 'continuous' (v = frequency). The h is 'Planck's constant'.

the 20th century physics functions fully in the Heraclitean spirit, as indicated by the summary and parallel thinking that follow.

"Heraclitus was the fi st who dared to announce the seemingly unprovable and logically inconceivable identificatio of opposite or contradictory concepts within a unity, as well as the hidden harmony which is more beautiful than that which is visible."

Contemporary physics, as we have seen, follows precisely the same dialectical path, all the way to the unificatio of two mutually exclusive contradictory concepts, 'particle' and 'wave'. Pauli describes the Heraclitean, direct, interdependent relationship between *unity* and *plurality* in physics as follows: "Phenomena in atomic physics possess a new *holistic* property, since they cannot be divided into *separate phenomena*, without essentially transforming each time the phenomenon as a whole".[49] For the two key founders of contemporary physics, Heisenberg and Pauli, "the ultimate goal was *unificatio* ".[50] "Looking back to the development of Greek philosophy from its beginning up to this point", Heisenberg notes, "one notices that it has been carried from the beginning by the tension between One and Many. For our senses the world consists of an infinit variety of things and events, colors and sounds. But in order to understand them we have to introduce some kind of order; order means: to recognize what is equal; order means some sort of unity. From this springs both the belief that there is one fundamental principle, and the difficult to derive from it and explain the infinit variety of things.... This leads to the polarity between 'Being' and 'Becoming' and finall to the solution of Heraclitus, *that change itself is the fundamental principle*". Furthermore, Heisenberg continues, "modern physics is in some way extremely near to the doctrines of Heraclitus. If one replaces the word 'fire by the word 'energy', he can regard Heraclitus's statements almost word by word as expression of our modern concepts... Energy may be regarded as the fundamental cause for all change in the world".[51]

"Heraclitus was the fi st to announce that the static view of nature is illusory and that the universe consists basically of processes manifested in an endless 'becoming', with 'due measure'."

Contemporary physics is in complete accord with this dynamic view. "The notion", notes D. Bohm, a widely known contemporary theoretical physicist, "that reality is to be understood as process is an ancient one, going back at least to Heraclitus, who said that everything fl ws. In more modern times, A.N. Whitehead was the firs to give this notion a systematic and extensive development ... I regard the essence of the notion of process as given by the statement: not only is everything changing, but *all is flux* That is to say, *what is* is the process of becoming itself, while all objects, events, entities, conditions, structures, etc., are forms that can be abstracted from this process. The best image of process is perhaps that of the fl wing

---

49 W. Pauli, *Naturwissenschaftlche und erkenntnistheoretische Aspekte der Ideen des Unbewusten*, 115.

50 W. Heisenberg, *Wandlungen in den Grundlagen der Naturwissenschaft*, XXVII.

51 W. Heisenberg, *Physik und Philosophie*, 44–45.

stream, whose substance is never the same".[52] Thus, twenty-f ve centuries later we return to the Heraclitean river image: "Heraclitus says somewhere that everything ceaselessly moves forward and nothing stays still, and he likens beings to the current of a river, saying you cannot step twice into the same river".[53]

Physicists would attempt to 'describe', insofar as this is possible, the new situation created by quantum mechanics: "According to the assumptions of contemporary physics", observes physicist H.-P. Dürr, "The particle does not exist anymore in the old sense. This means that there is no localized object *temporally identical with itself*. . .In terms of quantum mechanics, no temporally continuous objectifie world exists, but this world in some way *is created anew each moment*"[54] D. Bohm theorizes that "the new form of insight can perhaps best be call '*Undivided Wholeness in Flowing Movement*'. . .There is a universal flu that cannot be define explicitly but which can be known only implicitly . . . In the fl w, mind and matter are not separate substances. Rather, they are different aspects of one whole and unbroken movement".[55]

> *"Heraclitus would announce that "all things move eternally, but that evades our senses"*[56]*. Everything is involved in endless movement, conflic and change. "One cannot . . . lay hold of any mortal substance in one permanent state. Owing to the impetuosity and speed of change, it scatters and brings together again, or rather, not again, nor later, but at the same time it comes together and flow away, approaches and retreats."*[22B91]

This is the very spirit of contemporary chemistry. Macroscopically, matter appears inert and motionless; nevertheless, microscopically the atoms of which matter consists of are in ceaseless motion. Electrons, according to classical atomic theory, move around atomic nuclei with inconceivable speed. All the molecules that are made up of atoms are involved in sustained movement of *transfer, rotation, and oscillation.* Most important, these movements are not random, but observe specifi relations of 'ratios' of the number of oscillations per time unit (frequencies).

Chemical reactions are not static but are in dynamic equilibrium. This means that although nothing 'appears' to change in the state of equilibrium, nevertheless at a molecular level there are tremendous and continuous agitation and collisions, as the end (AB) formed by the forward reaction $A+B \rightarrow AB$ continuously decomposes at the same rate $AB \rightarrow A+B$. Thus what comes about is not the apparent one-way reaction $A+B \rightarrow AB$, but the dynamic, reversible equilibrium $A+B \rightleftharpoons AB$, which constitutes the contemporary chemical statement of the Heraclitean concept of nature as perpetual movement, conflic and change, always with 'due measure': "All things come to pass in accordance with this *Logos*".[22B1] This dynamic Heraclitean

---

[52] D. Bohm, *Wholeness and the Implicate Order*, 48–49.

[53] Plato, *Cratylus*, 402a.

[54] H.-P. Dürr, *Wissenschaft und Wirklichkeit, über die Beziehung zwischen dem Weltbild der Physik und der eigentlichen Wirklichkeit*, 37, 38.

[55] D. Bohm, *Wholeness and the Implicate Order*, 11.

[56] Aristotle, *The Physics*, 253b10.

'equilibrium' of bi-directional chemical reactions that beneath the seeming static surface controls and directs the hidden tension and collision, constitutes the very expression of life itself. '*The same and not the same*' could be one of the sayings of the 'enigmatic' Heraclitus. It is not that; instead it is the title of the book by R. Hoffman, Nobel laureate in chemistry (1981),[57] which describes the dialectic tension upon which the structure of contemporary chemistry is based, explaining that the opposites are in such a complex dynamic interrelation that they cannot be distinguished from one another.

"Heraclitus emphasizes the cyclic character of interrelations and change: "It is death of earth to become water, and death of water to become air, and of air to become fi e, and the reverse[22B76] *. . . On a circle beginning and end are common."*[22BB103]

Today we know numerous such *cyclic* processes, in *physics* (e.g. Carnot's cycle), in *chemistry* (e.g. catalytic reactions with organometallic compounds), in *biochemistry* (e.g. Krebs's or citric acid cycle, urea cycle, glyoxylate cycle, dicarboxylic acids cycle), in *biogeochemistry* (carbon cycle, nitrogen cycle, phosphorus cycle).

"Heraclitus discerned behind the seeming disorder and agitation a deeper decorum and order. "Invisible harmony is stronger than the visible",[22B54] *he would say. "The most beautiful cosmos seems to be a heap of disorder, scattered things."*[22B124]

Two and a half millennia later, physical science begins to grasp these enigmatic words of Heraclitus. Prigogine, Nobel laureate in chemistry (1977) and Stenger in their book *Order Out of Chaos* write: "For a long time turbulence was identifie with disorder and noise. Today we know that this is not the case. Indeed, while turbulent motion appears as irregular or chaotic on the macroscopic scale, it is, on the contrary, highly organized on the microscopic scale".[58]

"Heraclitus was criticized as 'obscure'."

Was Heraclitus's obscurity due to personal inability, or to the objective diffi culty in expressing the very nature of 'becoming' and the 'union of opposites'? The answer comes from contemporary physics: "Aristotle's logic," physicist H. Pietschmann observes, "forms the foundation of our attitude toward our life in the world and thus toward 'clarity'. Consequently, *quantum mechanics* as a seeming *contradiction* is basically unclear. It engenders a feeling of imperfection and creates uncertainty".[59] This feeling of.unsatisfaction led some of the main founders of 20th century physics -Planck, Schrödinger, Einstein, and others- to question quantum mechanics,[60] just as Aristotle had rejected the Heraclitian concepts in ancient times:

---

57 R. Hoffmann, *The Same and Not the Same.*

58 I. Prigogine and I. Stengers, *Order Out of Chaos*, 141.

59 H. Pietschmann, *Phänomenologie der Naturwissenschaft*, 206.

60 The two latter scientists, with the familiar thought experiements, 'Schrödinger's cat' and the 'EPR paradox', attempt to demonstrate that quantum mechanics are in conflic with the generally accepted assumptions of physical science. The experimental evidence of the 'EPR paradox', however, confirm quantum mechanics.

"This violent reaction on the recent developments of modern physics", Heisenberg notes, "... meant that one had not yet found the *correct language* with which to speak about the new situation, ... which can not be described with concepts of daily life or even of earlier physics.... The real problem behind all these controversial questions was the fact that no language existed in which one could speak consistently about the new situation".[61] How would it have been possible, then, for Heraclitus to have expressed himself with greater clarity?

"Heraclitus was criticized as 'enigmatic'[22A1(6)] *because he sought to express the principle of unity and the interaction of opposites through images, symbols and a polysemic, elliptical language."*

Is contemporary physics any less enigmatic? Indeed, it is not. As we have seen, today's ordinary and even specialized language does not suffic to explicate the Heraclitean dialectic synthesis of opposites. In this regard, Heisenberg writes: "The quantum theory is an excellent example of one's being able to understand a condition of things with absolute clarity; at the same time, however, recognizing that he can speak of it only through *images* and *parables*".[62] Thus the sole means of expression, as chosen by Heraclitus, is vindicated. Of course contemporary physicists have the advantage of withdrawing to a second line of defense not available to Heraclitus: "As soon as this vague and unsystematic use of the language leads into difficulties the physicist has to withdraw into the *mathematical model*".[63]

"Heraclitus was condemned by Aristotle and others as absurd since he blatantly ignored the basic logical laws of 'contradiction' and the 'excluded third'."

Is contemporary physics any less absurd -by Aristotle's definition As in the case of Heraclitus and his commentators, various theoretical physicists give differing interpretations of the quantum theory. "This is a clear indication", writes physicist H. Pietzchmann, "that here we have to do with a pure synthesis toward thinking in categories of Being that *exceed the bounds of rigid logic*.... The structural elements of matter, the elementary or fundamental particles are *particles* as well as *waves* (duality: particle-wave). Of course this is in contrary to the axioms of *logic*, for *discrete* and *continuous* are *mutually exclusive terms*. Nevertheless, experiments irrefutably prove the capability of both interference and location by collision".[64]

Heisenberg would arrive at the conclusion that "the mathematical scheme of the quantum theory can be interpreted as an *extension* or *modificatio* of classical logic. One fundamental principle of classical logic especially seems to require modifica tion. In classical logic it is assumed that, if a statement has any meaning at all, either the statement or the negation of the statement must be correct. Of 'here is a table' or 'here is not a table,' either the firs or the second statement must be correct. 'Tertium non datur,' a third possibility does not exist...In quantum theory this law 'tertium

[61] W. Heisenberg, *Physik und Philosophie*, 137–138, 143.

[62] W. Heisenberg, *Positivismus, Metaphysik und Religion*, 313.

[63] W. Heisenberg, *Philosophie und Physik*, 148

[64] H. Pietschmann, *Phänomenologie der Naturwissenschaft*, 208, 204.

non datur' is to be modifie ... *Classical logic* would then be contained as a kind of marginal case in *quantum logic*, but the latter would constitute the general logical pattern ... However, if one wishes to speak about the atomic particles themselves one must either use the mathematical scheme as the only *supplement* to natural language or one must combine it with a language that makes use of a *modifie logic* or *no-well-define logic at all*".[65]

> *"The visionary spirit of Heraclitus turned from the visible, static, material world, to 'relationships' to the 'Logos' that pervades the universe and man as its integral part."*

This is the tendency of contemporary physical science as well: If one can speak in our time of an image of nature within the natural sciences what is involved is no longer an image of nature, but a image of *our relationship* to nature. Descartes' distinction between *res cogitans* and *res extensa*, no longer functions as a point of departure for understanding in contemporary natural science. Within the range of science's vision above all stand the interwoven *relations* between man and nature. Natural science is no longer an objective observer before nature, but perceives itself as a part of this interaction between man and nature.

Epilogue

In Heraclitus we fin the crucial crossroads in the evolution of ancient Greek thought. From him Ionian 'physics' receives two powerful blows: To *static matter* he opposes *dynamic interrelation* as a perception of reality. Against the concept of the domination of *one material arche* over plurality, which ultimately springs from that arche, the Ephesian opposes balance and interchange between two ontologically equal factors, 'One-All', which engage in a hidden harmony which underlies tension. Heraclitus thus invites man to radically change his way of thinking. Instead of the seeming stability of the material world, he invites him to see the continual conflic of opposites which is expressed in a perpetual process of motion and alteration; and behind this collision of opposites to discern the hidden dynamic relation of 'due measure' and '*Logos*', of the interaction of all things in a harmonious 'balance'. This new, dialectical perception of reality poses for later thinkers a series of difficul problems regarding the maintaining of 'identity' of beings within perpetual 'becoming'. For Heraclitus, as Reinhardt would state, 'becoming' is actually what can be realized, since 'Being' is rendered possible only through 'becoming'.

"We have here", Popper observes, "the firs hint of what was soon to come: of the *general problem of change*, which became the central problem of Greek cosmology, and which ultimately led, with Leucippus and Democritus, to a *general theory of change* that was accepted by modern science almost up to the beginning of the twentieth century. . .This *general problem of change* is a philosophic problem; indeed in the hands of Parmenides and Zeno it almost turns into a logical one. *How is change*

[65] W. Heisenberg, *Physik und Philosophie*, 149, 150, 153.

possible – logically possible, that is? How can a thing *change* without losing *its identity*? If it remains the same, it does not change; yet if it loses its identity, then it is no longer that thing which has changed".[66] Heraclitus's theory of 'the union of all things' and the 'identity of opposites' appears itself to contain the seed of doubt that 'becoming' is realizable. "For change", Popper notes, "is the transition from one opposite to the other. Thus if in truth the opposites are identical, though they appear different, then *change itself might be only apparent*. If in truth, and for God, all things are one, there might, in truth, be no change".[67] Parmenides would focus on this syllogism with relentless consistency; and to him we shall now turn.

66 K. Popper, *Conjectures and Refutations*, 142.

67 K. Popper, *Conjectures and Refutations*, 145.

Parmenides of Elea (ca. 515–450 B.C.)

With Parmenides pure philosophy begins.
Georg W.F. Hegel

Being

'What is Being'? The question may seem superfluou – the response a mere tautology: 'Being is to be'. Or could it be that the question is not so simple as it firs appears? 'What, precisely, is the nature of Being'? A tree in the garden is a being, so is myself, the city, a poem by Edgar Allan Poe, Beethoven's Third Symphony. All these are beings insofar as they exist. That is self-evident and incontestable, so the question remains unanswered, 'what exactly does *Being* mean'? What is that which makes up the Being in the being, that which makes it be a being instead of non-being? The word *on* (being) in ancient Greek held a two-fold meaning: first that which 'is', the being itself, that which exists; and, second, that by virtue of which something exists – that which constitutes the 'Being' in the 'being', if it is a being. It is precisely the *second* sense -the beingness, 'to-be', which relates to the fundamental 'question' that according to Aristotle "was raised long ago, is still and always will be, and always baffle us- 'What is Being (to-be)?' "[1]

The meaning of the word 'Being' is at once both perfectly clear and completely obscure. It is clear that everybody knows how to distinguish between that which exists and that which does not exist. According to Heidegger, if man did not have this capacity to comprehend a being as being – that is, to conceive in advance his own *being*, then not a single word would exist, language itself would not exist, and consequently man would not be capable of being what he is: "For to be human means to be a sayer".[2] Nevertheless, the meaning of the word 'Being' at the same time remains obscure in that it is indefinable It cannot be define because the moment that we make the attempt to determine it, we render it an object, a 'being' like the other 'beings'. But 'Being' (to-be) is not 'being' as such, but that which provides, we can say, the possibility for something to-be a 'being'. For this reason

[1] Aristotle, *The Metaphysics*, 1028b, 3–4.

[2] M. Heidegger, *Introduction to Metaphysics*, 86. References to Heidegger during our analysis of Parmenides will be frequent. As the supreme contemporary 'philosopher of Being', Heidegger penetrated as few have been able the spirit of the Greek thinker who firs posed the question concerning 'Being'.

C.J. Vamvacas, *The Founders of Western Thought – The Presocratics*, Boston Studies in the Philosophy of Science 257, DOI 10.1007/978-1-4020-9791-1_10,

the firs Greek thinkers used the infinit ve form *einai* (to-be) without prefixin the article *to* (the), which would turn the term into a noun. We cannot comprehend the significanc of the word 'Being' also for an additional reason: 'Being' has a polysemic meaning. The statue of Aphrodite of Melos *is* a 'being'. Yet of what does its 'being' consist? Of the marble? Of the form? Of what it represents?

We return, therefore, to the initial question: 'What is Being'? According to Heidegger, this question implies the spiritual destiny of the West – the most vast, deep, fundamental and crucial of all questions, and consequently the understanding of 'Being', which remains "dark, confused, covered over and concealed, must be illuminated, disentangled and ripped away from concealment".[3] The firs European thinker who would explicitly pose the question about 'Being' and would attempt its unconcealment, its revelation, would lead to a 'decision', as Gigon observes, "which was to determine the course of Greek philosophy, whereby the problem of 'Being' would become in an irrevocable manner the property of the West".[4] This thinker was Parmenides.

Personality

Though Socrates and Plato pay tribute to Parmenides as 'revered' and 'enormously capable', they also note that it was extremely difficul for this thinker to be understood.[5] The indelible influenc of Parmenides on the course of western thought springs from his groundbreaking elaboration of the meaning of 'Being'. Parmenides would attempt to respond to the fundamental issue of the 'firs principle', the *arche*, in a different manner from that of Hesiod and the Ionian 'natural philosophers'. In Parmenides, the concept of the *arche* has lost its temporal ties and has now become a purely *ontological* issue. Believing like Heraclitus, that "the criterion [of truth] is reason ... and the senses are faulty",[28A1(22)] and positing that reason, independent of empirical knowledge, can grasp the essence of things, Parmenides would reject as illusory the entire world of the senses and consider as the *solely existing* the 'Being', which can be recognized by the *logos* and is identifie with it: "It is the same thing that can be *thought* and can *be*".[28B3] 'Being' is comprehensible, and only what is comprehensible exists. Parmenides is not only the firs thinker to introduce the fundamental ontological concept of 'Being' into European philosophy, but he is also the firs to construct a *purely theoretical* system through exclusively logical argumentation, without any reference whatsoever to the empirical world. Thus he claims for his theory the validity of absolute truth, because it refers to the one changeless, eternal, absolutely credible, *mental* 'Being', not to any entity of the empirical world which by nature is subject to change and decay. It becomes already apparent why the lucid, transparent thinking of Parmenides would constitute a crucial moment of transition in Presocratic philosophy, which indeed would have a profound impact on later European thought.

---

[3] M. Heidegger, *Introduction to Metaphysics*, 87.

[4] O. Gigon, *Der Ursprung der griechischen Philosophie*, 250.

[5] Plato, *Theaetetus*, 183e.

Life

He was born in Elea, lower Italy, around 515, B.C.Scion of an aristocratic family, he -like most of the Presocratics- would be committed to civic affairs in which he would distinguish himself as a "lawmaker[28A12]". His association with the Pythagorean, Ameinias[28A1(21)] converted him to 'peacefulness'[28A1(21)] and philosophical pursuits. His life thenceforth would be one of contemplation and would be etched in history as the 'Parmenidean Life'. It is certain that he was familiar with the works of Homer and Hesiod as well as those of the Ionian 'natural philosophers' while the extent of his knowledge of his contemporary Heraclitus's teachings is in doubt, and thus also in doubt is the firs impression that his teachings could constitute an attack on Heraclitean thought. He must have collaborated with the Pythagoreans, and it is very likely that he knew Xenophanes. According to Plato -as he notes in his namesake dialogue, *Parmenides*- he visited Athens in 450, B.C. , at the age of sixty-f ve, and greatly impressed the young Socrates. He would be the founder of the Eleatic School, the main exponents of which are Zeno of Elea and Melissus of Samos.

Writings

He would leave behind only one 'treatise', which like most of the works of the Presocratics would be titled *On Nature*. It is a hexameter poem, only 18 fragments of which survive today – a total of 153 lines. M. Heidegger observes: "What we still possess of Parmenides' didactic poem fit into one slim volume, one that discredits the presumed necessity of entire libraries of philosophical literature. Anyone today who is acquainted with the high standards of such a thinking discourse must lose all desire to write books". In his age, grammar, logic and scientifi thought in general as yet comprised a single, undifferentiated whole, which makes an approach to Parmenides's thinking extremely difficult Precise translation is impossible because the senses of many words are different from our contemporary meanings. Words like *einai* (to be, exist, happen), *logos* (reason, speech, word, etc.), *noein* (conceive, perceive by sight, observe, perceive by the mind, apprehend, think, consider, reflect presume), *eidenai* (I see with the mind's eye, i.e., I know), *doxa* (expectation, notion, mere opinion, judgment, conjecture, estimation, repute, honor, glory), retain as yet a primordial ambiguity, so that an interpretation of Parmenides's work can be neither perfect nor complete. It can consist only of a glimpse of Parmenidean thought. "The discussion of Parmenides", Heidegger would say, "is endless; not only because many of the extant fragments of his didactic poem remain obscure, but also because his language retains perpetual conceptual values".[6]

The poem is made up of three parts: the 'Prologue',[28B1] which was preserved as a whole; 'part two',[28B2–28B8,49] which refers to the essence of *truth*; and 'part three',[28B8,53–28B19] which treats *the opinions of mortals*, that are based on seeming and are distant from truth.

---

[6] M. Heidegger, *Moira (Parmenides Fr. VIII, 34–41)*, 52.

'Prologue'

The teachings of Parmenides offer a sudden opening up into a hitherto unsuspected realm of thought – an opening to which Parmenides would lend his 'Prologue'[28B1] -through metaphor- the quality of a divine revelation. What is involved here -in contrast to the style of the rest of the poem- is a dynamic opening, lucid and intensely vivid, written in the firs person: "Horses that carry me as far as my heart sped me on ... brought and set me on the far-famed road of the god [the Sun], which bears the man of knowledge over all cities". The chariot at once takes the broad road, and guided by the daughters' hands reaches the 'goddess'. The dramatic symbolism is intense in this sudden, rapid journey from Night to Day, from the darkness of ignorance and human wandering to the light of absolute knowledge, of 'well-rounded'[28B,11] truth.

It is significan that the journey to the supreme truth begins with a 'road' (*odos*) which is to say a *meth-od*, toward seeking knowledge that is accessible *exclusively* to the 'man of knowledge',[28B1,3] to the individual who *aspires* to travel far. This individual is already the sensitive, mature, inquiring person, cultivated and capable of receiving inspiration, so as to devote himself to and preserve the *logos* of Truth: "I, therefore, will speak", the goddess will tell him, "and you shall be attentive and receive the word you hear",[28B2,1] but also, "[you must] judge by reason [*logos*] the much-contested refutation spoken by me".[28B7,5–6] The word of the 'goddess' also constitutes the fina guarantee of the truth of what is told. In the poem's introduction Parmenides's position and message are already clear: *The truth is not revealed through the powers of the thinker himself, but is un-concealed and offered in her own light, to one who 'opens himself' to her and yearns to receive her.*

The goddess intends to reveal to the poet both, the absolute 'Truth', but also the empirical, subjective world of 'seeming'. Though this world bears no relation to 'Truth', it constitutes a plausible and necessary description of 'seeming', which alone is accessible to mortal man, whose steps carry him away from the unveiling of Truth. Thus, though "what seems to mortals"[28B1,30] lacks truth, it does have a meaning. This is the reason why Parmenides, speaking through the words of the goddess, considers it imperative to study exhaustively not only the 'Truth itself', but also 'mere opinion'. From this point to the end of the prologue, the goddess is the speaker and she 'teaches' the poet concerning: [a.1] the *absolute Truth* (which is 'Being' versus 'non-Being'); [a.2] the *attributes* of 'Being' which follow logically from its own definition [b.1] the *false conjectures* (or *mere opinions*) of mankind; [b.2] the seeming *cosmology* that arises from them.

Being–Non-being

"For the firs time in the Western mind, a thinker wonders how it is that Being exists and why it is impossible for us to conceive that 'nothing' can exist", Jaspers notes. "What is supremely self-evident is at once both highly enigmatic and completely clear. 'Being' is; 'nothing' is not; this constitutes for Parmenides an apocalypse of intelligibility achieved through the intellect itself".[7] Parmenides's fundamental achievement, however, lies not only in his scrutiny -for the firs time- of the meaning of Being, but also in the discovery of a formal logical method for the valid deductive production of conclusions. Pamenides would be the firs European thinker to apply today's four basic logical principles, which in his time, of course, were yet to be established. The major function of the logical construction of *syllogisms* in Parmenides's thought is shown by the fact that he begins with a basic alternative premise,[28B2] and proceeds to the logical exclusion of all premises except the one that conducts to truth.[B3–B7] Then -only when he has established- the "convincing truth",[28B1,30] the sole path to Being -does he-, again in a strict logical manner, elaborate its attributes.[B8]

Through the mouth of the goddess, the philosopher begins with two statements[8] that are linked by the *logical principle of contradiction*:

"Come now, I will tell (and do thou lay up my word when thou hast heard it), / the only paths of inquiry that are to be thought of:

- the one, that *is* and impossible *not to be*,
- the other, that *is not*, and must necessarily *not be*."[28B2,1–5]

These two contradictory statements, 'is' and 'is not' -[A] and [not A]- cannot *both* be valid or invalid. Necessarily the one is true and the other false, and this necessity is stressed by the phrases 'impossible' versus 'must necessarily'. For Parmenides the firs path constitutes "the unshaken heart of well-rounded truth",[28B1,29] but also "the path which is far from the footsteps of men".[28B1,27].

Men commit the error of not perceiving the absolute validity of the principle of contradiction, [A] or [not A], 'is' *or* 'is not', and by ignoring the disjunctive *or* they accept, or do not accept at the same time both. This is the track on which "ignorant mortals wander, two-headed",[28B6,4–5] believing "that to be and not to be are the same and not the same".[28B6,8–9] These opinions represent "what *seems* to mortals, in which [there] is no true conviction".[28B1,30]

Thus, Parmenides, combining the logical principle of *contradiction* with the *principle of the excluded third*, rejects any other possibility besides the fundamental disjunctive question "*is* or *is not*".[28B8,16] It must be judged and decided which of these two 'sole paths of inquiry' is that of 'being' and that of 'non-being'. "The one, that it is and that it is impossible for it not to be, is the path of persuasion (for she attends on truth); the other, that it is not, and it must necessarily not be, that I declare is a

---

[7] K. Jaspers, *Aus dem Ursprung dendende Metaphysider*, 33.

[8] We must not forget that we are considering the historical moment in which, on the basis of the colloquial Greek tongue, the various scientifi and philosophic terms were created and used for the firs time. The very words of Parmenides remain *unmatchable* in semantic density and lucidity. They convey in their immediacy and archetypal quality the seminal meaning of Parmenidean thought, which is incomparably superior to any interpretation. Line-by-line translation can only be partial and suggestive.

wholly indiscernible path; for you could not know what is not (that is impossible) nor declare it".[28B2,3–6] In approaching the question, Parmenides posits a decisive premise: "it is the same thing that can *be thought* and can *be*".[28B3] This premise is self-evident to everyone: When we think, we always think of 'something,' and never of 'nothing'. This 'something' expresses 'being' as opposed to 'non-being': "Apprehension and that, for the sake of which apprehension comes about, are the same".[28B8,34]

Every thought is possible and comes about through speech. And like 'to think', 'to speak' also always refers to 'something', never to the 'nameless'. On the basis of the *logical principle of sufficien reason*, combining the two premises, [a] 'Thought' and 'Being' are the same, and [b] 'I think'; therefore: [c] 'Being' 'is'. "It must *be*, what can be spoken and thought of".[28B6,1]

Parmenides would stress this fundamental conclusion through both affirmat ve and negative statements[28B8,35]: "For you could not know what is not (non-being), which is impossible; nor could you declare it".[28B2,7–8] "What is not, is inexpressible and inconceivable".[28B8,8–9] "The one [path] must be abandoned as unthinkable, unnamed, because it is not the true path".[28B8,17–18]

Partmenides's fina choice crystallizes as follows:

(a) "The track of 'is' *and* 'is not' ",[28B8,40] which ignores the basic principle of contradiction in formal logic, can only lead to conjectures, "in which there is no true conviction".[28B1,30]
(b) The path of 'is not', which he terms "a wholly indiscernible track",[28B2,6] is completely blocked, since the 'is not', the nothing, is both 'unthinkable' and 'unnamable'.[28B8,17]
(c) "There thus remains only one path of encompassing what 'is' ".[28B8,1–2] For I think (and speak); therefore what I think (and of which I speak), 'is'.[9]

Therefore the fina conclusion is:

"It is possible 'to be', but impossible for nothing 'to be'."[28B6,1–2]

This statement, in accordance with the firs *logical principle of identity*, constitutes a tautology: [A] = [A]. However, Parmenides's syllogisms have led to this tautology, which suggests that the statement "Being is" contains something truly *profound* and *archetypal*. The careful reader will have realized that all of the above syllogisms of Parmenides are *independent* of possible semantic suggestions of 'is' and are based exclusively on formal logical principles, which have a general, absolute and binding validity; because "this is the path of persuasion (for she attends on truth)".[28B2,4]

---

[9] This conclusion is comparable with Descartes' saying two thousand years later: 'Cogito, ergo sum'.

Attributes of Being

Only now, having define the logical horizon, does Parmenides turn to the description of the attributes, the 'signs', that characterize 'Being'. Setting out from the fundamental conclusion: "it is the same thing that can *be thought* and can *be*.";[28B3] "for it is possible for it to be, but impossible for nothing to be"[28B6,1–2] Parmenides proceeds to an un-concealment of the attributes of 'Being', based on [a] the *principle of self-sufficien reason*, and [b] the exclusion of conditions or processes that could presuppose 'nothing', i.e. 'non-being'. Thus, in a strictly logical, deductive way, he demonstrates a series of characteristic 'signs' or 'marks': "Being is unborn and imperishable, whole, unique, immovable and without end. It *was* not in the past, nor yet *shall* it be, since it now *is*, all together, one and continuous".[28B8,3–6]

Unborn, imperishable, eternal: 'Being' is unborn, and there are two proofs for this: If it had a *beginning*, it must have come either [a] from 'non-being', or [b] from 'being'. But it could not have come from 'non-being' for two reasons: [a1] for 'non-being' is 'nothing', inconceivable and non-existent, and [a2] even if 'nothing' did exist -which impossible- there would be no sufficien reason, '*obligation*', in 'non-being' to bring about 'being'. [b] 'Being' cannot, in turn, come from 'being', since that would presuppose another 'being' aside from 'Being' itself. In the same way the philosopher proves that Being does not have an *end*, since the end of 'Being' would mean its transformation to 'nothing'; but 'nothing' does not exist; therefore such a transition is impossible. The conclusion, thus, follows that 'Being' neither was, nor shall be, because "it now is, all together, one and continuous".[28B8,5] With amazing acuteness, Parmenides is the firs to endow eternity with an a-temporal dimension. The eternity of 'Being' resides in an enduring timeless 'now', without past, without future. For Heraclitus, reality "always was and is and shall be".[22B30] For Parmenides, "it neither was, nor shall be, because it is *now*".[28B8,5] Time is determined only by the processes of change, birth, and decay, which, however, are, as we have seen, non-existent for 'Being'. Parmenides's 'Being' resides outside time; it is an imperishable, perpetual, pure 'now' – a *continuous presence*.

Continuous, indivisible, unique: What holds above regarding *time* also applies to the *spatial* continuity of "being." If we accept for a moment that 'Being' in space is interrupted by 'something,' the later would be 'non-being'. (For we have accepted that our only alternatives are between 'being' and 'non-being'). But 'non-being' is inconceivable and non-existent, and as 'nothing' it cannot divide 'being'. Therefore, 'Being' is an indivisible unit; it is entire and continuous. If void existed, it would contain 'nothing' i.e. 'non-being', but since 'nothing is not', *void also cannot exist*. 'Being' is necessarily uniform ("it all equally is"[28B8,22]); it is homogeneous. For if it were heterogeneous, here it would be 'a' (e.g., more condensed) and there 'b' (e.g., more rarefied) But 'b' means 'not-a'. But 'not-a' (not-being) is inconceivable. Therefore, 'Being' is entirely undifferentiated.

Motionless, unchangeable: Movement means a shift of 'being' to a void space, where 'nothing'. But 'void' and 'nothing' are non-existent. Therefore, movement -and change in general- is impossible. For movement means the transition from one condition of 'being' 'a' to a condition 'b'; but 'b' means 'non-a' (non-being).

'Non-being', however, is inconceivable and non-existent, and the same holds true for 'b'. It follows that 'Being' remains immovable and qualitatively, temporally and spatially unchanging.

Within bounds, complete: Immovable and unchanging, 'Being' resides within the bounds of itself and its identity: "remaining the same in the same place it rests by itself and so remains firml where it is".[28B8,33] "For powerful 'Necessity' holds [Being] in the bounds of a chain that hems it in all around, because it is not allowed that what *is* should be incomplete. For it is not lacking; while 'not Being' would lack everything".[28B8,30] While the void is viewed as not rounded off, incomplete, the finit 'Being' is considered a plenum in its completion; it has no need for, or potential benefi from, movement or change. Only an eventual imperfection could cause movement or change. "Regnant Necessity"[28B8,30]demands that 'Being' be fully accomplished, "not without bounds, uncompleted".[28B8,29] The bound, the end (the Greek word *telos* signifie both the 'bound' and the 'end' and at the same time 'accomplishment', 'perfection') holds here the meaning of absolute completion and perfection. This perfection, characterized by "wholeness in all respects"[28B8,42] -uniform, symmetrical, equal in strength in all directions- is represented metaphorically by the form of a smoothly rounded globe, "a well-rounded sphere",[28B8,43] which, according to the Pythagoreans, constitutes the perfect shape, "the most beautiful form".[58C3,35]

Following the sole path of 'Being', Parmenides would dis-close the identifying 'signs' or attributes of 'Being', which are: *motionless, changeless, finit , perfect, uniform, unique, indivisible, unborn, imperishable, continuously present*. However, does the notion of such a reality -in which *becoming*, *motion* and *change* is unthinkable and non-existent- not constitute a scandal, in terms of the everyday empirical world of the senses?

The Opinions of Men

Parmenides's response is: That is precisely what I am the firs to demonstrate – following a purely abstract logical method, completely purged of any sense experience, constituting thereby, "the trustworthy account and thought concerning truth".[28B8,50–51] To this single road of 'Truth' he juxtaposes the 'mere opinion' (*doxa*) of mortals, in which "true certainty does not lie".[28B1,30] They follow the faulty method of inquiry, in which "mortals, knowing nothing, wander two-headed ... both deaf and blind, mazed, hordes with no judgment, who believe that to be and not to be are the same and not the same, and the track of everything is one that turns back upon itself".[28B6,4–5,7–9] Accordingly, the goddess advises: "Judge by reason (*logos*)", so that "thou keepest thy thought from this [false] method of inquiry and lettest not habit born of much experience force thee along this way".[28B7,2–3] However, the avoidance of this untrustworthy method of empirical inquiry presupposes the ability to recognize and pinpoint the cause of error, that is, the knowledge based on the *seeming* structure of the world.[28B1,31–32]

Keeping the goddess this promise, in the fina portion of the poem she presents precisely that faulty world of *seeming*, announcing and warning that "from this point forward, you will learn the mere opinions of mortals, listening to the deceitful pattern of my words".[28B8,51–52] Where does this crucial fallacy lie? Men confuse 'being' and 'non-being', faultily perceiving phenomena in a dualistic manner,[10] which is presented *cosmologically* by the primary pair of opposites, 'light' and 'darkness'.[28B8,53–59]

For the firs time in the history of Western thought such a clear distinction has been made between 'being' and 'seeming', between reality and the empirical world of the senses; and this distinction had a decisive influenc on intellectual developments thereafter. Parmenides announces that *doxa* (mere opinion based on seeming) must not be taken as reality; but at the same time he acknowledges that for man appearances are a *necessary* form of knowledge, since he is not capable of freeing himself from his "much experienced habit and nature"[28B7,3] and cannot conceive 'Being' with the eyes and ears of the mind, thus remaining mentally blind and deaf.[23B12] The sensory world is presented as a distorted image of 'Being' in which, however, exists a *semblance* of being. Thus, Parmenides characterizes the phenomenological world as "a *seeming* like cosmic order".[28B,60] The error of mankind consists of not accepting the uniqueness of 'Being' and not realizing the inviolable logical principle of the *excluded third* (the rule *either* 'is' *or* 'is not'), thereby accepting both and thus forming the fundamental pair of opposites, 'light' and 'darkness'. The fallacy of this con-fusion stems from *naming*, from the fact that "men have assigned a name to distinguish each one [of all these tangible, perishable things".[28B19,3]

The Cosmology of Seeming

The entire 'likely-seeming' cosmic order that arises out of man's 'deep empirical habit and nature' is based precisely on the archetypal pair of opposites, 'light' and 'darkness'. To that pair refer all other opposites such as hot-cold, soft-hard, light-heavy, rarefied-condensed male-female, and so on, so that in the fina analysis all natural phenomena go back to the initial pair.[28B9,3–4] These opposites are presented as supplementary, in the sense that to the extent darkness withdraws, light increases, and vice-versa.[28B8,57–58] Since Parmenides excludes any qualitative permutation between the two, when it comes to explaining the plurality and variability of phenomena, he will invent the *combination* of elements, introducing the concept of 'mixing' as well as the idea of a supreme power that initiates and controls the process. In this way he maintains the opposites unalloyed and explains the seeming variety and change in appearances by the differing *proportions of mixing*, elevating 'mixing' for the firs time into a fundamental cosmological principle.

---

[10] In regard to language, the philosopher applies the singular to *truth* and the plural to *mere opinions*.

The extant fragments are so incomplete that they do not provide a clear cosmological image.[28B10] The philosopher created a series of significan discoveries. Some scholars believe that he was the firs -not Pythagoras- to maintain that the earth is spherical and that the Evening and Morning star, i.e., Venus, are one and the same. He believed that the moon is spherical and lighted by the sun, and that its various phases of waxing and waning are caused by the way, in which its lighted hemisphere appears from earth.

The anonymous goddess rules cosmic becoming. She resides within, not outside or above, the *cosmos*: "and in the middle of these is the goddess who steers all things".[28B12,3] This constitutes a personificatio of the supreme power, 'Necessity', which imposes cosmic order, determining all things – matter and cosmic powers, birth and perishing, as well as the mixing of opposites. She functions as the archetypal cause of the various divine powers and as the conductor of souls: she is the cause of the gods . . . and she sometimes brings souls forth from invisibility to visibility, and sometimes the reverse.[28B13] Thus, death does not mean non-existence, but a reverse displacement from the visible to the invisible, from light to night. As with the Ionian 'natural philosophers', Parmenides's cosmology indivisibly integrates the natural, psychological and social world of mankind.

Our seeming world varies in appearance to us according to the proportions contributed in the mixing by the ever-changing opposites, which is actually subjective, since they are reflecte the mixing of those very opposites within us, leading to the forming of 'mere opinions' or impressions. Parmenides would be the firs thinker to develop a rounded theory of mental and sense perception. This theory is based on the age-old principle of 'like is attracted to like'. Sense impressions and their mental extensions depend on the given "blending of the parts"[28B16,2] within the individual.[11] This relationship is interdependent: the mind corresponds to the body, and the body reflect the mind. It is this *krasis* ('blending'; but also 'temperament') – which varies from individual to individual that determines the particular personal perception of things. Thus, Parmenides is able to develop a purely *natural* explanation of how individuals perceive -and by extension apprehend- the same object differently.

Parmenides's cosmology serves the purpose of showing that the world of appearances is based on the *faulty notions* of mankind concerning the existence of *opposites*, to which names were assigned so as to characterize empirical phenomena like birth, perishing, plurality and perpetual change, which are projected chronologically on past, present, and future. With these words he would conclude the fina extant passage of the poem: "Thus, in appearance these things come into being and now are, and having matured will come to an end in the future; and them men have assigned a name to distinguish each one".[28B19]

---

[11] In modern Greek the word *krasis* (blending) remains the same; in turn, the particular 'temperament' of the individual is expressed in the word *idiosynkrasia*, source of the same word 'idiosyncrasy' with similar, but narrower, meaning in English.

The Path of 'Truth' – The Track of 'Opinion'

Here one crucial question arise which may forever remain unanswered, since only hints can be presented as responses: For what purpose does Parmenides wish to present in detail -through the mouth of the goddess- a cosmology, which seems credible but according to part one of the poem does not correspond to reality? Many commentators maintain, finall , that the presentation of the world of human 'opinion' is necessary in that its complete refutation as illusory constitutes a revelation of the utmost significance "The relation between Part I and Part 2 of his poem" observes K. Popper, "is an *inversion* of the 'traditional style'. In other words, he used the 'traditional style' when he made his great discoveries, but he inverted it when he decided that the world of appearance was unreal, false and no more than an illusion or a nightmare – a dream not to be believed".[12] This view therefore maintains -in accordance with an initial suggestion by Nietzsche- that this cosmology is that of Parmenides himself, which he does not hesitate to juxtapose in the knowledge that it does not correspond to the ultimate Truth, which he had later disclosed. Popper, in an affirmatio of this hypothesis, ventures a bold parallel to modern physicists who, precisely like Parmenides, develop a theory *knowing* that it is at a remove from reality. While Newton's theory of gravity, for example, is based on the concept of 'force at a distance', he writes in a letter: "That one body can act upon another at a distance ... is to me so great an absurdity that ... no man can ever fall into it".[13] Einstein, considered his General Theory of Relativity "in reality ... valid only as a limiting case".[14] Concerning his theory of quantum mechanics, Dirac writes: "The difficultie being of a profound character, can be removed only by some drastic change in the foundations of the theory".[15] And though he himself stipulated the temporally dependent wave functions, Schrödinger believed -as did Parmenides- in a *timeless* ultimate reality.

What is 'Being'?

We can now return to the central question that we posed at the outset: 'What is Being'? As we have seen, the meaning of the word 'being' is general, abstract and polysemic. "(The word) 'Being' has many and varied significances"[16] Aristotle notes, depending on the sentence in which it appears. There are two basic contemporary meanings of 'being': the existential, 'being' as *existing* -referring to a subject's existence- e.g., 'the letter is in the vault'; and the *predicative*, -'being' as

---

[12] K. Popper, *The World of Parmenides*, 90.

[13] Letter of I. Newton to R. Bentley, 25 Feb. 1693; cited in Popper, *The World of Parmenides*, 204.

[14] A. Einstein, *The Meaning of Relativity* (Princeton, 1956); cited in Popper, *The World of Parmenides*, 123.

[15] P.A.M. Dirac, *The Principles of Quantum Mechanics*, 310.

[16] Aristotle, *The Metaphysics*, 1028a10.

link between subject and predicate- e.g., ‘it is white’ -together with a variety of uses according to circumstance-, like “ ‘it is happening’, ‘it is standing’, ‘it is native to’, ‘it consists of’, ‘it is located’, ‘it is present’, ‘it is somewhere’, ‘it is something’, ‘It is true’, etc”. However, beyond all these various meanings, which refer to a being as an entity, ‘that, which is’ (*the being, entia, das Seiende, étant*), there is also the *ontological* significanc regarding that which constitutes the *essence* of each being, ‘being itself’ -that by virtue of which all beings are as such- (*to be, esse, Sein, être*). To which sense of the word ‘being’ is Parmenides referring? Before attempting an answer, we must take the following into consideration.

In Parmenides’s time, the various meanings of the word ‘being’ just referred to remained largely undifferentiated. In Parmenides’s thought, we must accept that as yet there exists an individed *unity* of the various meanings, which were expressed both with the nominative ‘being’ and the verb ‘to be’. “The grandeur of Parmenides’s thought about Being is lost, if one strives to fil it with something that does not belong there”, Jaspers warns. “The meaning of the word does not, as a meaning, constitute the essence of Being”, Heidegger notes. “The word and its meaning are bound more originally to what is meant by them . . . Being itself relies on the word in a totally different and more essential sense than any being does”.[17] Thus, the semantic approach to Parmenides’s ‘Being’ demands extreme care and can never achieve perfect or fl wless expression. The most effective way to trace his semantic import is to follow the guidelines toward ‘Being’ which Parmenides himself sets forth, involving the ‘marks’, which necessarily arise from them.

We come upon the word ‘being’ eighty-six times in Parmenides’s poem. Of these, fifty-s ven involve the adjectival linking function and twenty-nine involve ‘being’ in relationship with something else: ‘being’ and ‘non-being’, ‘being’ and ‘opinion’, ‘being’ and ‘meaning’, ‘being’ and ‘what can be spoken and thought of’. These ‘marks’ confin and defin the semantic precincts of ‘being’.

‘Being’ and ‘Not Being’

We have seen that the path of ‘non being’ is inaccessible. Heidegger observes that here we have philosophy’s firs report concerning a *unifie* conception of ‘being’ and ‘non being’ (Nothing): “That ‘Nothing’ is not a being, however, by no means prevents it from belonging to Being in its own manner”.[18] With the delimitation of ‘Being’ against ‘non Being’, ‘Being,’ existence itself, ceases to be something self-evident. Existence is understood and experienced as the *antithesis* to ontological nothingness. Under the cold light of the logical principle of identity and of the excluded third, the merciless question ‘Being’ or ‘non Being’ (nothing) causes for the firs time the Being of beings -which up till now was self-evident- to be suspended before the possibility of ‘non Being,’ non-existence. As long as we are

---

17 M. Heidegger, *Introduction to Metaphysics*, 92–93.

18 M. Heidegger, *Introduction to Metaphysics*, 117.

within the framework of this formal *logical* principle of *contradiction*, 'Being' or 'non Being',[28B6,16] and of *identity*, 'Being is, non Being is not',[28B6,1–2] the content of 'Being' remains as yet completely *undifferentiated*. This general term 'Being' -as yet lacking concreteness- will begin to be differentiated the moment that Parmenides's thought is extended into the metaphysical and gnoseological sphere.

'Being' and 'Truth'

Within the above logical framework, Parmenides separates the concept of 'Being' from all illusory tangible experiences, lending to the term a pure rational with general and necessary validity, constituting the unshakable Truth, "the trustworthy account and thought concerning truth".[28B8,50–51] Just as the ultimate Truth is one, unique, and not subject to division, expansion, reduction, deterioration and dependence upon time, the same must necessarily hold for Being. This is a logical truth, completely self-contained and self-evident, like $2 + 2 = 4$, independent of time, space, change and sense experience; and for this reason it attains general validity, constituting incontestable certainty; it is "true conviction".[28B1,30] The property of 'Being' as 'true' retains the archetypal etymological sense of 'truth' as 'revealing', 'un-concealment'. *Lanthano* means not to be seen, not to be observed. *Litho* is from the same root, meaning *forget*. Thus, Parmenides's calling 'Being' *a-lithea* (truth) means that 'being' is *pro-phanes* (made manifest) in the sense of self-emerging out of forgetfulness and standing in *un-concealment*. The truth (*a-lithea*) is not a property of 'being'; it is its *very essence and identity.* According to Heidegger, this emerging, this stepping forth to appearance out of concealment, is *physis* itself. Thus, "*Physis* is Being itself, by virtue of which beings firs become and remain observable".[19] The archetypal meaning of the root *phy-* is 'emerging, coming to presence, coming-into-being'. 'Being' = 'Truth' = '*Physis*' – the self-revealing, the self-setting forth and standing in un-concealment, making itself manifest.[20] For Being to step forth and stand in appearing and become that which 'makes a being being' means: to reach and attain its 'limit', its *telos*,[21] in the sense of completion and coming into fulfillmen – to self-limit itself, to place itself and stand forth outside time, a continuous and constant presence. In a word, 'Being' is '*presence*'.

---

[19] M. Heidegger, *Introduction to Metaphysics*, 15.

[20] M. Heidegger, *Introduction to Metaphysics*, 15. The original sense of the root *phy* was not only 'being', 'existence' but also 'process of natural *development* or growth'. (C. Kahn, *Anaximander and the Origins of Greek Cosmology*, 201) Also, according to Heidegger, to be made manifest includes both 'becoming' and 'Being' in the strict sense of constant presence. Thus, according to this contemporary philosopher, a basic contrast does not exist here, as most believe, but congruence and complementarily between Heraclitean and Parmenidean thought.

[21] "This is the key", Heidegger writes, "to understanding the highest term that Aristotle used for Being: *entelecheia*, something holding- (or maintaining) itself-in-its-completion (or limit: *telos*)". (M. Heidegger, *Introduction to Metaphysics*, 63).

'Being' and 'opinions'

Parmenides warns, however, of another form of truth that man constructs for himself when he is on the accessible but misleading road of 'opinions' – "whatever mortals have laid down believing it to be true [real]".[28B8,39] 'Being', '*physis*', by self-appearing, is continually in danger of falling to the level of *seeming*, a mere semblance. *Doxa* is a view, an 'aspect' and by extension an 'assumption', an 'opinion'. The view that a being offers from itself lets itself be apprehended differently, depending on the time and the angle of approach, thus degrading itself to *doxa*, in the sense of mere conjecture. Wherever there is un-concealment and self-appearing there is also the converse possibility of re-concealment, of the covering-over and distortion of the view and reduction of it to a deceiving 'mere seeming', a semblance. This distortion is not restricted only in the view of the beings but also affects 'seeming' itself, which deceives as it shows itself as Being, as 'apparent truth', in which "mortals, knowing nothing, wander two-headed".[28B6,4] 'Being' as truth must clearly be one, unique, self-defined constant, timeless. Within the framework of these characteristic 'marks', *becoming* -as something which is neither any longer what it was, nor what it tends toward- belongs on the one hand to '*physis*', but assumes on the other an unstable view, becoming a 'seeming', since it is "coming into being and perishing, being and not being".[28B8,40] "Only by undergoing the struggle between Being and seeming", Heidegger observes, did the Greeks "wrest Being forth from beings, did they bring beings into constancy and un-concealment. The *thinking* and the '*Dasein*' (Being here) of the Greeks struggles over a decision between the great powers of Being and becoming, Being and seeming".[22] Thus, the Greeks would choose *apprehension* and *logos* to subdue their passion for truth.[23]

'Being' and 'Apprehending and Saying'

Noein does not mean 'thinking' but to 'perceive by the mind', 'apprehend'. It is *a direct mental perception* of a subject or a situation. It consists of an *intuitive act* by which one 'observes', 'discerns' beyond the sense world, and 'grasps' *true* reality. By definition then, that which is apprehended makes itself manifest, is existent. For to *noein*, to apprehend means to grasp 'something', not 'nothing', with the

---

22 M. Heidegger, *Introduction to Metaphysics*, 111, 122.

23 This titanic struggle between 'Being' and 'seeming' did not concern the firs Greek thinkers alone. We meet it in a peak expression also in Greek tragedy. Heidegger presents Sophocles's *Oedipus Rex* as a key example: "Oedipus, who at the beginning is ... in the brilliance of glory ... is hurled out of this seeming .... In the end, he is unconcealed in his Being as the murderer of his father and the defile of his mother. He must, step by step, place himself into an un-concealment that in the end he can endure only by gouging out his own eyes -that is, by placing himself outside all light, ... and then crying out, as a blind man, for all doors to be flun open so that such a man may become revealed to the people as the man who he *is*" (M. Heidegger, *Introduction to Metaphysics*, 112).

mind. That what is 'apprehensible' is 'graspable', and as the acceptor I am always the receiver of 'something', not 'nothing'. Just as 'seeing' always presupposes a 'view', an 'image' (we cannot say that we see in the dark), 'apprehending' presupposes a clear and complete perception of 'something', which for Parmenides is the disclosed Being. "It is the same thing that can be apprehended and can be".[28B3] 'Apprehension' captures the disclosed 'Being' and is identifie with it. 'Being' shows itself only through 'apprehension' and not through sense perception. Transcending sense awareness of the empirical world, which is characterized by apparent plurality and change, 'apprehension' invites the distant in time and space into the present: "Look at things", Parmenides states, "which though distant are securely present to the mind".[28B4,1] To recall into presence means to recall 'Being', for Being is precisely *presence*. That which I apprehend intellectually, is presence (Being); and, conversely, that which is present (Being) is manifest only by virtue of being comprehended by the mind. The one presupposes the other, and without the one the other is not existent or conceivable. Therefore, "apprehension and that for the sake of which apprehension comes about are the same".[28B8,34] Apprehension comes about for the sake of Being.

Furthermore, *noein* (apprehending) is manifested and expressed exclusively through *legein* (saying). Therefore, since 'apprehension' and 'Being' are one, "it must *be*, what can be *said* and *apprehended*".[28B6,1] To 'saying' and 'comprehension' the same ancient Greek concepts apply: *Legein* originally means 'gathering', 'collecting'[24] > 'congregating' > 'counting' > 'recounting' > 'saying' (speaking). These archetypal meanings, 'gathering', 'congregating', 'counting', evidently always refer -as in the case of 'apprehend'- to 'something', never to 'nothing'. For this reason, 'non Being' "cannot be said or perceived".[28B8,8] In Parmenides, then -as in Heraclitus- *logos* is the constant *gathering* of Being.

Heidegger would lend a profoundly existential[25] dimension to the saying of Parmenides, "It is the same thing that can be apprehended (*noein*) and can be".[28B3] 'Apprehending' is not identifie with 'Being' but they are the same in their *belonging together*. We would be mistaken to consider *noein* to mean 'thinking', for thinking is an act of subjectivity, which define 'being' as an 'object' – an approach which leads to a total *subjectificatio* of Being. This misconception -which began, as we shall soon see, with the Greeks themselves after Parmenides- dominates in the history of Western philosophy and makes an authentic understanding of Parmenides's saying difficult The saying actually means: "Apprehension and that for the sake of which apprehension happens are the same".[26] Apprehension takes place for the sake of

---

[24] It is significan that in modern Greek, *syllogi* means 'collection' but also 'intellectual activity'. In turn, *syllogizomai* means 'ordering impressions and judgments so as to reach a conclusion', and by extension, 'thinking', 'contemplating'.

[25] '*Existential*' refers to the 'ontological' condition of man, while '*existentiel*' to his 'ontical' condition. 'Ontological' refers to the capacity to *be* – that is, to man's relationship to the '*on*' (Being), while 'ontical' refers to *what* man is in his worldly experience.

[26] M. Heidegger, *Introduction to Metaphysics*, 148.

Being, not the reverse. In the same essence of *physis*, 'Being' and 'apprehension' belong together, so that with the coming-into-appearance of 'Being', 'apprehension' also necessarily occurs along with appearance. This concept leads to the conclusion that 'apprehension' is not a faculty of the human being that determines what 'Being' is, but, conversely, *the essence of Being determines the essence and the way toward being a human being.* "It must be, what can be spoken and apprehended".[28B6,1] "Apprehension is a happening (*Geschehen*) in which humanity itself happens, and in which humanity ... firs appears – that is [in the literal sense], itself comes into Being ... What is fulfille in this saying", Heidegger continues, "is nothing less than the knowing entrance-into-appearance of the human being as historical preserver of Being. This saying is the determination of Being-human that is definit ve for the West".[27]

The Eleatic School

The main followers of Parmenides in the Eleatic School, which was formed in the meantime, were two: *Melissus of Samos* and *Zeno of Elea.* The two would take diametrically opposed directions in support of the position of their master. Melissus would be directly involved with Parmenides's theories, with a view to their clarification Zeno, conversely, would be involved with the arguments of Parmenides's opponents, attempting to show through reductio ad adsurdum the illogical conclusions to which their unfounded arguments lead.

Melissus of Samos

Melissus of Samos (flourishe mid-5th century, B.C.), in an easily understood style, to analyze and explain Parmenides's doctrines. He too assumes that 'Being' is "eternal and unbounded and one and same throughout".[30B7(1)] "For if it came into being, before it came into being there must have been nothing; if there were nothing, nothing would ever come into being out of nothing".[30B1] Nevertheless, in contrast to the Paramenidean eternal *present*, Melissus lends the eternal a more understandable form, viewing it as 'continuous' with extension both into the past and the future: "It always was whatever it is, and always shall be".[30B1] He rejects, in turn, a spatially finit 'Being' and posits without proof that since Being is temporally infinite it must also be spatially boundless: "Precisely as it exists forever, thus it must be infinit in magnidtude".[30B3] On the basis of this arbitrary axiom, he conversely demonstrates Parmenides's position on the 'one': "To be infinite it must be one; for if there were two, they could not be infinit but would have limits in relation to [*or* be limited by] one another".[30B6] Moreover, since 'being' is 'one' it must also be changeless, "Because if it changes, necessarily it will no longer be the same, but lose its former being and become non-being; ... nor is it possible for a modificatio to take place,

[27] M. Heidegger, *Introduction to Metaphysics*, 150.

for the cosmic order which preceded is not lost, nor does it become an order which did not exist".[30B7(2–3)]

Melissus's concept of a 'singular' and 'unchangeable' Being inevitably leads to the conclusion that the empirical world is untrustworthy, since it presents -in contrast to true 'Being'- an image of plurality and change: "It is evident therefore that we are not seeing correctly, nor is our belief correct that things are many, . . . for nothing is higher than true being".[38B8(5)] One of Melissus's major conclusions, which would later influenc the thought of the Atomists, is that a *precondition* of every movement is the existence of *void*. Since, however, there is no void, movement also cannot take place: "Nor is there any void, for void is nothing, and nothing cannot be. Nor does it [i.e. what is] move, for it has no place to which it can withdraw, but is full. If there were void, it would withdraw into the void; but since there is not void, it has nowhere to withdraw to".[30B7(7)] Melissus's arguments aim at a fuller understanding of Parmenides's teachings, but they lack depth and the strict logical coherence of his mentor.

Zeno of Elea

Zeno of Elea (ca. 490–430, B.C.)was the student and close friend of his fellow citizen Parmenides. Concerning his life, little is known besides his heroic death, the result of his refusing to name his associates in a plot against the tyrant, Nearchus. Besides his treatise titled *Concerning nature*, he is believed to have written other works as well. "This study," Zeno mentions in Plato's dialogue, *Parmenides*, "is an effective defense of Parmenides's teachings against those who attempt to discommode him . . . It readily refutes their criticisms, aiming to demonstrate that their hypothesis that many [things] exist leads to a more ridiculous conclusion than [the hypothesis] that only One thing exists".[29A12] Zeno's argument is directed mainly against movement, against Pythagoras's plurality of discrete units, against Empedocles, and -according to one view- against Anaxagoras's infinit divisibility of matter. He skillfully, consciously, exploits -for the firs time- the *dialectical method* of arriving at contradictory conclusions on a position, thus *indirectly* demonstrating the fl ws of that position and, consequently, the correctness of its opposite. Aristotle would correctly call him 'discoverer'[29A1] and 'leader'[29A10]of 'dialectic'. "Through the art of his language", Plato states in *Phaedo*, "he made the same things appear to his listeners alike and not alike, one and many, immobile and moving".[29A13] The reference is precisely to Zeno's familiar 'paradoxes'.

The Paradoxes of Zeno

Plurality

Zeno is said to have stated that "if someone could explain to him what on earth a *unit* is, he would also be able to explain what the *pluraltiy* of things is".[29A16] His query is ingenious, contesting both the unit as the constituent of plurality and plurality itself.

If, Zeno says, we accept, contrary to the Eleatic position on a single and indivisible reality -that the latter is composed of many units- then we are inevitably led to contradictory conclusions, that is, that 'Being' must be [a] infinitel small, or [b] infinitel large: [a] Let plurality consist of units which -in order to be units- cannot be further *divided*. But what is indivisible is only that without 'size', 'thickness', or mass, that is, 'nothing'. In this case, then, units cannot produce the plurality of being, since no matter how many times 'nothing' is multiplied by itself, it remains 'nothing'. [b] If, on the other hand, we accept that plurality consists of units of *divisible size*, each unit must be divisible into other *divisible* sizes (for otherwise it would be 'nothing', as we saw above), and these into yet other sizes, and so on, ad infinitum resulting by the addition of an infinit number of these units in one, infinitel large being. "Thus, if many things exist, they must be both great and small, so great as to be infinit in size, so small as to have no size at all".[29B1]

He proceeds, as well, to a second contradictory conclusion: "[a] If being is a plurality, then things must be just as many as they are, no more and no less. In this case, they must be *finit* in number. [b] If being is a plurality, then the things that are must be *infinit* in number. Because between separate beings there will always be other beings, and between those others, still other beings, *infinit* in number". The contradictory conclusion thus is: "If being is a plurality, the things that are it must be at the same time *finit* and *infinit in number*".[28]

With the two above arguments, Zeno rejects both plurality and the units, of which it is composed, aiming at establishing the Parmenidean concept of a *single* and *indivisible* reality. According to the contemporary mathematical set theory and infinites imal calculus, the infinit number of elements in an infinit quantity (set), although they cannot be counted, they are exactly as many as they are. Consequently, from the mathematical viewpoint, plurality (set) can consist of exactly *as many* elements *as they are* and at the same time of elements *infinit* in number; thus, Zeno's 'paradox' is *mathematically* refuted.

Space

Using his familiar tactics of demonstrating that the positions of Parmenides's opponents lead to an impasse, Zeno would present this argument concerning space: The Parmenidean concept demands only one 'being' which is *identical* with space. If for the sake of argument we accept the opposite position regarding plurality and motion, then the moving 'being' ceases to be identifie with space, as it successfully occupies positions *within* space. If, however, we accept that 'beings' are located *within* space, since 'space' is 'being', it too, as being, must be located in another space, and that space into another, ad infinitum Thus, the idea of plurality and motion leads to the paradoxical conclusion that "space does not exist".[11.S.498.8]

---

[28] Some contemporary commentators conceive 'finite and 'infinite *arithmetically*, and others *geometrically*, with reference to the properties of the points of a line.

The Millet Seed

Another argument, which is directed against the infinitesimal at the same time demonstrating the unreliability of the senses, is the example of the millet seed: It is absurd that when a single millet seed falls, no noise is heard, while when a thousand -say- fall at once, a noise is heard. Since the sound of one millet seed is nothing, then any sum of the same 'nothing' will also be nothing; therefore, sound does not exist!

"There are four arguments of Zeno concerning motion", Aristotle states, "which cause difficultie to those who wish to solve these problems".[29B25] The famous four 'paradoxes' are:

First paradox: 'The dichotomy' (dividing in two)

"Motion does not exist, since an object must always arrive at the half-way stage before reaching the end".[29A25] That is, a man who wishes to reach the end of a distance, *a*, must firs cover half the distance, $a/2$, but before that he must cover half the distance of the half, $a/2^2$, and before that, half of the half of the half, $a/2^3$, and so on, ad infinitum Because there are *infinit* half points it is impossible for them to be covered in any *finit* time, and therefore movement is impossible.

Aristotle would attempt to reject the paradox of the above argument with this reasoning: "The length and time and any continuum can be considered *infinit* from two points of view: as regards *divisibility* and as regards *extent*. Thus, while a thing cannot reach things *quantitatively* infinit within *finit* time, it can nonetheless reach things that are infinit in *divisibility*. Because in this respect time itself is also infi nite. Thus, the infinit (in *divisibility*) distance can be traversed in infinit and not finit time, and contact with the infinit takes place in infinit and not in finit in number moments".[29] Aristotle juxtaposes to the infinit halves, $a/2$, $a/2^2$, $a/2^3$, $a/2^4$, and so forth, ... $a/2^v$ of a finit distance, a, infinit temporal halves $t/2$, $t/2^2$, $t/2^3$, $t/2^4$, and so forth, ... $t/2^v$ of finit time, t. A *finit* time, can be *infinitel* divided. Consequently, the distance infinit in divisibility can be covered in *finit* time. Nevertheless, Aristotle would later perceive that the problem remained (if stated in more general terms), i.e., to explain how an infinit number of acts -crossing one interval of an infinit number of points- can be serially completed. For this is precisely what Zeno maintains: that it is impossible for one to complete an *infinit* number of acts. Aristotle would attempt to speak to this issue through the distinction between *entelecheia* (actuality) and *dynamei* (potentiality), – the *actual* infinit (whose infinit exists at some point in time) and the *potential* infinit (whose infinit is spread over time). All the objections to the infinite according to Aristotle, are objections to the actual infinite "In a continuum there is an infinit number of halves, but only potentially, not actually".[30]

---

[29] Aristotle, *The Physics*, 233a23.

[30] Aristotle, *The Physics*, 263a27.

Second paradox: 'Achilles and the tortoise'

"The slower runner will never be overtaken by the faster".[29A26] If 'swift-footed' Achilles gives the tortoise a start, he will never catch it, for when he reaches the point, a, from which it started, the tortoise by that time will have gone further to point a′, and when he reaches that further point, a′, it will have gone further to a″, so forth, ad infinitum The time that is necessary to eliminate the difference of distance between them is infinite This second paradox is a simple variation on the first involving this time two moving objects instead of one.

These firs two paradoxes hold provided we accept the Pythagorean teaching concerning *indivisible units*, which is exactly what Zeno aims to refute. If we abandon this assumption, then -according to the contemporary mathematical theory on an infinit series- the 'paradox' ceases to be a paradox. The addition of such a geometrical series -like the race of Achilles and the tortoise- in which the parts of the series continually and regularly diminish, is define by a *single, concrete number*, which can never be exceeded by the sum of the parts of the series, however numerous they may be. In other words, it is possible *mathematically* to calculate the point at which Achilles will reach the tortoise.

Third paradox: 'The flyin arrow is stationary'[29A27]

This paradoxical conclusion derives from the following syllogism: (a) An object occupying a space equal to its own size must be motionless in that space (for it does not have additional space into which to move). (b) At every given instant (every 'now') anything that moves can only occupy a space equal to itself. Therefore at every 'present' moment of its movement in space, it is motionless. The flyin arrow remains motionless in each 'present' moment. This is true for the entire duration of its flight which is made up of the sum of moments. The conclusion is: "That which is moving does not move either in the place where it is or into the place it is not".[29B4]

This paradox holds if we accept -as does Zeno- that the instantaneous 'now' corresponds to *zero* time, whereby the covered distance in *zero* time is zero; consequently, movement *cannot exist*. G. Vlastos develops a pointed parallel between the temporal instant and the spatial point that leads to a similar paradox: "It is as if one asks, how can the bow be bent, since none of its points is bent".[31] The movement at a given instant, as we understand it in today's physics, does not refer to a temporal zero 'instant' but always to temporary *intervals* that tend to approach zero. The 'momentary speed' is determined by the ratio $\Delta S/\Delta t$, where Δt *tends* toward zero, so that the momentary speed $\Delta S/\Delta t$ approaches a limit that is *different* from zero; therefore movement does exist. Of course Zeno could not as yet distinguish between an instant *without* temporal duration and an instant *with* an *interval tending to zero*.

---

[31] G. Vlastos, *Simeiosi gia to Velos tou Zinona (Note on Zino's arrow)*, 310.

Fourth paradox: 'The Stadium'

"Half a time is equal to its double".[29A28] Assume that two rows of equal-sized objects B and C move in a stadium with equal velocity and in opposite directions, passing in front of a *stationary* row of same-sized objects, C, as in Fig. (i):

```
  AAAA          AAAA
BBBB →        BBBB →
   ← CCCC      ← CCCC
    (i)           (ii)
```

When they move one step to position (ii) in time t, the firs B has passed two C's and only one A, in relation to position (i). But objects of equal size (like A,B,C), moving at the same speed, demand *equal* time to pass the same of objects. Therefore, time t needed for the firs B to pass one A must be equal to the time t/2 needed to pass one C. Therefore, 'Half a time is equal to its double'.

Aristotle correctly observed that "the error in the paradox lies in the assumption that an object moving at a steady speed needs the same time to pass an object *in motion* and another object of the same size which is *at rest*".[29A28] This appears today so evident that one might wonder whether Zeno was actually so naïve as to miss the point. Many maintain that in Zeno's time the idea of *relative motion* was not yet understood, whereby Zeno's syllogism is justifiabl in that he was striving to prove on a theoretical basis that if we accept motion, this assumption leads to paradoxical results. Others, however, claim that Zeno's aim was not the concept of motion but the refutation of the Pythagorean concept of the *non-infinit divisibility* of matter and of time. In this case, Zeno shows in a brilliant and original way that if A (=B, =C) constitutes the minimal *indivisible* spatial unit, and t the minimal *indivisible temporal* unit, then the supposedly 'indivisible' A and t can be divided into A/2 and t/2.

Using the dialectical method with astounding acuteness, Zeno aimed at demonstrating that *reality is one, immovable, indivisible, continuous in space and time*, and that the assumption of *movement, plurality* and *empty space* leads to an insoluble logical impasse. His arguments would for the firs time focus the attention of thinkers on a new set of problems related to the infinitel small and the infinitel large, motion, infinit divisibility – problems which from then up to our own time are the object of mathematical and philosophic investigation. His dialectical method of thought would provide fertile ground for the atomic theory of Leucippus, the Aristotelian analysis of motion and infinit divisibility of matter, and later the mathematical infinitesima calculus. Zeno seems to be the firs who would relate through the concept of *motion* the disparate entities -according to Newtonian physics- of *space* and *time*.[32] Especially in the modern age, his thought would remain extremely timely, as shown in the discussions, among others, of Kant, Hegel, Bradley, Herbart, and Russell, who would characterize Zeno's thought as incalculably acute

---

[32] C.A. Brandis, *Griechisch-römische Philosophie*, I, 413, 415.

and profound, realizing "how little modern orthodox metaphysics has added to the achievements of the Greeks".[33]

The paradoxes of Zeno have remained a source of vital discussion for two and a half millennia, without our being certain even today that a generally accepted fina solution has been found. The difficult lies in the fact that every attempt to 'divide' a *continuum* into *discrete* separate points, or fin its 'synthesis' from them, leads inevitably to contradictions. The entire argument of Zeno refers for the firs time to the area of *infinitesimal* , which *by definitio* are *not measurable* and consequently evade empirical understanding. Infinitesima is any number that in infinitel small, yet greater then zero; zero itself as well as concrete numbers -however small they may be- are *not* infinitesimal Consequently, infinitesima is the number which is *greater* than zero and *smaller* than the smallest positive concrete standard number which one can conceive. The numerical difference between two concrete numbers is always *concrete* (standard) and never an infinitesima number; as a result the two limits of an infinitesima segment cannot be determined by *concrete* numbers. Hence, an infinitesima interval is not *measurable*. This inherent difficult renders the Zenonean 'paradoxes' unresolved – at least within the framework of *physical* empirical reality.

From the *mathematical* viewpoint, however, there have been many attempts to fin a solution, beginning with Aristotle. Although from the outset, the very *existence* of infinitesima numbers has been disputed, the Greek mathematicians applied them to solve geometrical problems such as quadratures (calculating the surface area enclosed by curved lines), cubic measure, and determination of tangents of curves. Eudoxus (3rd century, B.C.) was the firs who established the sum of a geometric series with diminishing terms, and Archimedes (3rd century, B.C.) the firs who calculated the summation of converging series, thus estimating the area of a circle. During the 17th century, Newton and Leibniz, at virually the same time but independently of one another, developed the *differential* and *integral calculus*. The existence, however, of infinitesima magnitudes would continue to puzzle both philosophers (Berkeley would call them 'ghosts') and mathematicians. This doubt would lead in the 19th century to the invention of the theory of limits. *Instantaneous velocity* is no longer define as the ration dS/dt but -according to the Weierstrass (1815–1897) method – as a *limit* which is approximated by ratios $\Delta s/\Delta t$ of infinitel small increments. This approach gives a 'mathematical' answer to some of Zeno's 'paradoxes', like that of the instantaneous velocity we saw above in the race of Achilles and the tortoise. At the same time, set theory and the area of transfinit mathematics contributed to the understanding of the 'plurality' paradox.

However, all these mathematical approaches deal with the problem by mathematical operations, without providing a solution for it. Recently, a new nonstandard

[33] B. Russell, *Our Knowledge of the External World*, 180.

analysis of infinitesima numbers, the Internal Set Theory (IST),[34] would attempt to present a different answer to Zeno's 'non-existence of motion' paradoxes.[35]

"Nonstandard analysis -in contrast to standard numbers, which do not include infinitesima numbers- does also include infinitesimals which it considers 'real' ",[36] thus making the infinitesima method precise for the firs time. From the firs moment, however, there were again doubts concerning the validity of this new mathematical method. "Zeno's paradoxes", S. Dillingham observes, "question the validity of our descriptions of physical reality. They are not simply mathematical puzzles and should not be considered solved unless there is reason to believe that space-time is accurately described by the mathematics used to formulate the solutions. Can one formulate all the known laws of physics using Internal Set Theory? Can any experiments be performed to determine whether infinitesima nonstandard points exist"?[37]

Two and a half thousand years after the ingenious arguments of Zeno, in spite of repeated 'mathematical' approaches and 'solutions' provided, his paradoxes continue to demonstrate contradiction and remain a logical 'scandal'. At the beginning of the twentieth century, "it is precisely this contradiction," H. Pietschmann notes, "which should be integrated by quantum mechanics as 'wave-particle dualism' in the description of matter in space and time".[38]

Heraclitus – Parmenides

"Parmenides", Heidegger notes, "shares Heraclitus's standpoint. And where else should these two Greek thinkers, the founders of all thinking, stand if not in the Being of beings"?[39] They lived at approximately the same time, but in diametrically opposed places. As to which thinker's teachings were the earlier and influence the other, opinion differs. As Heidegger observes, "A thinker is not dependent on another thinker, but -when he truly contemplates- leans toward what he should contemplate, Being. And only as long as he belongs to Being can he be open to the influence of that already contemplated by thinkers. Thus, it is the exclusive privilege of great thinkers to mutually influenc one another".[40]

---

[34] E. Nelson, *Internal Set Theory: A New Approach to Nonstandard Analysis*, 1165–1198.

[35] W.I. McLaughlin and S.L. Miller, *An Epistemological Use of Nonstandard Analysis to Answer Zeno's Objections against Motion*, 371–384.

[36] Up till the 19th century, mathematicians believed that the subject matter of their study was *objectively real*. Today this does not concern mathematicians. It suffice that their mathematical theory can be used in proofs.

[37] S.G. Dillingham, *Closing in on Zeno*, 5.

[38] H.Pietschmann, *Phänomenologie der Naturwissenschaft*, 42.

[39] M. Heidegger, *Introduction to Metaphysics*, 145.

[40] M. Heidegger, *Was heisst Denken*? 39.

For some Heraclitus was chronologically the first for others Parmenides. Many believe that the one did not know the work of the other. In ancient times they were considered intellectual opponents, since Parmenides penetrated 'Being', while Heraclitus probed 'becoming'. The distinction, however, is simplistic. At bottom both sought in *becoming* the *Being of beings*. Since they posed different questions and took different approaches, it is to be expected that they would develop contrasting views.

Both believed that the cosmos is composed of pairs of opposites. In Parmenides, however, the one member of the pair *excludes* the other, while in Heraclitus the one member constitutes the *precondition* for the other.

Both did not believe in the witness of the senses, but each for his own reason: Parmenides because he considered the senses responsible for the faulty perception of *change* and *motion*, in contrast to the true immobile and changeless Being; Heraclitus because he considered the senses responsible for the faulty perception of the *seeming stability* surrounding us, in contrast to hidden perpetual cosmic change and movement.

They are both concerned with the one 'Being', but they view 'plurality' differently. Parmenides *rejects* plurality as illusory, reflectin the "erring beliefs of mortals",[28B130] while Heraclitus *enrolls* plurality in a dynamic dialectical relationship with Being; unity is composed of plurality, and plurality springs from unity: "from all things one and from one all things".[22B10]

The language of both has retained its grandeur, with intense religious tones in Parmenides and mysticism in Heraclitus. Yet their style is totally personal. Parmenides chooses the archaic epic poetic tongue as his means of expression, while Heraclitus chooses the language of sage epigrams and aphorisms. For Parmenides words are conventional 'names', arbitrarily assigned by men to express illusory empirical contradictions, while for Heraclitus the ambiguity of words expresses the very essence of reality, which is the unity of opposites. The language of Parmenides is characterized by cold logic. He strives to admonish and to persuade. Conversely, the language of Heraclitus vibrates with passion. He strives to stimulate and to inspire.

In contrast to Parmenides, Heraclitus would be extensively concerned with the soul, life and death, which constitute the keystone for the conception of the universal law of the union of opposites.

For Parmenides, reality neither 'was' nor 'shall be' but endures concentrated in the 'now', while for Heraclitus reality 'was', 'is' and 'shall be' a *perpetual process*.

In spite of the above differences, Heraclitus and Parmenides are still conceptually very close to one another. By means of 'thought' and 'apprehension' -for the firs time surpassing tangible experience- they would attempt with superb vision and conviction to bring to light the singular un-concealment of beings. "Their disagreement is not of the type to mutually exclude one another in objective thought", Jaspers observes. "Rather, there are correspondences in their thought in which each contemplates that which is forever. The one examines Being as the logical identity and transcendent stillness of constant completion; the other as the logical dialectic and transcendent serenity of unshakeable Law. The one perceives the meaning in the identity through which contradictions are abolished; the other conceives the meaning in the contradiction resolved through the unity of opposites ... [Both] ceaselessly seek repose in the thought of the Being of beings".[41]

Influenc

The conclusions of Melissus and Zeno were frequently so extreme as to make refutation upon revision of Parmenidean concepts all the more urgent and necessary. As a result, the one-sided Parmenidean conception, which is in direct contrast to everyday experience of change and motion, could not prevail. Plato described the Eleatic thinkers as 'the partisans of the universe',[42] and Aristotle called them 'unscientific' [28A26] Common sense, unable to view the empirical world as illusory, would desperately seek a new theory that would preserve the world of the senses, without, however, ignoring -so far as possible- the Parmenidean position. This in the light that following Parmenides, the Ionian 'monistic' tendency to a single *arche* as source of a pluralistic world, could not be seriously maintained and was abandoned in favor of a 'polyarchy'. The later *Presocratic* thinkers -Empedocles, Anaxagoras and the Atomists- would attempt to achieve -as we will see- an integration of Parmenidean and Heraclitean concepts, limiting the unchanging, eternal Parmenidean 'Being' to their primal more-than-one 'elements' ('roots', 'sperms', 'atoms'), but at the same time accepting birth, perishing and change in beings of the world, according to the Heraclitean assumption of unity and separation.

Eleatic thought would influenc such *Sophists* as Gorgias, Protagoras, Cratylus and Xeniades, as well as the firs followers of Socrates, Antisthenes, Aristippus, and Euclid of Megara, who developed a variety of theories.

Plato would be significantl influence by Parmenides but would not embrace all his extreme views. In developing his theory of 'Forms' (*idea, eidos*), he would recognize him as 'our father Parmenides...the great one'. However, this tribute refers mainly to the initial *question* Parmenides posed and not to the philosopher's *response*. The latter, by the completion and enrichment of concepts during the course of the next two centuries, was excluded from Platonic thought. Thus, Plato confessed, "I fear that we may no longer grasp his words and thus remain far behind what he expressed in his contemplations".[43] The 'plausible narrative' of the Platonic cosmogony in *Timaeus* -the only dialogue treating of nature and its relation to man- is believed to have been derived from the 'likely-seeming cosmogony'[28B8.60] of Parmenides.

Plato would be deeply preoccupied with the ontological problem. He asks in the dialogue *The Sophist*: "What do you wish to designate when you pronounce the

---

41 K. Jaspers, *Aus dem Ursprung denkende Metaphysiker*, 43, 44.

42 Plato, *Theaetetus*, 181a

43 Plato, *Theaetetus*, 184a.

word Being"?[44] Reacting to the Parmenidean 'immovable' being, he would burst out: "But for heaven's sake, shall we let ourselves easily be persuaded that motion and life and soul and mind are really not present to absolute Being, that it neither lives nor thinks, but awful and holy, devoid of mind, is fi ed and immovable"?[45] Consequently, Plato would replace the single, unique Parmenidean 'Being' with the 'real existents' (*ontos onta*) which can be anything whatsoever that can be expressed with names, nouns or adjectives, but not with verbs. In this way he would lend his 'Forms', not only 'essence' -a philosophic term which is introduced for the firs time as the foundation of the existence of any being- but 'motion', 'life', 'soul', 'judgment'. In his dialogue, *Parmenides*, he would treat the Parmenidean position exhaustively, and in the dialogue, *The Sophist*, he would attempt to solve the problem which the position involves: If we assume that 'Being' is, and that 'non Being' is not, then -according to the Sophist's syllogism- we arrive at precisely the opposite conclusion, namely, that 'Being' is not and 'non Being' is. To avoid this absurd conclusion, Plato rejects the non-existence of 'non Being' as such. Instead, he attributes to 'non Being' *otherness* in relation to Being, assuming that it is a different sort of Being: "When we speak of 'non Being', we are not referring, as it seems, to the opposite of 'being' but only to something different".[46]

Plato's doctrine of 'Forms' marks the beginning of the great change from the Presocratic conception of 'Being'. The moment that *physis* (nature; the natural constitution of a person or thing; the essence) is viewed as *idea* (Form) -as the appearance of things- the Parmenidean identificatio Being – 'Physis' is transformed to 'Being – *idea*'. In this way, *idea* refers to the *exclusive* expression of 'being'. The consequence of this identificatio is that the *idea* is understood as the 'viewed', the 'what-it-is' (*ti estin*), i.e., *what* a thing actually *is*. Its 'essence' is thus shifted from the original concept of *idea*, which was the perception of Being coming-into-appearance, the 'that-Being' (*hoti estin*) i.e., the thing 'which *is*'. This change is decisive in regard to the subsequent distinction between 'essence' (*essentia*) and 'existence' (*existentia*). "The transformation of Being from *physis* to *idea*", Heidegger notes, "brings about one of the essential forms of movement within the history of the West, not just the history of Western art".[47] The *idea* is elevated to a 'prototype' image, as opposed to mere off-prints'. The latter tend toward equaling and resembling the archetype Form (*idea*). This means that the prime concept of Truth as self-emerging, un-concealed 'Being', is reduced to mere *homoiosis* (resemblance) and *mimesis* (imitation). 'Truth' depends now on the degree to which *likeness* is in accord with the Form (*idea*).

This concept of truth as *correctness* would become even clearer in *Aristotle*, where the 'true' is set in opposition to the 'false'. The primary goal of Parmenides, which is the experience of 'Truth' as the revealing and un-concealment of 'Be-

---

44 Plato, *The Sophist*, 244a.

45 Plato, *The Sophist*, 248e.

46 Plato, *The Sophist*, 257b.

47 M. Heidegger, *Introduction to Metaphysics*, 198, 192.

ing' from its concealment, is replaced by the effort to distinguish it from falsehood: "The truth is confirme when the subject and predicate are combined, and negated where they are divided; but the false has the contrary arrangement"; consequently, " 'falsity' and 'truth', according to Aristotle, resides not in *things* ... but in *thought*".[48] 'Truth' (*a-letheia*) is not *un-concealment* but an attribute of thought. While according to Parmenides "That which can be apprehended can be",[28B3] in Aristotle apprehending (*noein*) is transformed into a judgment concerning truth and falsehood. This conception, which would prove decisive in the development of subsequent European thinking, arises from the fact that *noein* has lost its Parmenidean meaning of apprehending and *identifying* with un-concealed 'Being' and now occupies a dominant *critical* position as 'thinking', as *logos* in the sense of logic regarding 'Being'. Aristotle's view of man was fatal for Western metaphysics.

Aristotle would reject the Parmenidean alternative "it 'is' or it 'is not' ", introducing "the intermediate state *genesis* (coming-to-be) between 'being' and 'non-being' and the developing subject, between what 'is' and what 'is not' ".[49] Change is for Aristotle the transition from *being potentially* to *being actually*, and in this respect the concept of being is *twofold.*

The logical methods which Parmenides followed unconsciously in developing his concepts would be developed by Aristotle into a perfect *logical* structure composed of ten 'categories' of being (*substance, quantity, quality, relation, place where, time when, position, state, acting, being acted upon*), which contain all the objects and events of the empirical world. Accepting the Parmenidean correspondence between 'being' and 'saying', Aristotle lends to these *logical* concepts *ontological* hypostasis. Just as Aristotle discerns many meanings of 'being',[50] by the same token he contradistinguishes diverse meanings of truth: "[Parmenides's] assumption is false inasmuch as he treats 'Being' as having only one meaning, whereas in reality it has several".[51] In *Metaphysics* and *Organon* Aristotle would develop an exhaustive theory of 'Being' and would emphasize even more strongly the Parmenidean split between the 'intellect' and the 'senses' to the degree that he would posit two sorts of souls, the mortal 'sense-perceiving soul' and the immortal 'intellectual soul'. Rejecting the split between 'Forms' (*ideas*) and 'sensations', he did not accept the Platonic view that sensations merely *participate* in 'Forms'. He believed that the Form (*idea*) -that is, the 'being'- of any tangible object has an *immediate* 'presence' within the tangible object and does not appear from having an hypostasis (real state of existence) that is different from the object. The Parmenidean theory of 'Being' is thus converted to an *organon* (instrument) of logic, which aims at definin truth through investigation of the various categories. Thus, 'categories', 'Form' (*idea*), 'essence' (*ousia*) would comprise the basis upon which Western thought would henceforth be evolved.

---

[48] Aristotle, *The Metaphysics*, 1027b20.

[49] Aristotle, *The Metaphysics*, 994a27–28.

[50] Aristotle, *The Metaphysics*, 1017a22.

[51] Aristotle, *The Physics*, 186a24.

Overview

Parmenides is without doubt one of the supreme thinkers of the ages. His stature is so great that some still wonder whether he was the great *philosopher* who founded *ontology*, or the firs *rational physicist* who deductively developed *cosmology*. These labels are, of course, anachronistic and bear no relation to the universal genius of Parmenides. A *separate* overview of the *cosmological* and the *ontological* dimensions of his work is not in order except to facilitate lucid exegesis.

Cosmology

Parmenides was the firs to oppose to empirical sense of perception the critical *rational* thought as a means toward findin absolute truth. He was the creator of a bold deductive system,[52] demonstrating by strict logical arguments that beneath the deceptive world of tangible appearances an ultimate, unchangeable theoretical reality is hidden. The concepts of stability and consistency he introduced to *physis* (nature) led subsequent generations of scientists to the pursuit and discovery of fundamental universal *constants* as well as natural laws of *conservation*, regarding, for example total energy and total momentum. He was the firs to apply as a scientifi investigative method the comparison and critical examination of two conflictin theories – those of reason and of experience. He was not only the creator of the theory of the 'continuous' structure of matter but indirectly also of the theory of 'discrete' structure, because it was precisely Leucippus's and Democritus's critical, empirical examination of Parmenides's conclusion concerning the non-existence of movement and change which led them to a partial revision of Parmenides's theory and to the development of the *atomic* theory. The confrontation between the two concepts 'continuous' and 'discrete' has preoccupied scientists and philosophers from that time till today; and it has proved to be extremely fertile in probing the problem of the ultimate structure of the universe. Aristotle was a supporter of the 'continuous' theory, which prevailed through the Middle Ages. Those remaining in agreement through the Renaissance were, among others, Descartes and Huygens. In contrast, other scientists like Gassendi and Newton brought the *atomic* theory back on stage, and the two concepts have remained a fundamental problem of physics up to Einstein, Schrödinger, and the contemporary quantum fiel theory.[53]

As we have seen, strictly applying the deductive method, Parmenides arrived at the absurd conclusion that in reality the world is an *immobile* and *unchanging* 'now'. Yet, is this conclusion actually as absurd as it seems, given the views of modern physics two and a half thousand years later? In daily life, time is divided into three

---

[52] The logical method that moves from the general to the specific versus the *inductive* method that moves from the specifi to the general.

[53] For a brief summary, see '*Overview*' of Democritus below.

sections: 'past', 'present' and 'future', the 'now' of our conscious awareness gliding steadily onward. The question is, whether this *flo of time* is an *objective* aspect of the world or, as A. Einstein wrote, 'the past, present and future are only *illusions*, even if stubborn ones'.

The proponents of a continuous state of flu maintain that the 'arrow of time' pointing toward the future is explained by the second law of thermodynamics, which states that entropy (the amount of disorder within a system) increases with time, denoting, thus, an *asymmetry* between past and future. Yet it seems that for such a conclusion no one can really rely on the second law, because its explanation, according to L. Boltzmann, is *probabilistic*, and thus, *symmetric* in time. "The arrow of time", theoretical physicist Paul Davies argues, "denotes an asymmetry of the world *in* time, not an asymmetry or flu *of* time. The label 'past' and 'future' may legitimately be applied to temporal directions, just as 'up' and 'down' may be applied to spatial directions, but talk of *the* past or *the* future is as meaningless as referring to the up or the down".[54] Einstein tied objective natural time with three-dimensional space, creating a uniform and continuous *four-dimensional spacetime*, which expresses cohesive universal reality. This concept contains 'in a single stroke' the entire history of the *cosmos*. Movement and change are nothing more than a sort of 'apparent' view – Parmenides's 'likely seeming' ordering of the world, experienced merely from the observer *perceiving subjectively*, while the *objective* four-dimensional reality remains unifie and unaltered. The attempt to combine Einstein's general theory of relativity with the quantum theory by canonical quantization, produces the Wheeler-DeWitt equation without a time variable, indicating that the universe should be 'frozen in time'.

"The objective world simply *is*; it does not *happen*," observes H. Weyl, one of the authoritative global mathematicians of the 20th century. "Only to the gaze of my consciousness...does a section of this world come to life as a fleetin image in space, which continuously changes in time".[55] The Parmenidean concept of a unique, immovable and unchanging 'Being' is revived in a variety of forms in contemporary scientifi thought. Einstein himself accepted the description of his theory as 'Parmenidean'.[56] Popper concludes: "Historians of science or philosophy who are reluctant to attribute to a great thinker like Parmenides a doctrine so severely unempirical as the illusionary character of the world of change ... may perhaps be less reluctant when they see that great scientists, such as Boltzmann, Minkowski, Weyl, Scrödinger, Gödel and above all Einstein, have seen things in a similar way to Parmenides, and have expressed themselves in strangely similar terms".[57]

---

[54] P. Davies, *That Mysterious Flow*, 27

[55] H. Weyl, *Philosophy of Mathematics and Natural Science*, 116.

[56] K. Popper, *The World of Parmenides*, 165ff.

[57] K. Popper, *The World of Parmenides*, 172.

Ontology

"The great begins great, sustains itself only through the free recurrence of greatness, and if it is great, also comes to an end in greatness", Heidegger notes. "So it is with the philosophy of the Greeks. It came to an end in greatness with Aristotle".[58] While Plato and Aristotle are considered the peak of Greek thought, at the same time they mark the 'end' of the 'great beginning' of the thought of an Anaximander, a Heraclitus, a Parmenides. "The philosophy of the Greeks attains dominance in the West not on the basis of its originary inception [the Presocratics] but on the basis of the inceptive end,"[59] Heidegger concludes. With the separation of 'Being' from 'physis' and its identificatio with 'idea', the Parmenidean self-emerging and un-concealment of 'Being' is lost. The Logos is transformed into a logical tool which, with the help of 'categories', adjudicates between truth and falsehood. No longer un-concealment, the Truth is now limited to the concept of 'correctness'.

The moment that noein (apprehending) is understood as 'thinking', which by means of 'logic' determines truth, the subject/object split has taken place. Henceforth, European metaphysics would consider truth as the agreement between 'thought' and 'thing' (adaequatio rei ad intellectum). 'Being' is simply considered an 'object' of depictions, which refers to a subject. The term 'Form' (idea) within the framework of Christian philosophy loses its objective character, and in this subjective sense is established in the course of European philosophy through such major exponents as Descartes, Leibniz, Kant, Fichte, Hegel, Schelling, and Nietzsche.

In the 20th century philosophy restates the pressing metaphysical question concerning 'Being'. *What is Metaphysics? The Essence of Truth. Being and Time*, these are some of the titles of Heidegger's works which deeply mark intellectual developments in the 20th century. Heidegger would come back to the Parmenidean question, 'What is Being'. Returning to the Presocratics, who were the firs to grasp the deepest meaning of 'Being', he would distinguish between 'Being' (Sein) and 'beings' (das Seiende). The object of ontology, according to Heidegger, must not be -as it has been up to today- the ontic 'being of beings' but the ontological 'Being as such'. Returning to the fundamental saying of Parmenides, "It is the same [thing] that can be apprehended and can be"[28B3] -which was expressed two and a half thousand years ago- Heidegger would consider 'the same' as the third term from which 'Being' springs and is linked to human 'apprehending'. Heidegger translates Parmenides's 'the same' as the 'Event' (Ereignis) in which 'apprehending' and 'Being' belong together as self-sameness, the 'Event' itself beyond 'Being' and 'apprehending', in a realm that is more mystical than philosophical. In order to rearticulate the question concerning 'Being', man must firs reorient himself and defin his relation to 'Being'..."According to Heidegger", Malevitsis notes, "we do not have on the one hand man and on the other Being; these two cannot be understood separately. Classical metaphysical thought has created this split, and the ending of metaphysics

[58] M. Heidegger, *Introduction to Metaphysics*, 17.

[59] M. Heidegger, *Introduction to Metaphysics*, 202.

must also bring an end to that concept. Being is neither a concept nor an object; one cannot understand it because it is one's very depths. Man himself, too, is not the 'rational animal' of metaphysics. Being belongs to man and man to Being. Being invites man and man responds to Being. Being is this invitation and man is this response".[60]

Parmenides would be the firs Greek thinker to direct European thought toward abstract thinking, separating it from tangible experience. From Parmenides, as we have seen, originate the probative methods of thinking which later philosophers will apply as a pure syllogistic vehicle, leading to the science of Logic and Dialectic. From him arose the fundamental question concerning 'Being' upon which Metaphysics would be based, and -from the 17th century onward- Ontology. From him begins the contradistinction between 'absolute Truth' and 'empirical knowledge', which constitutes the crucial point of departure in Western philosophy and forms the foundations of Ontology.

Nonetheless, Parmenidean thought itself contains in a tragic way the seeds of the unattainable. When we think, we inevitably think with reference to differences and relationships, setting up contrasts and assigning names. From the moment, however, that we introduce 'differentiation' and 'naming' into our thought we have ceased to apprehend Parmenidean 'Being' and already fin ourselves in the realm of 'non being'. "In the beginning of speculative philosophy", Jaspers notes, "lies already the impossible. Philosophy fails and vanishes when it reaches its truth; for it attempts expression at the cost of again deserting its truth the moment it is acquired".[61]

---

[60] C. Malevitsis, *M. Heidegger, Introduction to Metaphysics* (Greek edition), 269

[61] K. Jaspers, *Aus dem Ursprung denkende Metaphysiker*, 39.

Empedocles of Acragas (ca. 494–434 B.C.)

He was the most multi-faceted figu e of ancient Greek philosophy: With him departs the age of myth, tragedy, and orgiastic rites, while at the same time in him appears the new Greek – a democratic statesman, orator, reformer, allegorist, scientist. In him the two epochs clash . . .

Friedrich Nietzsche

Personality

He was to be likened to Faust. Nietzsche saw him as "oscillating between physician and magus, between poet and demagogue, between god and mortal, between scientist and artist, between statesman and priest, between Pythagoras and Democritus".[1] Renan limned him epigrammatically as "a cross between Pythagoras and Democritus, Newton and Cagliostro".[2] He was born in Acragas (today's Agrigenti) in southern Sicily. As an offspring of an aristocratic family he would participate passionately in the political affairs of his community, but contrary to what one would expect, he committed himself to the democratic cause. The success of his struggle against tyranny would be so great that his fellow citizens honored him with royal powers, which he disdained, preferring to devote himself to the investigation of nature. Soon he would be a renowned physician, poet and speaker, as well as a seer who would lead the people on the road to salvation. Moving among them in a grand and striking manner, he wore a gold wreath on his head, a purple robe with a golden girdle, and bronze scandals.[31B112] We must not forget that in that age medicine was as yet closely tied to both philosophy and sorcery. The intensely mystical character of Empedocles's language and work bears witness to the Orphic influence which must have come through the Pythagoreans. He was not only a 'superb doctor'[31A1(58)]-considered to be the founder of the Italian school of medicine-but also a distinguished speaker. Aristotle himself calls him the inventor of the art of oratory.[31A1(57)]

Many legends grew up around his work, suggesting both rare genius and supernatural powers. These achievements would lead the populace "to bow down before

[1] F. Nietzsche, *Philologica* III, 201.

[2] E. Renan's portrait is hyperbolic. As Guthrie properly reminds us, even Isaac Newton -with his interests in alchemy and the prophecies of Daniel- exhibited mystical tendencies and thus has been called "the firs modern scientist and the last of the mages" (W.K.C. Guthrie, *A History of Greek Philosophy: The Presocratic Tradition from Parmenides to Democritus*, II, 123). The same characterization could apply to Empedocles.

C.J. Vamvacas, *The Founders of Western Thought – The Presocratics*, Boston Studies in the Philosophy of Science 257, DOI 10.1007/978-1-4020-9791-1_11,

him and worship him as a god".[31A1(70)] Empedocles himself embraced that role. In the opening lines of his *Purification* , he would announce: "I, an immortal god, no longer a mortal, go about among you all, honoured as is meet, crowned with fillet and blooming garlands".[31B112,4–8; 31B113,2] However, side-by-side with this arrogance and condescension there exist -as is characteristic of a deeply religious temperament- extreme humility and contrition: "I too am no one, a fugitive and a wanderer from the gods, having put my trust in raving strife".[31B115]

His death, like his life, is shrouded in the mantle of legend. One version has it that, hounded by political foes and opponents, he would fle to the Peloponnese and there breathe his last. An other legend has the philosopher on the peak of Mt. Etna, hurling himself into the fier crater, "confirmin the hearsay of his deification [31A1(69)]. Later, the volcano would erupt, sending up one of his bronze scandals. . .This legend would inspire romantic spirits in Europe at the beginning of the 19th century. It would comprise the core of the dramatic poem, *Empedocles on Etna* by Matthew Arnold and J.C.F. Hölderlin's tragedy, *The Death of Empedocles*. In his unfinishe play composed of three versions, Hölderlin lends psychological, historical and political dimensions to Empedocles's action.

Writings

The work of Empedocles reflect his many-sided and integrated spirit. According to tradition, he must have written one prose work, *Medicine*, political essays, many tragedies, poems, hymns, epigrams, and the two known works which were most likely posthumously titled, *On nature* and *Purification* . These are the only writings of which today around four hundred and fift lines survive from an estimated total of three to f ve thousand. Empedocles would be the last of the Presocratics who -emulating his teacher Parmenides- would compose in epic hexameter verse. His style is personal and lacks the strict logical development of a Parmenides.

There are significan differences between the two surviving works.[3] *On nature* represents a philosophy of nature, which attempts on a rational basis to combine Parmenides's doctrines with human experience. It is written in the second person and is addressed to the philosopher's favorite pupil, Pausanias,[31B1] explaining the writer's physical theory in a confidentia manner.

In contrast, *Purification* , which presents the moral principles of spiritual life, is addressed to the poet's fellow citizens.[31B112] Here the style changes utterly. It becomes mystical, meditative, fille with intense religious illumination, as it displays the fall of man and the stages through which he must pass to achieve expiation, rehabilitation. The differences between the two works are so marked that many would maintain that Empedocles wrote them during two very distant moments of his life under the influenc of differing experiences and situations. Others, however,

[3] Few of the fragments belong with certainty to the one or the other poem. Most have been assigned to the two works by later scholars.

would reply that we also encounter this seeming *duality* in other great thinkers, for example the *mathematici* and *acusmatici* of Pythagoras, or the *Critique of Pure Reason* and the *Critique of Practical Reason* of Kant.

There is, as we shall see, a common underpinning, so that the two works can be considered not as opposites but as a *complementary* approach to a *unifie* cosmic vision. Such a complementarity and interdependency in the ancient Greece of Apollo and Dionysus is not only possible but to be expected. "Indeed it is here above all", Guthrie notes, "in the union of rational thought with mystical exaltation, that Empedocles sums up and personifie the spirit of his age and race".[4]

The Natural World

Immediately following Parmenides the Greeks found themselves at an intellectual impasse, since the rigid Parmenidean 'logic' had delivered the fina blow to Ionian 'monism', and standing in clear conflic with Heraclitean 'becoming', had entirely rejected the sense world as 'illusory'. Empedocles's pragmatic view could not support the absolute Parmenidean denial of empirical reality. Faced with a logical dead-end, he is clearly outraged by those who "boast that they have found the whole",[31B2,6] and he begs the gods to protect him from their madness.[31B3,1–2] Of course, he would not consider sense experience as entirely true, since human perceptions are inconsistent and subjective. However, while recognizing the fallibility of the senses, he believed that we should not utterly deny tangible experience as one criterion of truth, for we need aid "wherever there is a path for understanding".[31B3,12;31B3,9] In an evident allusion to the Parmenidean invitation, "From this point learn the opinions of mortals, listening to the deceitful pattern of my words",[28B8,51–62] he would consciously oppose his own command: "Hear thou the *undeceiving* order of my discourse".[31B17,26] What is the *undeceived* vision of Empedocles? It is a conception, which ingeniously combines the *eternal stability* of Parmenides and the *perpetual movement and change* of Heraclitus and of empirical experience. In order to achieve this synthesis, Empedocles would selectively reject or accept certain basic positions of foregoing systems.

He would reject the 'monistic' conceptions of the three Milesian 'natural philosophers' regarding *one* unique cosmic *arche* (Thales: *water*, Anaximander: *the infinit* ; Anaximenes: *air*) as the source of all things. He would deny, in turn, the simplistic conception of the *auto-kinesis* of beings, organic or inorganic -since everything was considered animate from the start- the universe is besouled ("all things are animated"[11A23]). He would, nevertheless, accept the basic opposite pairs which the Ionians used to explain the plurality of the world, such as hot-cold, dry-wet, and so forth[12A16].

[4] W.K.C. Guthrie, *A History of Greek Philosophy. The Presocratic tradition from Parmenides to Democritus*, II, 125.

The latter notion, which coincided with the medical view of the time that "health is maintained through the balance of the powers of wet, dry, cold, hot, bitter, sweet, and so on",[24B4] was supported both by the Pythagorean teaching that the elements are four ('fire water, earth, air'[58B1a]) and the Heraclitean view that "in death earth becomes water, and in death water becomes air, and air becomes fire and conversely".[22B76]

Empedocles would also agree with Heraclitus that everything is involved in perpetual change and motion, as confirme by the human senses. He does not, however, accept that the vast variety of tangible things constitutes *ultimate* reality.

He agrees word for word with Parmenides that being is uncreated, immortal and indestructible.[31B12] He also agrees that non being, the void, does not exist.[31B14] He would not accept, however, the two other basic Parmenidean attributes of being: that it is [a] singular, and [b] immobile.

Through the above inspired synthesis, Empedocles would free himself from the bonds of merciless Pamenidean reasoning and restore the tangible world of motion and change to the position assigned it by common sense. The new cosmological system, which emerges, can be described in broad outline as follows:

- The Parmenidean attributes 'uncreated' and 'imperishable' no longer refer to a single Being, but to four completely equal 'uncreated'[31B7] 'roots' – "four are the roots of everything",[31B6] *earth, water, air, fi e*. These four elements constitute the *ultimate realities*.
- The innumerable variety of sense phenomena, such as change, generation, and perishing are due exclusively to the *mingling* of the root elements in various concrete *numerical proportions*: "I shall tell thee: there is no birth of any mortal things, nor any end in baneful death, but only mingling and separation of what is mingled".[31B8,1–3]
- 'Mingling' and 'separation' as well as movement come about under the influenc of two external opposite powers: 'Love' (*Attraction*) and 'Strife' (*Repulsion*): "They never cease to alternate continually, now all coming together into one through Love, and now again each one drawing apart by Strife's hatred".[31B17,6–8]

The implication of these positions is that all the transient, transmuted forms of the empirical world come about in reality through the combination in varying proportions of the four *eternal and immutable* root elements. This would be the new 'combined' or 'pluralistic' cosmology of Empedocles, which would break ground for the immediately upcoming systems of Anaxagoras and Democritus.

The 'Roots'

Although, as we have seen, the basic elements earth, air, water and fire had figure in the cosmologies of previous Presocratics, in Empedocles they would assume for the firs time total *equality* amongst themselves, as well as the attribute of *ultimate* reality. To be sure, Empedocles did not use the term *element*, which was introduced later, but instead *roots*, foundations.

The four 'roots' are 'equal and coeval'.[31B17,27] They are equal and stable not only qualitatively but quantitatively.[31B17,30–33] While they are equal, "each has another competence and its own character".[31B17,28] Empedocles would attribute to the four roots all the properties of the Parmenidean Being, except immobility. They are 'uncreated',[31B7] unchanging,[31B17] forever imperishable, homogeneous and compact.[31B13] Void does not exist.[31B12,1–2;31B14] That which might be considered a void is, in reality, air. One of the historic achievements of Empedocles is the establishment of the non-existence of void, not exclusively on theoretical grounds, but by the *experimental* replacement of an equal volume of air by water, and vice-versa.[31B100,8–21]

Becoming

Having define the properties of the four 'roots', Empedocles could now readily explain *plurality*, *change*, *coming to be*, and *perishing* in the empirical world. These come about through the *syntheses* of the four root elements in a variety of *discrete* arithmetical proportions. Thus the philosopher refers *qualitative* differences of matter coming about to its *quantitative* composition. Blood is formed, for example, from equal parts of the four roots, earth (E), water (W), air (A), fir (F); the nerves from F and E with a double portion of W; the bones from two parts W and E and four parts F, and so on. Analogous to contemporary chemical terminology, we can readily write out the corresponding molecular formulas: blood = EWAF; nerves = FEW_2; bones = $W_2E_2F_4$, and so on. Basically, then, generation and attrition do not exist; becoming is a phenomenon due exclusively to the restructuring – that is, the mingling and separating out of the four 'roots' in differing proportions.[31B8,1–3] Empedocles's conception is categorical: Being -that is, the four 'roots' – is ungenerated and indestructible. "For they [the four roots] endure forever stable, and they are not [in a process of] becoming".[5] Generation and disintegration refer *solely* to the variety of compounds of the four elements among themselves, and men mistakenly consider that the ultimate Being is subject to alteration and perishing.[31B11] Empedocles, "does not deny creation, except from nothing, nor attrition, except complete – leading, that is, to nothing";[31B11] However, he admits that phenomena can sweep one away into faulty language and description: "When the roots combine to form animals or plants, men say these are born, and when the roots scatter again men call it death, wrongly, but I comply with custom myself".[31B9]

Forces

"Empedocles deserves to go down in history", Guthrie notes, "as the firs European to introduce into science the idea of a force operating on matter".[6] Indeed this is

---

[5] Aristotle, *The Metaphysics*, 984a9.

[6] W.K.C.Guthrie, *A History of Greek Philosophy. The Presocratic Tradition from Parmenides to Democritus*, II, 159.

so. From the moment that he rejects Parmenidean *uniqueness* and *immobility* of being and posits that the four root elements are subject to *mingling* and *separation*, he is obligated to interpret *how* the *mingling* and *separation* come about. The old Ionian notion that living matter is auto-kinetic can no longer be seriously entertained. Empedocles would, therefore, resort for the firs time to an *exterior* cause of motion – to a pair of forces, *attraction-repulsion*.[7] These two opposing forces are responsible for all the phenomena and events of the *cosmos*. 'Love' (attraction; friendship; affinity or 'Aphrodite', 'Harmony', 'Joy', 'Affection'), and 'Strife' (repulsion; conflict or: 'Aggressiveness'; 'Rage') are divine and immortal powers.[31B16] The two forces are invisible, equal, and isotropic, i.e., extending in space with like force in all directions, "equal in length and breadth".[31B20] Although Empedocles speaks of two forces, 'equal in every respect'[31B17,19] with the four 'roots', he does not seem quite to regard them as material bodies. He perceives them, rather, as operating within an extended spatial field isotropic, homogeneous, dynamic.

Their influenc is not upon material things only, but on living creatures as well; it is, moreover, not only *mechanical* but *psychological* and even *moral*. "For if one follows up and comprehends the statements of Empedocles in accord with their true meaning and not with his obscure language", Aristotle observes, "he will fin that love is truly the cause of good and strife the cause of evil, and thus he will be able to conclude correctly that Empedocles, who indeed was the firs to do so, spoke of evil and good as firs principles".[8] Introducing for the firs time this pair of opposing forces, 'Love' (Attraction), 'Strife' (Repulsion), Empedocles shows a key characteristic of his genius, for he states not only a *natural law* but simultaneously a *moral principle*.[9] 'Love' and 'Strife' exercise influenc both on the natural world, as attraction and repulsion, and upon living things, as love and friendship, or hate and aggressiveness.[31B21,7–8;31B17,23–24;31B22,5–9] These same opposing powers act on both a cosmogonic and cosmologic scale, on the macrocosmic and microcosmic level, and in material and biological fields 'Attraction' unites dissimilars and on the other hand separates similars. Conversely, 'repulsion' separates dissimilars and consequently unites similars.

The 'Cosmic Cycle'

"The *cosmos* itself is alternately created and destroyed, and again it is created and again it perishes, and these opposed states continue to infinit . Thus Empedocles teaches that Love and Strife alternately prevail: Love combines everything into one, destroying the world which came about through Strife, and from that creates the Sphere, and Strife again separates out the elements and creates a world like this",[31A52] in a never ending cosmic cycle. Empedocles describes in broad

---

[7] Aristotle, *The Metaphysics*, 985a29.

[8] Aristotle, *The Metaphysics*, 985a4.

[9] Aristotle, *The Metaphysics*, 984b38.

terms[31B17,6–13] a universe forever *pulsing* between two extreme conditions: that of the *One*, a compact, symmetrical *Sphere*, compactly bonded within itself, sustained by Love; and that of fragmented, disordered and differentiated *plurality*, which is the work of Strife. In each transition from the one extreme stage to the other, the world of matter and living things is formed and subsequently destroyed.

The extant fragments are so incomplete as to give rise to diverse interpretive approaches to this basic 'cosmic cycle of Empedocles', which can be divided into four phases:

(a) Complete *unity* – the *Sphere*, where Love (attraction) prevails.
(b) Gradual strengthening of Strife (repulsion), and withdrawal of Love (attraction), the intermediary stages being the creation and perishing of the world.
(c) Complete *plurality* in which Strife (repulsion) prevails.
(d) Reverse gradual strengthening of Love (attraction) and withdrawal of Strife (repulsion the intermediary stages again being the creation and perishing of the world; and so forth and so on, in perpetual cyclical alternation.

Cosmogony – Cosmology

W.K.C. Guthrie observes, "the motion originated by Strife, in keeping with his character, is one of separation pure and simple. Its further results are *accidental* consequences of the separation".[10] Thus, "in effect the formation of a cosmos is the undesigned and purposeless result of a clash between material substances, each driven in certain ways by its own internal impulse. That is the essence of this type of cosmology, against which Plato and Aristotle directed their heaviest batteries in the interests of teleology".[11] Empedocles speaks, of course, of 'chance' events, but he does not view chance as we do today; nor does he introduce a definitio – a lack which will later be criticized: "Empedocles . . . appears to refer to chance in regard to smaller things, but he never explains what [it is]".[12] *Tyche* (chance), *Anange* (necessity), *Physis* (nature; constitution), are ancient Greek words frequently encountered together, as in 'necessary chance', 'by chance, according to necessity', 'by nature and chance', suggesting, as Guthrie puts it, that "*Physis* . . . is a natural necessity inherent in each separate thing, or substance, not a law of interaction between them. With each thing moving as its *physis* (constitution) dictates, the clashes between them will be fortuitous though caused by necessity".[13]

---

[10] W.K.C. Guthrie, *A History of Greek Philosophy. The Presocratic tradition from Parmenides to Democritus*, II, 186n.4.

[11] W.K.C. Guthrie, *A History of Greek Philosophy. The Presocratic tradition from Pamenides to Democritus*, II, 164.

[12] Simplicius, *Aristotelis Physicorum Commentaria*, 331.15.

[13] W.K.C. Guthrie, *A History of Greek Philosophy. The Presocratic tradition from Parmenides to Democritus*, II, 164.

It is significan that picturing the creation and evolution of the *cosmos*, Empedocles gives fir a primal position].[14] Fire itself -and by extension, heat- play a central role in the physical-chemical processes of Empedocles's cosmogony. He correctly perceives evaporation and drying as the effect of heat,[31A69] and on the basis of this principle he interprets a number of natural phenomena of solidificatio and crystallization.[31A51;31A60;43B56;43A69] The effect of fir is not limited to material things alone, but is also extended to living things.[31B73;31B62,1–2] It is apparent that Empedocles perceives in *fi e/heat* not only a material element, but -in today's terms- a form of energy that not only plays a primal role in the *formation* of the universe, but also constitutes the main cause in the *initiation* of cosmogony itself, by the *asymmetry* and the resulting imbalance created during the firs disruption of the Sphere, as we shall see immediately below.

Having define in broad terms the primary role of *chance according to necessity*, or of *nature and chance*, and the special significanc of *fi e/heat*, we can now turn to the cosmogony and cosmology of Empedocles.[15]

Under the initial repulsive impact of Strife, the compact, symmetrical and homogeneous Sphere breaks apart: "Air was firs separated off from the primal mixture and spread around it in a circle. After air, fir burst out, and having nowhere else to go ran out under the solid mass of air".[31A30] By this separation, the disrupted Sphere's symmetry is destroyed and the imbalance of masses thus created initiates the beginning of a revolving, continually accelerating movement, a 'vortex'.[31A30] The same whirling motion creates the "two hemispheres revolving round the earth, one entirely of fir and the other a mixture of air and a little fire This, he supposes, is the night".[31A30] As the revolving motion of the whirlwind continually increases, the length of the day respectively decreases, so as to reach today around twelve hours: "When the human race was firs generated out of the earth, the day was of the length that ten months are now, owing to the slowness of the sun's advance. As time went on, it became the length of seven months".[31A75] The same motion of the whirlwind holds the earth suspended and keeps the heavenly bodies in their places, preventing them from falling on earth. Empedocles would explain this centrifugal property of the *cosmic* whirlwind by applying an *earthly* observation: "The movement of the sky as it turns with increasing speed blocks the earth from shifting, as happens with water in a container. When the container swirls in a circle, and frequently the water find itself below the bronze bottom, it does not pour out, though it would be natural for it to do so. The cause is the same".[31A67]

Empedocles pictures the shape of the earth as spherical-plane, as it extends to the side more than to the top, in the shape of an egg – a metaphor probably derived from Orphic rites. He believed that at firs the celestial North Pole was at the Zenith, but later the polar axis of the universe was tilted.[31A58] The composition of the sun is

---

14 Aristotle, *The Metaphysics*, 985a31.

15 The extant fragments are so elliptical and irregular that they permit differing and multiple conflictin interpretations by later commentators.

not precisely defined [31A49] The moon was formed separately of the air cut off by the fire which solidifie like hail. It has its light from the sun".[31A30] The stars are "fier , made from the fier element, which the air contained within itself and squeezed out at the firs separation".[31A53] "The fi ed stars are imbedded in the crystal [vault of heaven], while the planets move freely".[31A54] Regarding the solar eclipse, Empedocles would provide the correct interpretation: the moon "cuts off [the sun's] rays as he passes above her, and darkens as much of the earth as is the breadth of the pale-eyed moon".[31B42]

Creation of Life

Under the influenc of Strife, the firs separation of the four elements in turn according to their (specific weight into *earth* at the center, *water* on its surface, *f re* and *aether* in heaven, is not complete. Fiery pockets continue to exist in the depths of the earth; water and steam gushes out of its depths and evaporates in the atmosphere. As yet we fin ourselves at the stage of intense conflic between the forces of Strife/repulsion and Love/attraction. Under the influenc of Love, the firs organic compounds form from different elements, according to concrete, simple numerical ratios. Earth, fire water and air constitute the *raw materials*, fire-hea is the 'forming condition', and *Cypris* (Love) is the 'binding power': "So then *Cypris*, after she had moistened the earth with water, as she fashioned the forms of living things gave them to swift fir to harden".[31B73] Life on earth develops in four phases, from lower to higher organisms, not by chance or according to a divine teleological plan, but according to the dictates of *natural selection*: "The firs generations of plants and animals were not entire, but divided, with parts not grown together; the second, in which the parts grew together, were like creatures of fantasy; the third was the generation of the whole-natured; the fourth was no longer engendered from the elements like earth and water but from each other, when for some their nourishment became thick and for others the beauty of women excited the seminal motion. The kinds of all living creatures were distributed according to the character of their mixture: some had a more natural inclination to water, as many as had a preponderance of fir fl w up into the air, the heavier made for the earth".[31A72]

During the firs phase of life creation the "parts of things wandered separately". [31B58] "On the earth many heads sprang up without necks, arms wandered bare, bereft of shoulders, and eyes strayed alone in need of foreheads".[31B57] This random wandering and movement would lead to coincidental combinations creating at random a mass of fantastic shapes: "As they [the elements] were mingled, myriad kinds of mortal creatures were brought forth, endowed with all sorts of shapes, a wonder to behold".[31B35,16–17] A variety of monstrous creatures would arise, "shambling with countless hands, with faces and breasts both front and back, oxen with the heads of men, and conversely ... creatures in human form with the heads of oxen, and mixtures of men and partly of women's nature, fitte with shadowed [private] parts".[31B61] Such images of chimeras appear in the art of the Upper Paleolithic Period[16] and are *shamanistic* in origin.[17]

Of these deformed beings the ones that will eventually survive by *natural selection* are those that possess a constitutions enabling them to face the challenges of existence: "As many of these parts as were fitte together in such a way as to ensure their preservation became animals and survived, because they fulfille mutual needs – the teeth tearing and softening food, the stomach digesting it, and the liver converting it into blood. The human head, when it meets the body, ensures the preservation of the whole, but being inappropriate to the ox-body it leads to its disappearance".[31B61] The moment that this harmonious blending of the parts of the organism secure its survival, the third phase of life evolution begins, which is the creation of 'complete' beings – organisms, that is, that are completely functional but as yet without difference in gender. With the ensuing strengthening of Strife, we arrive at the fourth phase, which constitutes the fina differentiation and separation of the genders into male and female. We now fin ourselves in today's world of dominant discord and conflic (Srife), coming about by weakened Love and concord.

Knowledge

For Empedocles microcosm and macrocosm constitute a unified indivisible whole. Consistent in his theory of the four 'roots' and the two opposing forces, attraction-repulsion, 'Love'-'Strife', he incorporates into it his views regarding the processes of sensation, thought and knowledge. Empedocles's conceptions revolve around three basic concepts: (a) Sensation and thought are based on the same principle: 'like is both perceived and thought of by like'. (b) This principle is manifested through direct physical contact. (c) Knowledge, though confine within bounds, is useful.

(a) Empedocles believes that sensation and cognition come from the attraction between *things that are alike*. To understand the concept of likeness, we must firs see what it is that make a thing A, or B, or C. According to Empedocles, the *identity* of each thing derives from the concrete numerical *ratio* among the elements of which the thing is composed.[31A78]
    Each natural body is thus characterized by a concrete composition of elements in a specifi ratio. The person appropriately feels, perceives, thinks, feels joy and sorrow according to the degree that his constituents are *like* those of the outer world. Thus microcosm and macrocosm are attuned in such a way, as Brun puts it, that "the person, a true constituent of the *cosmos*, participates in the universe".[18] Sensation and thought, joy and sorrow, are activities and functions linked to the same basic principle of attraction of similars:[31B109] "Knowledge of like by like".[31B109] "Pleasure is induced by what is similar in respect to its

[16] R. Leakey, *The Origin of Humankind*, 116.

[17] E.R. Dodds, *The Greeks and the Irrational*, 145

[18] J. Brun, *Les présocratiques*, 97.

parts and their mixture, and pain by the opposite”.[31A86(9)] For Empedocles, then, feeling and thinking are *corporeal* not *mental* capacities.

(b) Sensation and cognition of like by like come about exclusively through *physical contact*. This is effected through each body’s outfl w of special ‘effluences’ which are received by the properly sized ‘pores’ of other beings. The giving off of effluence is an inborn characteristic of all entities.[31B89] Sensation and cognition presuppose a geometrical symmetry between the various outfl wing ‘effluences and the respective ‘pores’ of the recipients. Wherever there is no such correspondence, sensation and cognition would not be possible.[31A86(7)] Through this system of ‘effluences and ‘pores’ Empedocles would explain the functioning of all of the senses: *vision*;[19,31A86(7);31A86(8);31B109a] hearing;[31A86(9)] smell;[31B101;31B102] *taste;* touch.[31A86(9)]

(c) Empedocles would realize that the above purely natural processes that are based on interaction between similars and are the cause of sensation and thought, of joy and sadness, cannot guarantee man a complete picture of reality: “Narrow are the powers that are spread through the body”.[31B2] Even though our perception of reality is doomed to remain fragmentary and limited, Empedocles does not lose heart, nor does he condemn the entire world of the senses as ‘illusory’, as does his teacher Parmenides. On the contrary, he takes a positive stance in the face of this given situation. With full knowledge of the limitations imposed on us, we must extend our knowledge of the world surrounding us.[31B3] There is an interaction between perceived depictions and experiences, on the one hand; and, on the other, a person’s character, which is being formed accordingly.[31B106] Thus, “As much as men change their nature, so much it also befalls them to think different thoughts”.[31B108] The enrichment of man by knowledge and the widening of his intellectual horizons demand determination and his steady, active participation in this process.[31B110,1–8] Finally, the achievement of knowledge can lead man to the taming of nature and the control of her powers for his benefit[31B111] The attraction of similars, the affinit of man with surrounding nature, brings not only understanding of the latter but also aids in perfect control of her. Knowledge for Empedocles no longer constitutes only the means for discovery of truth; it is also the tool for mastering the powers of nature, so that the conditions of human life can be improved.

‘Purifications

‘Purification refers to catharsis – catharsis from the guilt, miasma, or sin. As the title suggests, this poetic work of Empedocles relates to an entirely different sphere than that of his study *On Nature*. The style and content change entirely. The theme here is not the *natural world*, but the *fate of the soul*. Souls, originating from an initial state of supreme unity, love, blessedness and innocence, fall through the influenc of Hate-Strife, into the sin of bloodshed and destruc-

---

[19] Plato, *Meno*, 76d.

tion of one another. The consequence of this fall and its resulting punishment is a tortured wandering of souls in hostile regions within a series of reincarnations. Gradually, however, to the degree that they retrieve the sense of their divine origin and observe certain rules of purification they reach the stage of total liberation from the cycle of reincarnations, regaining the company of the gods, "sharing hearth and table with the other immortals, free from human sorrows, unwearied".[31B147]

What relationship may one fin between this deeply mystical and religious work of Empedocles and the purely rational, scientifi presentation of his ideas in *On Nature*? We will examine this issue in the conclusion, having firs discussed the basic elements of *Purification* .

Introduction

"Friends. . .I bid you hail. I an immortal god, no longer a mortal, go about among you all, honored as is meet, crowned with fillet and blooming garlands".[31B112] From Empedocles's firs lines, we sense the change in expressive tone from that of *On Nature*. He no longer addresses his one, faithful, initiate friend, but instead the throng of 'mortals'. His attitude is condescending, since he believes that he has already attained the *fina stage* in the cycle of reincarnations, being at the same time seer, bard and physician.[31B146;31B113,2] He is convinced that he alone is in a position to teach truth to the wayward, but he doubts that they are capable of the requisite understanding and belief.[31B114]

At the same time, however, Empedocles does not forget his initial fall as well as the anguished and painful wandering of his soul's reincarnations. Thus, on the one hand we fin Empedocles imperious and arrogant, but on the other, humble and repentant. Thus, in contrast to "I, an immortal god, no longer a mortal, go about you all, honored as is meet",[31B112,4–5] we have his confession: "of these I too am now one, a fugitive from the gods and a wanderer, who put my trust in raving strife".[31B115,13] In contrast to the proud, "why am I hot against these men, as if it were some great thing to surpass mortals doomed to destruction",[31B113] we have his humble statement, "alas, that the pitiless day did not destroy me before I thought to bring the impious food to my lips".[31B139] Pompous pride and extreme humility exist side-by-side within the same personality. "Pride in possession of certain truth", Guthrie notes, "is the mark of the seer who believes himself divinely inspired; consciousness of the fallibility and inadequacy of the human faculties belongs rather to the *scientifi* mind. What we are witnessing in these men (in Empedocles far more fully than in Heraclitus) is the gradual transformation of the *seer* into the *philosopher*. That is what makes the essential unity of their thought so difficul to comprehend, while at the same time it lends them an absorbing interest".[20]

[20] W.K.C. Guthrie, *A History of Greek Philosophy. The Presocratic tradition from Parmenides to Democritus*, II, 248.

The Primal Stage of Bliss

During the initial stage of innocence, antagonism and violence [personifie in the god *Aris* (Mars)] are unknown, since this phase is governed wholly by *Cypris*, the goddess of Love. Souls (*daemons*) exist in a state of beatitude and blessedness. All offerings are *bloodless*, as they are performed with water, honey, wine, or olive oil. All things are in harmony, which brings plentiful crops throughout the year.[31B128;31B130;31B78]

The Fall

There comes a time that under the influenc of irresistible necessity, strife, hate and discord will infiltrate leading to slaughter and rampant bloodshed. This is the mark of the fall of souls into the earthly realm, into a cycle of painful reincarnations as punishment and propitiation for this 'primordial' sin. Empedocles's lines describing this fall are overwhelmingly tragic.[31B115,1–12] Empedocles himself powerfully describes this anguished wandering through successive reincarnations in "some of the strangest and most moving religious poetry which has come down to us from antiquity", as Dodds puts it.[21] "Alas, that the pitiless day did not destroy me before I thought to bring the impious food to my lips[31B139] . . . I, a fugitive from the gods and a wanderer, who put my trust in raving strife[31B115,13–14] . . . Before this have I been a boy and a girl, a bush and a bird, and a dumb fis of the sea[31B117] . . . I wept and wailed when I saw the unaccustomed place.[31B118] From what high rank and what a height of bliss, [we come to] a roofed-in cave, a joyless place [where] Murder and Anger and tribes of other spirits of death, and parching diseases and wastings and the works of flu wander in darkness over the meadow of Doom."[31B,119–121]

Guilt

In the ancient Greek spirit, the primordial feeling of guilt, which we also fin in other peoples, has its roots in the Orphic myth of Dionysus's dismembering by the Titans. It is precisely the violation of the Orphic command, 'Shun murder', which for Empedocles is the cause of souls' fall from their state of concord and innocence.[31B135] The profound contrition and instinctive fear toward shedding blood is indissolubly interwoven with belief in reincarnation, since the animal one kills for food could be the seat of a fallen human soul that had 'changed form'.[31B137;31B136]

E.R. Dodds observes "Morally, reincarnation offered a more satisfactory solution to the Late Archaic problem of divine justice than did inherited guilt or post-mortem punishment in another world.... The post-mortem punishment did not explain why the gods tolerated so much human suffering, and in particular the unmerited;

---

[21] E.R. Dodds, *The Greeks and the Irrational*, 153.

suffering of the innocent.[22] Reincarnation did. On that view, no human soul was innocent; all were paying, in various degrees, for crimes of varying atrocity committed in former lives. . .Only in this way, and on this cosmic-time scale, could justice in its full archaic sense -the justice of the law that 'the Doer shall suffer'- be completely realized for every soul".[23]

Purificatio

The torturous path of the soul towards fina salvation is determined by strict rules of abstinence that can be summed up in the saying, "fast from evil".[31B144] During the gradual purgation of the soul in a series of reincarnations, there is a hierarchy of lives, beginning with plants, continuing with animals, then on to ordinary human beings, and, finall to the highest category of humanity, which constitutes the threshold to ultimate apotheosis: "At the end they become prophets, bards, physicians, and princes among men on earth. Thence they arise as gods highest in honor".[31B146;31B147]

Daemons

Empedocles does not use the word 'soul' but, instead, the word *daemon*. 'Soul' denoted for Empedocles and his contemporaries the condition of the *living* body, which gives up its last warm breath in death, expelled into the air.[31B2,4] In contrast, the *daemon* (spirit) is the holy portion of the body, the indissoluble 'ego', detachable from the body, that undergoes and experiences all the successive reincarnations so as finall to unite with the divine. "The function of the *daemon*", Dodds observes, "is to be the carrier of man's potential divinity".[24]

The divine: The divine cannot be reached through the senses.[31B133] The reason is that the senses, being corporeal faculties, possess nothing in common with the divine.[31B134] Empedocles's lofty god is an incorporeal, invisible and indescribable holy intellect, a 'sacred mind' (*phrin hieri*). Since the word *phrin* not only signifie 'mind' but is etymologically related to the word *phrontis* (care), a wider meaning should be attributed to it, suggesting not only intellect but feelings, especially of love. Empedocles appears to suggest this in the statement that through Love, men think 'kindly thoughts'.[31B17,23]

Empedocles would also attribute divinity to other entities: to the four 'roots', the two opposing forces, the fallen daemons, and the traditional Olympian gods. The latter, however, have ceased to possess immortality; they are simply "long-

---

22 We must not forget that as a physician Empedocles witness every day the pain and discomforts of the sick.

23 E.R. Dodds, *The Greeks and the Irrational*, 150–151.

24 E.R. Dodds, *The Greeks and the irrational*, 153.

lived".[21B21,12] In these cases, the meaning of the divine derives less from religious feeling than from the uniqueness and permanence that characterize these natural elements and powers. In religious terms, Empedocles's lofty divinity is that 'sacred mind', which can be identifie with the Sphere of Love.

The Relationship of the Two Poems

We can now return to the issue of the apparent lack of relationship between the two poems. Of course, this could be due to *Purification* probably being written later under different circumstances and with different priorities. Still, the utterly different thematic development of the two poems remains baffling Indeed, as Dodds points out, "scholars have been astonished that a man capable of the acute observation and constructive thought which appear in Empedocles' poem *On* Nature should also have written the *Purification* and represented himself as a magician". This could be explained by the view of Empedocles as "a very old type of personality, the shaman, who combines the still undifferentiated functions of magician and naturalist, poet and philosopher, preacher, healer and public counselor. After him these functions fell apart; philosophers henceforth were to be neither poets nor magicians ... But men like Empedocles and Pythagoras may well have exercised all these functions. It was not a question of 'synthesizing' these wide domains of practical and theoretical knowledge; in their quality as Men of God they practiced with confidenc in all of them; the 'synthesis' was personal, not logical".[25]

As C. Kahn observes, "The two poems of Empedocles are fundamentally compatible with one another. His conjunction of natural philosophy and mystic religion is effected by his own notion of Love or *Harmonia* as a complex reality, at once physical and spiritual, inherent in the symmetrical mixture of unlike compounds but capable of existing apart form them as an independent, imperishable entity.... This is Empedocles' own version of the great discovery of his age: the recognition of the Rational or Spiritual as a distinct and dominant element in Nature. Love cannot be seen by the eyes, but only by the mind (*nous*) or, more precisely, by the Love which is in us. For 'by Love we see love'.[31B109,3] The element by which we communicate with the divine, creative force in the universe will fittingl stand for the divine principle embedded in man".[26]

The new fragments of the recently restored 'Papyrus of Empedocles'[27] in Strasburg appear to confir the view that both of the works of Empedocles, 'On Nature' and 'Purifications are indeed based on a unifie system.

---

[25] E.R. Dodds, *The Greeks and the irrational*, 145–146.

[26] C. Kahn, *Religion and Natural Philosophy in Empedocles' Doctrine of the Soul*, 448, 446.

[27] A. Martin, O. Primavesi, *L' Empedocle de Strasbourg*.

Overview

From the work of Empedocles arises his many-faceted personality – a personality, which would influenc many domains of later western thought. A gifted speaker -he was considered, as we saw, the founder of art of oratory- he would fin a worthy successor in the person of his pupil, Gorgias, who was distinguished as an orator in Athens. His disciples Acron and Pausanias are considered the founders of the Sicilian school of medicine. Empedocles's influenc on the thought of Plato is projected in clear relief in the latter's *Timaeus*. Plato would embrace Empedocles's theory of the four root elements as 'primal' and 'unalloyed', and he would render them as geometrical figures replacing them with the four regular 'Platonic' polyhedrons: *f re* = regular tetrahedron, *earth* = regular hexahedron or cube, *air* = regular octahedron, *water* = regular icosahedron [the fift figure the dodecahedron, Plato considers responsible for the decorum of the universe]. Lending the root elements as geometrical figures he explains their joining on the basis of stereometry. It appears, in turn, that he accepts Love as the bonding force, Strife ('necessity' in *Timaeus*) as the principle of division, the attraction of like to like, as well as features of Empedocles's physiology. Later philosophers like Nietzsche and Schopenhauer would bend with admiration over the work of Empedocles. And Freud would discern in the two opposing forces fundamental concepts for his own theory (love-death). Empedocles's life and its tragic end would inspire poets and dramatists like Hölderlin, Arnold, Guérin, Brock.

Without doubt, Empedocles is a unique case of scientifi genius. In spite of the limited and as yet untested vocabulary at his disposal, in spite of the impossibility of microscopic and telescopic observation, in spite of the lack of any mathematical background, he speaks with unparalleled visionary power and strict logical consistency of a complete and cohesive natural world which is in amazing *qualitative* agreement -taking historical perspectives into account- with the view of the natural microcosm and macrocosm reached by the physical sciences around two and a half millennia later.

Sense Experience

Empedocles acknowledges the weaknesses of sense experience as a means of attaining knowledge, but he considers it a necessary approach towards knowledge. "According to Empedocles, the criterion of truth is not the senses but sound logic".[31B2,1] Nevertheless, he continues: "Come now, observe with all thy powers how each thing is clear". *Thought* and *sense experience* in conjunction reveal how 'each thing is clear'.[31B3,9–12] This response is confined however, by the limits of human intelligence. "Truth is not completely ungraspable but can be perceived within the bounds of human intellect".[31B2,8–9]

Supporting the necessary cooperation between sense experience and thought for the investigation, not the acquisition, of ultimate truth, and rejecting the two extremes -on the one hand, absolute privileging of mind, which would later develop in the classical *rationalism* of Descartes, Spinoza, and Leibniz, and on the other, the absolute privileging of sense experience, which would later develop in the classical *empiricism* of Francis Bacon, Locke, Berkeley, and Hume- Empedocles would be the firs to point to the road that would be taken by contemporary science.

Experiment

The ancient Greeks -as we observed in the 'Introduction'- have been criticized for their lack of experimentation. In Empedocles's *clepsydra*,[28] however, we have the firs extant record of a scientifi experiment aimed at demonstrating the *corporeality* of air.

Most of the Presocratics formulated cosmogonies the nature of which excludes experimental proof. Empedocles conceived a cosmogony of a perpetually *oscillating* universe based on certain fundamental *concepts* like that of *symmetry* and the *balance* between the opposites 'attraction-repulsion', the 'one-plurality'. Does this make him naïve, anachronistic and less scientifi than today's scientists? The answer comes from a similar example provided by one of today's most renowned theoretical physicists, S.W. Hawking, in his study, *A Brief History of Time*: "Black holes are one of only a fairly small number of cases in the history of science in which a theory was developed in great detail as a mathematical model *before* there was any evidence from observations that it was correct. Indeed, this used to be the main argument of opponents of black holes: how could one believe in objects for which the only evidence was calculations based on the dubious theory of general relativity"?[29] Fortunately, as regards black holes, observations confirmin the theory were made less than forty years later. Empedocles had to wait, as we shall see below, two thousand four hundred years for the vindication of his theory.

Practical Applications

Empedocles is not solely a theoretician. For him the quest for knowledge is not an end in itself, but is also a means of controlling nature for the betterment of mankind. Perhaps some of his visions of the practical potential of knowledge may have seemed then -and even now- inapplicable and utopian: "Thou shalt learn all drugs that exist to ward off ills and old age. . .[and] shalt stay the power of the unwearied winds. . .and, if thou wish, bring back their breath again. After dark rain

---

[28] Water-clock (*clepsydra*): A small vessel for reckoning time. It was a water-butt full of water with a narrow neck and pores in the bottom from which the water slowly dripped. The length of time elapsed was determined by the amount of water lost. Others, however, have questioned whether what Empedocles describes actually constitutes an experiment in the contemporary sense.

[29] S.W. Hawking, *A Brief History of Time*, 92.

thou shalt cause a seasonable drought for men, and after summer's drought bring on the streams that nourish the trees...Thou shalt bring back from Hades the strength of a man who has died".[31B111]

Because of such ideas as these, many would mistakenly attribute to him the identity of magus and mystic. But aren't such aspirations -indeed, some that are even more wondrous- expressed by contemporary science? Theoretical physicist P. Davies writes: "One can speculate that technology will continue to advance dramatically over the coming millennia; if so, it is tempting to believe that human beings, or their descendents, will gain control over ever-larger physical systems and may eventually be in a position to avert disasters even on an astronomical scale".[30]

Determinism

As we have seen, Empedocles believed that a system of laws underlies nature. He would express the concept through various names –'care', 'broad oaths', 'necessary chance' – without limiting the idea to one precise definition

Later thinkers would attack this lack of clarity, but the unfairness of this critique is evidenced today by physicists' inability to provide an adequate explanation of the nature of this deterministic principle. In the work referred to above, Hawking concedes that "In effect, we have redefine the task of science to be the discovery of laws that will enable us to predict events up to the limits set by the uncertainly principle. The question remains, however: How or why were the laws and the initial state of the universe chosen"?[31]

The Universality of Natural Laws

Empedocles's point of departure is a principle that is for him simple and self-evident: the laws of nature that underpin the *microcosm* are the *same* as those which hold in the *macrocosm*. Thus, on the basis of the four elements and the two forces, he proceeds to the development of a *unifie* theory encompassing both the constitution of matter (*microcosm*) and the origin and evolution of the universe (*macrocosm*).

This presupposition, introduced into Western thought with clarity and consistency by Empedocles and the other Presocratics, was not at all self-evident to later thinkers. It would only become generally accepted in the fina decades of the 20th century, when it was confirme for the firs time that the natural laws that determine the creation and evolution of the universe (*macrocosm*) can indeed be identifie with those that determine the behavior of the elementary particles (*microcosm*). In their book *From Quarks to Cosmos*, two outstanding physicists, L.M. Lederman (Nobel Prize in Physics, 1988) and D.N. Schramm, describe as follows this startling

[30] P. Davies, *The Last Three Minutes*, 11.

[31] S.W. Hawking, *A Brief History of Time*, 173.

vindication of the Presocratics and Empedocles in particular: "This merging [between particle physics and early universe cosmology] is relatively recent and only recently intense. However, at the dawn of science, about 2,500 years ago in the Greek cities along the Aegean, inner and outer space were both essential and interconnected domains. The Milesian philosophers [i.e. the Presocratics] believed in overarching principles of poetic simplicity that gave coherence and order (*cosmos*) to an apparently complex world. The atoms and the voids, the consideration of primary substances and the structure of the heavenly vault, were a single subject in those times. During the past few hundred years, the subjects separated, constructed their very different instruments, and made their separate discoveries. Some overlaps persisted. . .But the real convergence began in the 1970s. In the closing years of the twentieth century, we fin again a *unity of knowledge* like that envisioned by *our ancestors*".[32]

Forces

Empedocles, as we have seen, would be the firs to introduce the concept of a pair of opposing *forces*, which act on the material world. He would be the first in turn, to attribute the composition of all material bodies to four elemental 'roots'.

This Empedoclean distinction between matter and force still applies today. And it applies in a way that is amazingly close to Empedocles's visionary theory of the cosmos. "Empedocles was the first" Sambursky observes, "who, in his own poetic style of expression, described the effects of causes in the natural world and identifie those effects with forces. The intuition of Empedocles that led him to the acceptance of the simultaneous existence of forces of attraction and repulsion elicits the admiration and amazement of contemporary physicists".[33]

Empedocles's ideas 'came true' in contemporary physics in regard to another point: As we have seen, viewing the forces as 'equal to and like' the four 'root' elements, he appears to attribute to them, too, *corpuscular* properties. Such a corpuscular property in forces was completely inconceivable till recently. But from the middle of the 20th century, physical theories have been developed which demonstrated the existence of 'force-carrying particles' – a total of thirteen boson force carriers (the photon; the particles W^+, W^- Z^o; eight gluons and one graviton). The existence of particles carrying the electroweak force W^+, W^-, Z^0 was proved *experimentally* in 1983 by C. Rubbia and S. Van der Meer, who received in 1984 the Nobel Prize in Physics for this discovery.

It appears, moreover, that Empedocles is attributing to his two forces properties of a *fiel* , when he states that they are equal in power in all directions. Physics today attributes to vacuum precisely this property of *homogeneity* and *isotropy*. It was recently claimed that Empedocles viewed Love/attraction not as an independent

[32] L.M. Lederman and D.N, Schramm, *From Quarks to Cosmos*, 10.

[33] S. Sambursky, *Das physikalische Weltbild der Antike*, 35.

physical entity, but as *immanent* in the four 'roots', thus establishing the unity, homogeneity and conjunction among them: "Love in their midst. . .as she whirls around amid the others".[31B17,20,25] This new interpretation closely approaches contemporary physical concept of forces.

Light

Empedocles would proceed to certain astonishing commentaries on the nature of light. First of all, he considered that light has a *limiting* velocity. This was confirme *experimentally* more than two thousand years later, in 1676, by the Danish astronomer, O.C. Roemer. Furthermore, Empedocles used ambiguous language in describing the emission of light: "the light streams forth, being corporeal, from the illuminating body".[31A57] 'Streams forth' suggests a continuous stream of light, while 'corporeal' expresses a corpuscular quality.

One can perceive here the firs glimpse of the *dual* nature of light, of the *continuous* and the *discrete*, of *wave* and *corpuscle* at the same time, which was the conclusion reached by contemporary physics only at the beginning of the 20th century.

Chemistry

If Lavoisier (1743–1794) is considered today the 'founder of modern chemistry', in the sense that he was the firs to introduce *quantitative* measurement, then Empedocles can be viewed as the 'father of the science of chemistry' in general, since he was the firs to establish its basic principle, which holds up to our time. When Lavoisier experimentally demonstrated the imperishability of matter regardless of its chemical transformations, in essence he confirme -having achieved a concrete quantitative formulation- Empedocles's principle: "Fools, ... who suppose that what formerly was not can come into being or that anything can die and perish wholly".[31B11]

Furthermore, when J. Dalton (1766–1844), working on the basis of Lavoisier's quantitative measurements, introduced the 'atomic theory' in chemistry, showing that "chemical analysis and synthesis go no further than to the separation of particles one from another and to their reunion; [so that] no new creation or destruction of matter is within the reach of chemical agency",[34] in essence he reconfirme the saying of Empedocles, "There is no birth of any mortal things, nor any end in baneful death, but only *mingling* and *separation* of what is mingled".[31B8,1–3] The difference lies in Empedocles's positing his principle within the framework of a general philosophic theory of the universe -which remained experimentally unproved in his age- while more than two millennia later the same theory returns, confirme by experimental data. In the meantime, the theory, supported by atomists Leucippus

---

[34] J. Dalton, *New Systems of Chemical Philosophy*, 93.

and Democritus and refuted by Aristotle, vanishing during the middle ages, would come back as a simple conjecture by Galilei, Boyle, Newton, and Lomonosoff, only to be reestablished by the quantitative finding of Lavoisier, J. Proust, and J. Richter. Comparing Empedocles's (E) theory of 'roots', Dalton's (D) 'atomic theory', as well as Lavoisier's (L) views, we fin impressive similarities:

E: *Matter constitutes of extremely small, indivisible particles, the 'roots'.*
D: Matter constitutes of by extremely small, indivisible particles, the 'atoms'.
E: *The 'roots' of an element are the same in all respects.*
D: The 'atoms' of the same 'simple body' (element) have the same mass and are the same in every respect.
E: *The 'roots' of different elements are of equal value, 'but each is master in a different province and each has its own* character'.[31B17,28]
D: The 'atoms' of different elements differ as to their mass and to their chemical properties.
E: *The compound substances are composed of two or more [same or] different 'roots'.*
D: 'Compound bodies' (molecules) constitute of two or more, same or different 'atoms'.
E: *The 'roots' are arranged during mixing and separation, forming various substances.*
D: 'Chemical analysis and synthesis go no further than to the separation of particles (atoms) one from another and to their reunion'.
E: *The 'roots' are unborn and indestructible, 'for ever remaining the* same'.[31B17,35]
D: Atoms are neither created nor destroyed; they remain always the same.
E: *The different substances arise from the combination of the 'roots' in various ways, according to define numerical proportions.*
L The relative *quantitative proportion* of each element in a certain chemical compound is always the same, independent of the origin and the way the chemical compound has been synthesized.

It is significan that while Lavoisier refers to chemical 'empirical formulas' (that is, the *ratio* of the number of atoms composing the molecule), Empedocles has taken a step beyond, since he is already referring to 'molecular formulas' (that is, not to simple *ratios* but to the precise *number* of each 'root' element that constitutes the compound). For example, regarding bones, although he could have simplifie the relation 2:2:4 to 1:1:2 and have given the 'emperical formula' WEF_2 (W = water, E = earth, F = fire) he gives the 'molecular formula' $W_2E_2F_4$; that is, he has already grasped the concept of the 'molecule' and the *actual number* of atoms that constitute it.[35]

[35] We know today, for example, that the compounds ethylene ($CH_2{=}CH_2$) and 1-butene ($CH_2{=}CH{-}CH_2{-}CH_3$) have the same *empirical formula*, CH_2, but different *molecular formulas*, C_2H_4 and C_4H_8, respectively, and thus they are different chemical compounds. Apparently

Empedocles is, then, the firs to introduce the general principle of chemistry that *qualitative* differences -i.e. physical and chemical differences in the properties of the vast number of materials surrounding us- depend on the *quantitative* proportions with which the 'roots' (today's atoms) combine, while the 'roots' (atoms) themselves remain unchanged.[36] He appears to view the roots as small, *indivisible* particles, 'fragments', 'parts'. These particles constitute the structural elements of the material world "as a wall is composed of bricks and stones. All of us and all the heavenly bodies are composed of the same elements; not, however, fused together, but as small particles fitte and adjoined side by side".[31A43]

This description comes surprisingly close to today's image of *molecules* composed of atoms. Yet Empedocles goes even further: His idea that one particle binds to another only when the "pores of the particles are symmetrical to each other" and that "the cavities of the one 'fit the protrusions of the other" is to be found in modern biochemistry: The capacity of proteins to distinguish among different molecules lies in their adoption of well-define tertiary structures that enable them to recognize and bind molecules and even catalyze reactions. Specifi receptors mediate the response of cells to extracellular signals. On the surface of the target cell, or in its cytoplasm, is a *receptor protein* that has a 'binding site' with high affinit for a particular signaling substance (a *hormone*, *pheromone*, or a *neurotransmitter*), which 'fits the binding site of the protein, initiating a sequence of reactions that changes the function of the cell.

Empedocles' theory of reproduction is based on the assumption that "within the male and female there exists something like a *symbolon* [one half of a cooperative whole], and wholeness does not come only from the one or the other".[31B63] We know today that both the sperm cells and the egg cells of a given species have *one* chromosome of each type. Each gamete thus contributes *equally* (ignoring for the moment the different sex chromosomes) to the fertilized egg, which has *twice* as many chromosomes as either gamete alone. Each single chromosome of the sperm or the egg cell represents thus Empedocles' *symbolon* [one half of a cooperative whole].

Empedocles's 'roots' must not be equated with the 'atoms' of Democritus's atomic theory, which will follow a few decades later. Aristotle makes the fin distinction, stating that "when fissio will stop, the fina particle will be either *a-tomo (in-divisible)* or *divisible*, but will never be divided, as Empedocles liked to claim".[31A43a] In this case, too, Empedocles's thinking is seminal and would be vindicated twenty-two centuries later: Indeed, the 'roots', the water, the air and the earth, are divisible, but in his time no division of them was possible. In the 19th century they were divided into their constituent molecules, hydrogen (H_2), oxygen (O_2), nitrogen (N_2), etc. – which again could not be divided further. Later appeared

---

Empedocles understood this crucial difference, because otherwise he would propose the simplifie formula WEF_2 instead of $W_2E_2F_4$ for bones.

[36] Aristotle could not accept Empedocles's advanced idea and would reject it, maintaining that each mingling is accompanied by a qualitative change in its ingredients.

the atoms hydrogen (H), oxygen (O), nitrogen (N) and many other elements (around 100 in all), which, as experimental techniques improved, were again split into electrons, protons and neutrons. Contemporary physics has shown that the latter (except the electron) are *divisible* into twelve fermions (six quarks and six leptons), and it is not out of the question that these too may eventually prove to be further divisible.

Cosmogony

Had this study been written but a few decades earlier, Empedocles's cosmogony – with the sudden disruption of the 'Sphere', the fragmentation of the 'one' into a plurality, the return to the hyperdense state, and the perpetual repetition of this process – would have seemed merely a fanciful myth. However, in the second half of the twentieth century, man's conception of the universe has changed so radically that today one can view Empedocles as the precursor of today's scientifi theories about the origin and the ultimate fate of the universe.

Prior to the twentieth century the view prevailed that the universe is what it appears to be: *static and unchangeable*. The great revolution in cosmology would take place in 1929, when E. Hubble proved that the universe is continually *expanding*. Unquestionably, this fact constituted one of the greatest discoveries of the century. Reversing the course of expansion in time, one arrives for the firs time at the conclusion that the universe must once have been in a *hyperdense* state, which exploded and then underwent expansion. The universe, then, must have had a beginning, and this beginning was initiated by the 'Big Bang'. This revolutionary theory was firs looked upon with doubt. For example, in 1948 astrophysicists T. Gold, H. Bondi and F. Hoyle countered with the Steady State theory of a universe, which has neither beginning nor end. This opposing position was refuted in 1965, when A. Penzias and R. Wilson discovered the cosmological microwave background radiation, firs predicted by G. Gamow, that can not originate from a static universe, but only from a universe which was once immeasurably dense and hot. This discovery (for which Penzias and Wilson were awarded the Nobel prize in 1978), along with the observation that the universe contains light chemical elements, established the 'Big Bang' theory in the 1970's, confirmin that the universe did have a beginning.

The moment that one accepts a beginning for the universe, the question spontaneously arises as to its *end*. Astrophysicists have not yet arrived at an answer. Among various unconfirme theories there are three basic scenarios that answer to the preconditions put forward by physicist-mathematician A. Friedmann:

(a) The universe will expand to a temporal limit and then begin to contract back to its initial hyperdense, hyperthermal state.
(b) Continuing to expand, after a point in time the universe will reach a steady speed, which will be maintained to infinit .
(c) The universe will continue to expand infinitel , with decreasing speed, which, however, will never reach zero.

Which of the above models will prevail depends on the present *expansion rate* of the universe and upon its *average density*. For the present these two parameters cannot be measured with the required accuracy, so that the issue of the ultimate future of our cosmos remains unresolved. If the firs conjecture proves true, then we will have a universe endlessly alternating between 'Big Bang' *expansion* and 'Big Crunch' *contraction*.

Let us look more closely at this scientifi confirmatio of Empedocles's astounding intuitive conception concerning the creation and evolution of the *cosmos*:

- As beginning Empedocles proposes a *spherical state* of *highest density*, homogeneity and symmetry: "The rounded sphere stays fast in the dense covering of harmony".[31B27,3–4]

Contemporary astrophysics assumes a similar initial state of highest density. According to the ideas initiated by J. Hartle and S. Hawking, "the universe, at those *extreme densities* where its quantum attributes become overwhelming, behaves like a four dimensional *ball*".[37]

- According to Empedocles, the Sphere, being "all things [in] one[38]...exists without qualities, since nothing within it possesses any property – either of fir or any of the other elements, each element having lost its characteristic feature".[31A41]

Contemporary science agrees. At the highest energies ($> 10^{19}$ GeV) and at extreme conditions of temperature ($> 10^{32}$ K) and density ($> 10^{92}$ gr/cm^3) prevailing in the very firs instants of the universe ($t < 10^{-43}$), the individual properties of the elementary particles have disappeared. "At this energy," Hawking observes, "the different spin-1/2 matter particles, like quarks and electrons, would also all be essentially the same, thus achieving another unification" [39]

- Empedocles's Sphere represents the greatest possible symmetry and isostropy, being "equal to itself on every side".[31B29,3]

Today's established isotropy of the cosmological microwave background radiation constitutes an indication of *homogeneity* and *isotropy* in the universe during its initial phase: "The starting stage of the universe must have been very highly ordered", Barrow observes in his book, *The Beginning of the Universe*, "and hence extremely special and perhaps governed by some grand principle of symmetry or economy".[40]

- Empedocles characterizes the Sphere as 'altogether without limit'.[31B28,1] We do not know whether he meant 'huge' or 'unbounded'.

In today's astrophysics, these two possibilities still hold for the universe: S. Weinberg conceives that 'all space' may mean "either all of an infinit universe, or

---

[37] J. Barrow, *The Origin of the Universe*, 116.

[38] Aristotle, *The Metaphysics*, 1000b1.

[39] S.W. Hawking, *A Brief History of Time*, 74.

[40] J. Barrow, *The Origin of the Universe*, 27.

all of a finit universe which curves back on itself like the surface of a sphere".[41] In the second case, as Hawking notes, "The universe is not infinit in space, but neither does space have any boundary. Gravity is so strong that space in bent around onto itself, making it rather like the surface of the earth. . .When one combines general relativity with the uncertainty principle of quantum mechanics, it is possible for both space and time to be finit without any edges or boundaries"[42] – in other words "altogether without limits".[31B28,1]

- Empedocles would speak of "a rounded sphere rejoicing in his circular solitude".[31B27,3]

The contemporary astrophysicist resorts to precisely such language -a "hyperdense period of absolute solitude"[43]- as, proceeding temporally backwards to the moment of the creation of the universe, one find the *causal region* shrinking continually, till the moment of the Big Bang. "The causal region of each particle contains that particle only . . .Although all space is concentrated at one point and the density of matter is infinite" astrophysicist Xanthopoulos notes, "the Big Bang is a stage of *absolute solitude*".[44]

- This hyperdense state of absolute symmetry and solitude will cease, according to Empedocles, the moment that the force of *repulsion* (Strife) penetrates the Sphere. This provokes its disruption and the dissociation of the 'one' into 'plurality'.[31B31]

Contemporary astrophysics agrees that the universe was created at a particular moment around fiftee billion years ago through the Big Bang, an explosion of a hyperdense and hyperthermal state.[45]

- Empedocles, as we have seen, theorizes the cause of the creation of the universe to be the *repelling* influenc of Strife: "If Strife (repulsion) did not exist in things, everything would have been one".[46]

---

41 S. Weinberg, *The First Three Minutes*, 2.

42 S.W. Hawking, *A Brief History of Time*, 44.

43 V. Xanthopoulos, *Peri asteron kai sympanton (On Stars and Universes)*, 12.

44 V. Xanthopoulos, *Peri asteron kai sympanton (On Stars and Universes)*, 67.

45 Indeed, it is believed that the four interactions were united immediately following the Big Bang at energies greater than 10^{19} GeV. Below 10^{19} GeV, gravitational interaction *separates out*; below 10^{15} GeV (10^{27} K, 10^{-33} sec) strong interaction *separates out*; and below 100 GeV (10^{16} K, 10^{-10} sec), the two last forces, the electromagnetic and the weak *separate out* from one another. In the opposite direction the experimental *unificatio* of the last two forces has been achieved. The Grand Unifie Theories (GUTs) have been developed regarding the unificatio also with the strong force; and it is the vision of physicists to reach someday a Theory of Everything (TOE), with the fina unificatio with gravitational interaction. The latter has met obstacles both *theoretical* (it demands the unification unattained to this date, of 'quantum mechanics' and the 'general theory of relativity') and *experimental* (it demands energies of universal dimensions of the order of 10^{19} GeV).

46 Aristotle, *The Metaphysics*, 1000b1.

Today's astrophysics agrees and goes farther, theorizing that even though the action of gravity itself is one of attraction, immediately after the Big Bang, for an extremely short time interval, it became a force of repulsion! Indeed, during the 1980s, aiming at an explanation of the observed homogeneity of the universe, A. Guth developed the 'inflationar theory', according to which between 10^{-35} sec. to 10^{-33} sec. from the Big Bang, the universe underwent inflationar expansion, that is, the expansion accelerated as the gravity temporarily acted repulsively. Afterward, it returned once more to a deaccelerating expansion rate under gravitational attraction, remaining so up to the present.

- Empedocles attributes a decisive role to fire not only during creation but also in the formation of the universe.[47] Fire-heat is considered a creative, dynamic element that, among the rest, causes the "beginning of motion".[31A30]

Today we know that *temperature* is the expression of the mean *kinetic energy* of particles. In the astrophysics of our time, heat and light rays, in turn, decisively affect the entire evolution of the universe.

- In Empedocles we have the separating out and the concentration of fire "After air, fir burst out and having nowhere to go ran out under the solid mass of air".[31A30]

In today's cosmogony there is also a decisive point in time when there is a decoupling of radiation from matter: The temperature drop in the initial stage of the universe provides the conditions for the creation and combination of the fundamental particles. When the temperature drops from 10^{32}K (10^{19} GeV/10^{-43} sec) to 10^{13} K (1 GeV/10^{-6} sec), quarks condense into baryons and mesons. At the temperature of 10^{9} K (100 KeV/3 min) nuclear reactions produce deuterium, helium-3, helium-4, and lithium ions. At the temperature of 10^{4} K (1 eV/10^{5} years) plasma of nuclei and electrons begin to condense into neutral atoms. At this time, the photons cease to interact with matter; they decouple, radiation *separates* from matter, and the universe becomes transparent. "Matter, not radiation, now dominates the universe: matter begins condensing under its own gravity into galaxies, stars, etc."[48]

- According to Empedocles, after the separation, the embryonic *condensations* that are created by *gravity* and *rotation* form the heavenly bodies.[31A30]

Rotational motion would play a significan role in the formation of the galaxies and planets in the later cosmogonies. It suffice to cite the model of Kant and Laplace in the 18th century and the theory of whirlwinds developed in 1943 by C.F. von Weizsäcker.

- Empedocles's cosmogony begins at a state of supreme *symmetry* and *order* -"the sphere was everywhere equal in itself"[31B29,3]- only to end up in a condition of "terrifi disorder and discord".[31B27]

---

47 Aristotle, *The Metaphysics*, 985a31.

48 L.M. Lederman and D.N. Shramm, *From Quarks to Cosmos*, 153.

Contemporary cosmogony also theorizes that the universe began in a state of greatest *symmetry* at the time of the Big Bang, and as it expanded this symmetry broke down, not only spatially, as regions of the least homogeneity were created, but also in another physical sense as a consequence of which massless gauge particles -W^+, W^-, Z^0 acquired mass at lower energies. S. Weinberg (Nobel Prize in Physics, 1979) writes: "As the temperature fell below 3,000 million million degrees (3×10^{15} K) a *symmetry* was lost – not its spatial homogeneity ... but the symmetry between the weak and the electromagnetic interactions".[49] Beginning in a supreme state of *order*, represented by the smallest 'entropy', the universe moves toward increasing disintegration and disorder, with a corresponding rise in entropy. Entropy would continually increase till it would reach its highest possible value, after which it would be impossible for any changes whatsoever to take place. No objects would exist that would constitute any order at all – such as stars, planets, or life.

- If we correctly understand Empedocles's line, "When great Strife waxed strong in the limbs and leaped to power, as the time appeared which is fi ed for them [Strife and Love] by a broad oath",[31B30] it means that time unfolds by necessity only through the conflic of the forces of attraction-repulsion, which are the cause of all movement. As long as the clash of Strife-Love is absent, at the stage of the unmoving and homogeneous sphere, time does not appear.

Many physicists seem to have reached precisely that conclusion. If there are no events, time also does not exist. "One may say", Hawking states, "that time had a beginning at the Big Bang, in the sense that earlier times simply would not be defined"[50] Others, however, are not in agreement: "Time precedes existence", I. Prigogine maintains. "Time does not have a beginning; it precedes the existence of our universe".[51]

- According to Empedocles, at some moment in the temporal course of the universe from greatest order to extreme disorder, the proper conditions arise for the creation of living things[31B61] (see page (175)).

This Empedoclean theory of the creation and survival of species is considered the firs clear suggestion of concepts related to the Darwinian theory 'concerning the creation of species through natural selection', published in 1859, according to which in addition to the law of inherited traits, the variety of bodily forms is related to the adaptation of each organism to conditions of maintenance and survival.

- According to Empedocles, the fina state of the disintegrating universe will be, as we have seen, a time of "fearful disorder and discord",[31B27;31B21,7] when "the elements of all things, [are]unmixed, unloving and solitary, not admitting combination or association but avoiding and shunning one another and moving with their own stubborn motions".[31B27]

---

49 S. Weinberg, *The First Three Minutes*, 134.

50 S.W. Hawking, *A Brief History of Time*, 9.

51 I. Prigogine, *La fi des Certitudes*, 215.

This is an ending amazingly similar to that foreseen by contemporary astrophysics: "The universe of the very far future", P. Davies concludes in his book, *The Last Three Minutes*, "would be an inconceivably dilute soup of photons, neutrinos, and a dwindling number of electrons and positrons, all slowly moving farther and farther apart. As far as we know, no further basic physical processes would ever occur. No significan event would come about to interrupt the bleak sterility of the universe".[52]

What will occur thereafter, contemporary physics is not as yet in a position to determine. The universe may continue to expand to infinit , or begin to contract. This depends, as we have seen, on the present *rate of expansion* and on the average *density* of the universe – two values, which today have not been measured with suffi cient accuracy. The critical density ρ_c, which is determined by the rate of expansion, is estimated to 4.5×10^{-30} gr/cm^3 or around 2.7 protons per cubic meter. If the actual cosmic density is smaller than the critical density, then the gravitational attraction would be too weak to reverse the present force of expansion, and the universe would remain *open*. If, conversely, the cosmic density is greater than the critical density, then the gravitational attraction will some moment halt the expansion of the universe and will, in turn, bring on its contraction. This development would lead to a *closed* universe. Today the *observed* average density is much smaller than the critical density, ρ_c. In 1998, however, the giant Super-Kamiokande detector, in the heart of Mount Ikenoyama in Japan, demonstrated that the neutrino metamorphoses in flight strongly suggesting that neutrinos have mass.[53] If neutrinos prove to be sufficientl massive, then their presence in the universe increases its average density. In turn, many believe, that beyond the visible celestial matter of the galaxies, there is additional non-radiant matter which could supplement the 'missing' mass, raising the total cosmic density beyond ρ_c, leading thus to a 'closed' universe.

- Empedocles makes a strange suggestion regarding this extracelestial matter: "According to Empedocles, there is one *cosmos* only, but the *cosmos* is not the whole [universe], but a small part of the whole: the rest is 'idle matter' ".[31A47]

The ultimate outcome would depend, then, on whether the force of repulsion or of attraction will ultimately prevail. "Gravity would play out a strange battle", Davis writes. "The expanding universe attempts to pull every object farther apart from its neighbors, but the mutual gravitational attractions oppose this and try to bring bodies together".[54]

- Empedocles is stating precisely the same idea in his own poetic style: "As much as [Strife-repulsion] continued to run forth, ever pursuing it there was a gentle immortal stream of blameless Love-attraction".[31B35.12]

Thus, if we accept the eventuality that in the future the universe will begin to contract, this is totally in agreement with Empedocles's theory of reverse development

---

52 P. Davies, *The Last Three Minutes*, 135.

53 E. Kearns, T. Kajita, and Y. Totsuka, *Detecting Massive Neutrinos*.

54 P. Davis, *The Last Three Minutes*, 114.

from extreme disorder and dispersion back to the supreme unity and compactness of a new 'Sphere'. During this reverse course a world of living things is once more created, later to be destroyed, as the universe proceeds under the influenc of all-powerful Love-Attraction to the unificatio of all things into the 'one', the hyperdense Sphere.

Today's physics describes this 'Big Crunch' as follows: "In the fina moments, gravity becomes the all-dominant force, mercilessly crushing matter and space ... Larger and larger regions of space are compressed into smaller and smaller volumes. According to conventional theory, the implosion becomes infinitel powerful, crushing all matter out of existence and obliterating every physical thing, including space and time themselves, at a space-time singularity. This is the end".[55] The fina words from contemporary physics seem to cast light for the firs time on Empedocles's enigmatic saying, "Love-attraction is not the cause of Being, for drawing all things into One, it obliterates all things".[56]

"This model of cosmic evolution", observes astonomer J. Barrow, "suggests the ancient idea of the cyclic universe – one that undergoes a never-ending sequence of rebirths, each time rising phoenix-like from the ashes of its previous demise. According to this view, we are living in one expanding cycle of an infinitel old oscillating universe with an infinit future. All the planets, stars and galaxies would be destroyed each time the universe plunged down to a 'Big Crunch' and bounced back into a state of expansion".[57]

This is precisely the model conceived by the visionary spirit of Empedocles twenty-f ve centuries ago: "The same cosmos comes to be and perishes alternately, and again rises and perishes, and this succession goes on for ever. . .Love-attraction and Strife-repulsion gain the ascendancy in turn: Love-attraction brings all things together into one, destroys the cosmos created by Strife-repulsion and makes of it the Sphere, whereas Strife-repulsion separates the elements again and creates a world like this.[31A52] And these things never cease from continual shifting, at one time all coming together, through Love-attraction, into one, at another each borne apart from the others through Strife-repulsion. [So, in so far as they have learnt to grow into one from many,] and again, when the one is sundered, are once more many, thus far they come into being and they have no lasting life; but in so far as they never cease from continual interchange of places, thus far are they ever changeless in the cycle".[31B17,6–13]

Rarely in the history of the Spirit and Thought does one encounter so many centuries later such a *retroactive* scientifi confirmatio of a man's *visionary* conceptions. Empedocles was unquestionably a great and unparalleled personality. He was the last Greek thinker who as yet could grasp the cosmos as a *unifie whole*, unimpeded by philosophic, scientific or religious blinders such as will appear thereafter. His genius arises from the fact that he was able to probe all these domains, which are

---

[55] P. Davis, *The Last Three Minutes*, 168–169.

[56] Aristotle, *The Metaphysics*, 1000b11.

[57] J. Barrow, *The Origin of the Universe*, 29–30.

separate in our day, without fragmenting or isolating them, but -on the contrary- creating a unifie framework and foundation that sustained their interrelation. "Some love the Greeks for their rational achievement, their classical sense of form, proportion, symmetry and order" Guthrie writes. "Others exalt the romantic, Dionysiac strain of *enthusiasmos* in which reason abdicates and man feels the ecstatic joy of possession by the god. We may be capable of responding to one side only of this contrast, but if we could do justice to the Hellenic spirit we must recognize that it included both; and we are helped to understand this by the knowledge that they were combined not only in one people but in one man, Empedocles of Acragas".[58]

[58] W.K.C. Guthrie, *A History of Greek Philosophy. The Presocratic tradition from Parmenides to Democritus*, II, 265.

Anaxagoras of Clazomenae (ca. 500–428 B.C.)

The progress of thought is leading toward a non-mechanistic concept of reality; the universe is gradually coming to be described as a great mind rather than a great machine.
James Jeans

Personality

If Empedocles is the last great Greek thinker in whom the Apollonian and Dionysian elements, the practical and the religious, the rational and the mystical, as yet remain unified Anaxagoras is the firs genuine representative of the scientific-philosophi spirit in today's sense. To the question of why it is better to be born than not to be born, he would answer, "In order to study the heavens and the order of the entire universe".[59A30] To this philosophic inclination, he would passionately devote himself, making it the aim of his life and freeing it of all social, political, practical and religious overtones. He would slight worldly interests, and to the objection of his fellow citizens that he was not concerned with politics and did not honor his homeland, he would answer, pointing to the heavens, "I honor her indeed and am deeply concerned for her".[59A1(7)] Anaxagoras would be the firs Western thinker to lend a wholly independent and self-contained dimension to scientifi investigation and thought, separating them from all mystical tendencies, which as yet remained widespread in his time.

Life

Born in 500, B.C., in Clazomenae, Ionia, he was a contemporary of Empedocles and Leucippus. Although he was of a well-to-do family, he left his property to relatives in order to go to Athens, the glowing intellectual capital of his time, 'the Greece of Greece', as it was styled by Thucydides. He would remain in Athens thirty years, teaching and associating with the intellectual leaders of the city – especially Pericles. It was the friendship with Pericles, which would be the cause of his later being forced to leave Athens. So as to attack Pericles through Anaxagoras, Pericles's political adversaries passed a bill[59A17] and on the basis of this vote, Anaxagoras was accused of impiety, because "he called the sun a fier stone",[59A15(12)] in an age when the Athenians still believed the sun to be a god. Anaxagoras finall was aided by Pericles to escape to Lampsacus on the Hellespont. There he would continue his teaching, founding a school of philosophy, where he would be succeeded after death

C.J. Vamvacas, *The Founders of Western Thought – The Presocratics*, Boston Studies in the Philosophy of Science 257, DOI 10.1007/978-1-4020-9791-1_12,

by his student Archelaus of Athens, who would become the mentor of Socrates. Though a stranger, he was greatly honored by the Lampsacenes, and according to his wish the day of his death was declared a holiday for children. Upon his tomb was inscribed: 'Here lies Anaxagoras, who ascertained to the highest degree the ultimate truth of the universe'.[59A1(15)]

Writings

The period immediately following the victorious Persian wars (492–479, B.C.) were Athens's 'Golden Age'. In this cultural capital of the Hellenes – 'the soul and the Magistrate's Hall' of Greece – where tragedy, comedy and the plastic arts were already at a peak, creating masterpieces that remain peerless to our day, Anaxagoras would be the *fi st* to introduce philosophy. It would also be inevitable that he would be the firs philosopher in history to stand trial and finall be forced to flee

In his main treatise, *On nature*, "written in an attractive and dignifie style",[59A1(6)] he would deal with the firs principles, matter and motion. He considered "as the *arche* of everything mind and matter; mind as creator and matter as product".[59A42(1)] Being the firs to introduce *mind* as the cause of motion, Anaxagoras would himself be called 'the Mind'.

Matter

It appears that what most concerned Anaxagoras was not the structure of the universe so much as the possibility of *change* within the restrictive limits of the Parmenidean principle that Being is uncreated and imperishable. In order to explain change, it would be necessary to develop a theory about the constitution of things in the tangible world. Anaxagoras would support this theory on the basis of the following principles:

- "Nothing comes to be from, or perishes into, non-existence".[59B17]

This is the pervading Parmenidean principle that could not be set aside by any later Greek thinker.

- "In everything there is a portion of everything except Mind".[59B11]

Anaxagoras would arrive at that conclusion as an answer to his basic question, "How can hair come out of not-hair, and fles out of not-flesh" [59B10] This would be possible, according to Anaxagoras, only if "everything has a portion of everything".[59B6] On the basis of this reasoning,[59A46] Anaxagoras would reject Eleatic monism, would refuse to limit existence to the four roots of Empedocles, and affir that 'being' is a common characteristic of *all* things. The infinit variety of all *material* things, as well as their *properties* (for example, cold-hot, white-black)[1] are contained within each entity. In this way, Anaxagoras explains every sort of change. Each thing contains a portion of all the other things, and all of the properties -as we call them today- so that every *change* comes about from the predominant manifestation of an already preexistent constituent, not from new creation out of nothing. Two additional basic statements round out this position. The one concerns *infinit divisibility*:

- "Of the small there is no smallest, but always a smaller (for what is cannot not-be), but similarly there is always a larger than the large".[59B3]

Matter is *infinitel divisible*, yet however minute the part may be, it continues to contain portions of all things, since "everything has a portion of everything".[59B6] Each particle, whether animate or inorganic, independent of its *discrete* size, is composed of *spermata* (seeds) that contain *infinit* portions of all other things, so that the given particle can be transmuted into any other particle. There is an infinit number of 'seeds', which are not understood as ultimately *discrete* particles, but as 'parts that are like the whole', *homoeomeri*",[59A46] according to Aristotle; just as a drop of orange juice, which -however small- contains all the constituents in the same proportions present in the whole glass of orange juice.

Two 'seeds' can be considered to be not only *alike* in that both contain portions of all things, but also *different*, to the degree that the quantitative relation of their parts differs. This leads us to the next Anaxagorean principle – that of the *dominant portion*:

- "Furthermore, nothing is like anything else, but what is most in each thing, that evidently is and was".[59B12]

Everything, regardless of size, is composed of portions of all the innumerable existing things, but in *differing proportions*. The *differentiation* is explained by the *quantitative predominance*[59A41; 59B5] of one of the constituent parts of a thing, which also characterizes it. For example, "What appears to be gold is the thing that contains a lot of gold, although it contains also all other things".[59A41]

Mind

The pioneer doctrines of Anaxagoras followed two basic axes: (a) The development of a theory concerning the structure of *matter*; and (b) The introduction -in contradistinction to inorganic matter- of an intelligible principle, *mind*, as the original

[1] A clear distinction does not as yet exist between *material* entities and opposing *properties*. They are considered equal and exist in everything: "This being so, we must suppose that there are many things of all kinds in all the things that are being mingled, and seeds of all things with every sort of shapes and colours and fl vours".[59B4]

moving and controlling force. For the firs time, Anaxagoras introduces mind as the acting, controlling principle of cosmological becoming.[2] Anaxagorean *mind*:

- is incorporeal and independent of a primal *material* state of existence,
- constitutes the firs cause of the initial impulse[59A45] that leads to the separation of matter,[59B13]
- holds the order of the universe as its ultimate goal,[3; 59A46; 59B12]
- exists in all living things[59B12] as soul, and in man also as reason,[59A100] and,
- although its qualities relate it to the divine, is not explicitly identifie with god.

The above attributes open new horizons for later Western philosophical and scientifi thought.

Mind exercises a dual control: (a) "Mind *controls* (i.e., *continuously*, even today) everything that has life, both greater and smaller",[59B12] and (b) "[Mind] *controlled* (i.e., only once) the whole revolution, to make it at the *beginning*".[59B12] Thus, mind acts on the basis of a predetermined purpose: to impose order on the universe, to endow it with form, to make it a decorum, a '*cosmos*'. Attributing to mind this specifi function of '*cosmos*-shaping,' Anaxagoras would be the firs to introduce the *teleological*[4] theory of the universe. The Anaxagorean mind is, then, not the 'firs mover' alone, but also the *teleological* principle which lends purpose and meaning to the initial movement. "Mind is the acting cause that ordered and arranged all things".[59A46]

The attributes of 'mind' are: 'Infinite [59B12] both temporally ("it is eternal"[59B14]) and spatially, since it is found with all things "infinit both in number and in smallness".[59B1] "Self-governing".[59B12] Pure: "it is not mingled with anything";[59B12] for if it were mixed, "the things mixed in it would have prevented it from controlling anything as it can when alone and by itself".[59B12] All alone unto itself: "alone and by itself".[59B12] "It is the fines and purest of all things",[59B12] a description which does not as yet precisely imply the immaterial -a concept as yet unknown- but sets it in clear contradistinction from all things material, though it retains a spatial dimension. Omniscient; it "has all judgment of everything".[59B12] Omnipotent; it "possesses the greatest power".[59B12] Uniform: "mind is alike throughout".[59B12] "It is nowhere in the extant fragments", Guthrie notes, "called god, but this may be accidental and it is impossible that Anaxagoras should not have thought of it as divine".[5]

[2] Aristotle, *The Metaphysics*, 984b17.

[3] Aristotle, *The Metaphysics*, 984b15.

[4] From the Greek: *telos* (=end, fina cause) and *logos* (=teaching, doctrine).

[5] W.K.C. Guthrie, *A History of Greek Philosophy. The Presocratic tradition from Parmenides to Democritus*, II, 279.

Cosmogony

In the thought of Anaxagoras, the creation and evolution of the universe -in contrast to Empedocles's 'cosmic cycle'- is characterized by a *linear* direction within time. "All things being together, motionless, infinit period of time, mind introduced motion and separated them".[6] The primal condition was a compact, inert mixture of all things: "All things were together, infinit both in number[59B4] and in smallness".[59B1] "The infinit of ingredients", Guthrie observes, "in the same unequal proportions in any portion of matter, however small, is the cornerstone of his cosmology, for it ensures that even in the original fusion, however large or small a portion may be, there is the lack of homogeneity which makes possible later emergence of distinctions in perceptible bulk. The acceptability of the infinit regress (if one likes to call it that) is the great discovery which enables him to parry, and even destroy, the weapons of Zeno".[7]

To this amorphous, motionless material mass, 'mind' at some moment gives the initiating motion, which leads to separation and *differentiation*. For Anaxagoras this initial motion of the universe is no longer viewed as instigated by an assumed inherent property of matter, but instead is affected by an *exterior* factor – the intellectual force of 'mind'. This firs movement serves a single *teleological* end: the 'orderly arrangement of the world'. The initial thrust of 'mind' provokes a 'rotary motion', a vortex, which steadily increases and extends to larger areas. 'Mind' provides only the *fi st* initiating step.[59B12] In the sequel it withdraws from the process, and all the ensuing dynamic evolution now obeys its own physical laws.[59B13; 59B9]

Cosmology

In a spirit probing, inductive, rational and unbiased by religious assumptions, Anaxagoras would attempt to provide a description of the universe and natural phenomena. His cosmology, though influence by the theory of Anaximenes, contains a plethora of new scientifi conceptions,[8; 59A42(3–5); 59A42(5); 59A90; 59A42(6–9); 59A42(10); 59A82; 59A1(9); 59A42(11–12); 59A85; 59B19] which would be considered advanced for his time and -for Athenian adherents to popular religion- impious.

Given this strictly rational approach, based on both penetrating observation and inductive reasoning, it is not surprising that the as yet conventionally religious and superstitious Athenians would consider and later condemn Anaxagoras as an atheist.

---

6 Aristotle, *The Physics*, 250b24.

7 W.K.C. Guthrie, *A History of Greek Philosophy. The Presocratic tradition from Parmenides to Democritus*, II, 298.

8 W.K.C. Guthrie, *A History of Greek Philosophy. The Presocratic tradition from Parmenides to Democritus*, II, 309.

Creation of Life

"Animals [he says] originally arose in the moisture, but later from one another.[59A42(12)] The air contains the seeds of all things and it is these seeds which, when carried down with the rain, give rise to plants".[59A117] This theory of Anaxagoras, in accordance with which animals were firs created on Earth by seeds originating in moist air, perhaps constitutes a distant echo of the popular archetypal conception that rain falling from the *sky* fertilized the *Earth*, thus creating life. Even the plant is "an earthbound animal".[59A116]

"Everything that has life, both greater and smaller, all these Mind controls".[59B12] Mind constitutes the essence of life; it exists in all living things, since it is the very factor that differentiates them from the soulless material world, which lacks mind. As we have seen, it seems that in the fina analysis mind is identifie with soul.[59A100] The mind's wisdom is not found in all men not because they do not possess intellect but because they do not always exercize it.[59A101a] "In all these things (like strength and swiftness) we fall short of animals, and only through using our own experience and memory and wisdom and artistry can we bring in harvest and milk and transport what we gathered".[59B21b]

He attributes man's mental superiority to the physique, and mainly to the fact that he possesses *hands*. The Anaxagorian position is confirme by today's science: "The hand, in particular gave the hominids pre-eminence in evolution, and consequently was continuously perfected, with the course of the neural machinery".[9]

Physiology

The ingenious theory of sensation constitutes yet another example Anaxagoras's brilliance. Produced by the opposites, sensation comes about through a *stimulation* or disturbance: "Anaxagoras maintains that sensation is by opposites, for like is not affected by like. And he attempts to list details in regard to each sensation. Each perception is accompanied by pain".[59!92(27–29)] Even plants feel pleasure and pain in relation to the growth and falling of their leaves.[59A117] In order to be felt, the difference between opposites must be significant as Anaxagoras demonstrated in this experiment: "If we take two colors, black and white, and pour one into the other drop by drop, our sight will not be able to pick out the gradual alterations although they exist in reality".[59B12] This *experiment* also constitutes the firs clear evidence of the limitation of the human capacity of discrimination beyond certain natural boundaries in the microcosm. Thus Anaxagoras arrives at the unforced conclusion that "owing to the weakness [of the senses] we are unable to discern the truth".[59B21] The mind *alone* "knows all things",[59B12] knows the true nature and essence of all things. Man perceives only that *which predominates*. Those things that are accessible to human perception do not constitute the ultimate truth, but

---

[9] H.J. Eccles, *Evolution of the Brain, Creation of the Self*, xiii.

only a *seeming appearance* of the unseen reality: "Phenomena are a glimpse of the unseen".[59B21a]

Overview

Introducing the principle of *mind*, in contradistinction to *matter*, Anaxagoras is the firs to instigate *dualism*, which would thenceforth constitute one of the major issues of western thought. Dismissing the least trace of mysticism, Anaxagoras approaches the problem of the creation and evolution of the universe with lucid rationalism. In his view of the cosmos, he would be the firs to introduce the intellectual element; and in this respect he is justly characterized as the founder of 'the metaphysics of mind' [Geistmetaphysik].[10]

In contrast to the other Presocratics, Anaxagoras is "the firs among all . . . to posit neither 'chance' nor 'necessity' as the *arche* of the harmonious ordering of the universe, but instead pure, unmixed 'mind' ".[59A15] Anaxagoras would not accept the view of the other Presocratics that the primal differentiation of matter is the result of some mechanical process or qualitative transformation. According to him, the firs motion and separation of archetypal material substance, would be initiated by an exterior intellectual force -Mind- which functioned on the basis of a *predetermined purpose*: to *differentiate* the hitherto chaotic, homogeneous mass, and to make an 'orderly arrangement' in the universe. Assuming a purpose, a *telos*, in the creation and evolution of the universe, Anaxagoras is the firs to introduce the concept of a 'teleological *arche*' into Western thought. To be more precise, Anaxagoras's principle is not 'teleological' but 'eutaxiological',[11] and given that this factor is mental and lies beyond the perceived world, it is a 'transcendent, impersonal, eutaxiological *arche*'. According to Anaxagoras, mind planned and imposed this cosmic order – without conceiving this order as serving some other ultimate purpose; and after the initial impulse let the entire cosmological evolution unfold in accord with purely physical causative laws. "The fact that [Anaxagoras] does not involve mind in relation to the causes of various events, but uses only natural explanations, is indeed the correct method for the study of nature".[12]

This innovative 'teleological' view introduced by Anaxagoras would lead Western philosophical and scientifi thought to fertile insights which -from that time to the present- oscillate between two poles, the *deterministic* (everything is settled by earlier natural causes) and the *teleological* (everything is governed by an ultimate

---

[10] J. Mansfeld, *Die Vorsokratiker*, 493.

[11] The distinction between 'teleological' and 'eutaxiological', firs made by the American philosopher L. Hicks, is that on the one hand 'teleological' implies that the existing cosmic order serves an *ultimate end* (e.g., the development of the universe in the best possible world), while on the other 'eutaxiological' is not concerned with some ultimate end, but simply posits that the order (*eutaxia*) of the universe is neither a matter of chance nor of natural causes, but is due to a *predetermined plan* based on an intellectual principle. (L.E. Hicks, *A Critique of Design Arguments*).

[12] Simplicius, *Aristotelis Physicorum Commentaria*, 177.10.

end, served also by natural causes). In the second case, there is a *purpose* in the world that, depending on its nature, leads to three basic principles:

(a) The '*eutaxiological principle*': A controlling purpose exists which determines cosmic order.
(b) The '*transcendent teleological principle', personal or impersonal:* Cosmological development is predetermined by an ultimate purpose, directed from beyond the perceived world, an incorporeal -personal or intellectual- principle.
(c) The '*immanent teleological principle*': The ultimate purpose is not dictated from outside, but inheres in matter.

One of the firs thinkers to reject the 'eutaxiological' *arche* of Anaxagoras would be Democritus, who would formulate a strictly deterministic natural theory. Plato would be the founder of 'transcendental teleology': There is an objectively existent kingdom of 'forms' (*ideas*) to which beings refer by 'participation' (*metochi*) and souls by 'recollection' (*anamnesis*). The kingdom of 'forms' constitutes an impersonal, transcendent teleological principle. In his cosmogony, Plato posits that aiming at the *ideas,* the Creator fashioned the world 'according to mind and necessity'. The teleological order derives from 'mind', while the mechanical causes, 'necessity', derives from the blind resistance of matter.

Conversely, Aristotle would introduce the 'immanent teleological principle'. According to him, the cosmological teleological tendency is not conscious and transcendent, but inherent in matter itself. In the beginning, matter is simply in a state of potentiality, '*dynamei on*', 'that which is potentially'. There is in matter an inherent force, '*entelechia*' striving to reach actuality, rendering matter '*energeia on*', 'that which is actually', by lending it 'form', the *eidos*, toward which it aims. Aristotle would become the main proponent of the teleological cause of the *cosmos*,[13] which -thrusting the Presocratics' causative principle into obscurity- would prevail for around two thousand years. Within the framework of the scholastics' synthesis of Aristotelian philosophy and Christian doctrine, cosmological purpose would be linked with divine omniscience (Duns Scotus, William of Occam); and in his so-called fift 'teleological proof' of God, Thomas Aquinas deduces the existence of God from the existence of natural order.

In the Renaissance attitudes changed. Francis Bacon would unconditionally condemn Aristotelianism and any sort of teleology, considering involvement with 'fina causes' as sterile and fruitless. Descartes would agree, believing that the *eutaxia*, the harmonious orderliness of the universe, constitutes an inherent property of inorganic matter, which was provided by the divine at the time of creation and thenceforth was maintained by strictly natural causation. Newton returns to the 'eutaxiological' principle of Anaxagoras, but posits God in the place of 'mind': Primordial order was created by God and, on the basis of natural laws, has been maintained by God to the

---

[13] Aristotle would include the 'ultimate cause' in his four principles; the other three are: the 'essence of Being' ('what is being'), the 'ultimate matter', and the 'motive cause'.

present. In contrast, Leibniz would continue the Aristotelian teleological tradition, asserting that at the time of creation God introduced order into the world and thenceforth this universal harmony has been maintained, even if its elements may not be causally interrelated. Cosmic harmony constitutes a new proof of the existence of God. Of all possible worlds, the choice of this one is due to its being 'the best of all possible worlds'. While Darwin rejects any hint of a teleological interpretation, A. Wallace, co-founder of the theory of natural selection, was convinced that human intelligence can only be explained by the direct intervention of Cosmic Intelligence: "A Christian agreeing to man's evolution by Natural Selection has to add that man has spiritual attributes of good and evil that are not a result of evolution, but are of supernatural origin".[14]

Kant would limit the teleological principle exclusively to the organic world, maintaining that the inorganic realm is completely explained by Newton's mechanical laws. Nevertheless, for Kant purpose does not inhere in the organic world but derives from man himself, and thus does not correspond to objective truth, but constitutes a subjective construction. German idealism is permeated by teleological concepts. "Because of the absolute identificatio of the spirit (*Geist*) within us and nature outside us", writes Schelling, "...the system of nature ... is also the system of our spirit".[15] Moreover, "all of the actions of the mind are directed to express the infinit within the finite The *purpose* of all these actions is self-consciousness".[16] According to Hegel, the entire teleological process is constituted by the unfolding of a plan of the loftiest *purpose*, the self-fulfillmen of the 'highest idea'.

In the 20th century, the question of the existence of a 'eutaxiological' or 'teleological' principle has been a continuing deep concern of philosophy and science. The development of new scientifi disciplines such as Cybernetics, Informatics, and Genetics has led to a partial limitation and reorientation toward teleological principles. On the basis of new evidence, F. Tennant would reformulate the teleological argument for the existence of God. Many others, however (like J. Monod and B. Russell), would maintain that scientifi knowledge cannot be generated within the framework of a teleological principle: "The world which science presents for our belief", Russell observes, "is even more purposeless, more void of meaning ... [Man's] origin, his growth, his hopes and fears, his loves and his beliefs, are but the outcome of accidental collocations of atoms".[17] The development of genetics would relegate to the sidelines the neo-vitalist teleological concepts of Bergson (*elan vital*) and Driesch (*entelechy*). Marxism -more specificall , dialectical materialism-maintains that any seeming purpose in nature can readily be explained as a result of physical causes. Conversely, other thinkers maintain that 'teleological' or 'eutaxiological' axioms lead to valid scientifi predictions. In the area of molecular biology there are researchers who believe that a strict reductionist mechanistic methodology

---

[14] Cited in D. Lack, *Evolutionary Theory and Christian Belief. The Unresolved Conflic* , 115.

[15] F. Schelling, *Ideen zu einer Philosophie der Natur*, II, 56, 39.

[16] F. Schelling, *Abh. Werke*, I, 382ff.

[17] B. Russell, *Why I am not a Christian*, 107.

is not capable of interpreting all the complex biological phenomena. These require a *synthesis* of various approaches -physics and chemistry combined with teleological thinking, perhaps- for their full interpretation.

The 'eutoxiological' principle of Anaxagoras involves two key points that are central to his system of thought:

i. Although all the attributes that defin mind refer to the divine,[18] Anaxagoras does not take the fina step, pronouncing mind divine. It is amazing that although the thinker lived at a time in which his surroundings were as yet overfl wing with allusions to the Olympian gods and mystical ceremonies, he would resist explicitly identifying the cosmological intellectual power with god, thus retaining a strictly natural, causal framework for cosmogony and cosmology. His spirit consequently appears more 'scientific than that of many later thinkers. In any case, Anaxagoras paves the way that would be followed by many scientists since that time: an approach to the divine, not through religious tradition but through a deep awareness -in a purely rational manner- of the amazing order underlying the natural world, a profound harmony which refers to a creator.
ii. In turn, Anaxagoras would provide a brilliant answer to the profound question that would trouble many later commentators – that is, how is it possible for the human intellect to grasp cosmic order: "What peculiar privilege has this little agitation of brain which we call thought", Hume asks, "that we must thus make it the model of the whole Universe"?[19] Darwin expresses the same doubt in dramatic words, acknowledging "the impossibility of conceiving the immense and wonderful universe, including man . . . as the result of blind chance or necessity. When thus reflecting I feel compelled to look to a First Cause having an intelligent mind in some degree analogous to that of a man and I deserve to be called a Theist. But then arises the doubt, can the mind of man, which has, as I fully believe, been developed from a mind as low as that possessed by the lowest animal, be trusted when it draws such grand conclusions"?[20] Contemporary philosopher T. Kuhn would summarize this issue in the key question: "What must nature, including man, be like in order that science be possible at all?. . .What must the world be like in order that man may know it? [That problem] is as old as science itself, and remains unanswered".[21]

Anaxagoras would be the firs to grasp this crucial problem and provide an ingenious solution. In order for science to be possible, for man to be capable of comprehending the orderliness and the harmony of the universe, it is necessary that both, man's reason and the orderliness of the universe have a common source – the same

---

[18] For this reason, Nestle considers him the founder of 'philosophic theism'. (W. Nestle, *Die Vorsokratiker*, 47).

[19] Cited in J.D. Barrow and R.J. Tripler, *The Anthropic Cosmological Principle*, 70.

[20] F. Darwin, *The Life and Letters of Charles Darwin*, I, 282.

[21] T. Kuhn, *The Structure of Scientifi Revolutions*, 173.

creator and guide: The Mind. Thus, Anaxagoras establishes an inherent intellectual connection between human thought and natural order, which is precisely that which makes the latter comprehensible to the former and which, in turn, renders science possible.

A number of topics in contemporary scientifi research, which are somewhat vague -and are thus variously interpreted- bear resemblances to the Anaxagorean perception of a transcendent relation between *human thought* and a *cosmic intellectual* principle. During the fina decades of the 20th century a number of scholars in several field developed the 'Anthropic Cosmological Theory', which offers "a means of relating Mind and observership directly to the phenomena traditionally within the compass of physical science .... Meaning is important, is even central. It is not only that man is adapted to the universe. The universe is adapted to man ... A *life-giving factor* lies at the center of the whole machinery and *design of the world*".[22] The last phrase of theoretical physicist J. Wheeler resembles the view expressed by Anaxagoras two thousand four hundred years ago: "As in *living* things, so throughout nature, mind is the cause of the whole order of the *cosmos*".[23]

We can refer, in turn, to the positions of two distinguished scholars from completely different scientifi fields neurologist, J.C. Eccles, and theoretical physicist, R. Penrose. Eccles maintains that "materialistic solutions fail to account for our experienced uniqueness". Thus, neither "genetic uniqueness with its fantastically impossible lottery ($1{:}10^{15,000}$)" nor "environmental differentiations which do not *determine* one's uniqueness, but merely modify it", can provide an adequate explanation. Eccles feels bound "to attribute the uniqueness of the Ego or soul to a super-natural, spiritual creation".[24] Thus, he adds, "we have to recognize that we are spiritual beings with souls existing in a spiritual world as well as material beings with bodies and brains existing in a material world".[25]

Penrose, though a mathematician and physicist-cosmologist, emphasizes that "a scientifi world-view which does not profoundly come to terms with the problem of conscious minds can have no serious pretensions of completeness. Consciousness is part of our universe, so any physical theory which makes no proper place for it falls fundamentally short of providing a genuine description of the world".[26] He accepts the Anaxagorean position that the same mind that exists in man as reason is that which endows the universe with order: "Whatever it is that controls or describes the mind must indeed be an integral part of the same grand scheme which governs, also, all the *material* attributes of our universe".[27] The writer is thus led to the conclusion that "a unity with the workings of Nature is potentially present within all of us,

[22] J.D. Barrow and F.J. Tripler, *The Anthropic Cosmological Principle,* 1, vii.

[23] Aristotle, *The Metaphysics,* 984b15.

[24] J.C. Eccles, *Evolution of the Brain. Creation of the Self,* 237.

[25] J.C. Eccles, *Evolution of the Brain. Creation of the Self,* 241.

[26] R. Penrose, *Shadows of the Mind,* 8.

[27] R. Penrose, *Shadows of the Mind,* 214.

and is revealed in our very faculties of conscious comprehension and sensitivity, at whatever level they may be operating".[28]

By today's standards, Anaxagoras undoubtedly possessed a scientifi genius which, in contrast to the other Presocratics, was not involved in social, ethical, or religious doctrines, or matters of soul or death. He aspired to interpret cosmic and human nature through a single concept, developing a unifie theory. He worked in strictly rational terms and recognized that through the senses we cannot reach ultimate reality, since phenomena are simply "a sight of the unseen"[59B21a] being. On the basis of observation, analogy, and the principle of symmetry, he developed theories on matters of astronomy, meteorology, physics, biology, and physiology. He resorted, moreover, to experimentation[59A68, 59A69] to prove some of his views. His entire approach is informed by a rational spirit which justifie his being called 'the most natural'[59B21] investigator of nature among the Ionian Presocratics.

Anaxagoras was not an innovative scientifi thinker only. He was firs and foremost a *Greek mind.* He envisioned that the aim of life is *theoria* – observation, rational contemplation, and the "*freedom* that springs therefrom".[59A29] It is signifi cant that he does not say theory and the *knowledge* that springs therefrom', as one of today's theorists would have put it. Nor does he say 'theory and the *practical applications* that spring therefrom', as one of today's researchers in applied science would have put it. Instead, Anaxagoras stresses the *theoria* and the *freedom* that springs therefrom.

The ultimate value of the ancient Greek is *freedom.* Just a few decades before, with unparalleled courage, the Greeks had given their lives for the sake of freedom in the unequal figh against the Persians. Beyond political freedom there is also the intellectual freedom from any irrational or mystical conceptions or biases. Where deviation from rationality leads would be shown by Anaxagoras's beloved student, Euripides, in his tragedies. "Euripides", writes Kitto, "like most Greeks, is a rationalist in that he believes reason, not belief or formula or magic, to be the guide of life; but he sees too that we have in us, besides reason, non-rational emotions which are necessary but may run wild, thwarting our reason and bringing calamity. In the last analysis Euripides' tragic hero is mankind".[29] Anaxagoras was not a tragic poet. True to his vision, through his teachings toward a rational interpretation of the world, he aims at the liberation of the human spirit from any traces of the demonic that blur our reasoned perception of nature. Human *happiness* itself derives precisely from this political and intellectual *freedom.* The other great friend and student of Anaxagoras, Pericles, would sum it up epigrammatically in his *Epitaph*: 'Freedom is happiness'.[30]

---

28 R. Penrose, *Shadows of the Mind*, 420.

29 H.D.F. Kitto, *Greek Tragedy*, 194–195.

30 Thucydides, *History of the Peloponnesian War* II, 43.

Democritus of Abdera (ca. 460–360 B.C.)

Nothing occurs at random, but everything for a reason and by necessity.

Leucippus

Truth is in the depths.

Democritus

Personality – Life

In contrast to Anaxagoras, who was reputed to be dour of countenance, Democritus would be dubbed the 'Laughing Philosopher' "because of his inclination to laugh at human vanity".[68A2] But he would also be styled 'the Wise one',[68A2] since he was the last -and one of the most significant of the great Presocratics. He was born around 460, B.C., of well-to-do parents, in Abdera, Thrace. An intense and restless spirit, he would strive even as a young man to gain knowledge of other civilizations, investing his inheritance in visits to Egypt, Babylon, Persia, and possibly Ethiopia and India – lands where "he gleaned wisdom from people of all walks of life".[68A2] He was the pupil of Leucippus – a decisive factor in the development of his atomic theory. Of modest and dignifie demeanor, he would return finall to his place of birth to devote himself to teaching and writing. Of serene and sunny disposition, he encountered life with a patient and cheerful spirit. It is said that "among the wise who seek to quench their rage, Heraclitus did so with tears, Democritus with humor ... He found each meeting with a new acquaintance to be an occasion for merriment".[68A21]

It is said that *contact* with things gives us *knowledge*, whereas *distance* from them gives us *wisdom*. According to an unconfirme story, Democritus "blinded himself because he believed that his thoughts and the insights of his spirit pondering the laws of nature would be deeper and more precise, were he freed from the grip of the visual and the impediments imposed by sight".[68A23; 68A22]

He died in extreme old age, probably beyond the age of one hundred. Democritus's brilliant intellect was steeped in the widest knowledge from all the domains of thought: mathematics, physics, grammar, logic, ethics, theology, aesthetics, history, education, art. He devoted his life entirely to study and research, and "he remarked that he would prefer discovering the cause of something to being King of the Persians".[68B118] He was, as Zeller and Nestle put it, a "universal intellect that encompassed all the knowledge of his time in his philosophy, and in this regard can be compared with Aristotle alone".[1]

[1] E. Zeller and W. Nestle, *Grundriss der Geschichte der Griechischen Philosophie*, 82.

C.J. Vamvacas, *The Founders of Western Thought – The Presocratics*, Boston Studies in the Philosophy of Science 257, DOI 10.1007/978-1-4020-9791-1_13,

Writings

He would be honored as 'all-knowing' Democritus, who broke through the limits of all the preceding Presocratics. Indeed, the extent and diversity of his work can only be compared to that of Aristotle. Truly prolific he would be judged the firs Greek encyclopedic mind by Marx and Engels. Later commentators would classify his work into thirteen thematic tetralogies in f ve parts: *Ethical* (two tetralogies), *Physical* (four), *Mathematical* (three), *Musical* (two), *Technical* (two), as well as nine unclassifie works. The works of the Physical section, which are also the most numerous (sixteen), contain the treatises, *Small World-System* and *Great World-System*, portions of which are attributed to Leucippus.

His writings are, by general consent, literary masterpieces.[68A34] From the mass of his works only approximately three hundred fragments have survived today, most of which are gnomic practical sayings concerning the prudent and ethical life. Unfortunately nothing has survived from his original and extensive treatise, *On Natural Things*, where, among other subjects, he elaborates upon his atomic theory. Thus, to discover his scientifi teachings we are indebted exclusively to later scholars, mainly Sextus, Theophrastus, Eudemus, Alexander Aphrodisiensis, Simplicius, and Ioannis Philoponus, who are recognized as relatively the most trustworthy commentators,[2] and Aristotle. "Democritus", Nestle notes, "was an Aristotle before Aristotle, and one could ask himself, what the historical evolution of Greek philosophy would have been, had his works survived like those of Plato and Aristotle. In any case, in him Greek thought had already reached both a peak and a turning point".[3]

Leucippus

The work of Democritus is solidly linked to the teachings of Leucippus, who was Democritus's 'teacher'[67A2] and will always be referred jointly with Democritus. Upon the city's devastation by the Persians, he fle to Elea in southern Italy and was apprentice to Zeno, intertwining the philosophy of Parmenides and Melissus.[67A8] Finally, he would settle in Abdera, Thrace, where he would found his own philosophic school. It is impossible today to evaluate separately the contributions of Leucippus and Democritus to the development of atomic theory. This is the reason that the writings of all the atomists of that period were gathered under the concise title, *Corpus* Democriteum.[67B2]

"It is as if natural philosophy in the person of its most recent representative would make a last, staggering attempt", writes Nestle in his lucid style, "to achieve its great goal, a unifie interpretation of the cosmos, lest it be carried away in the floo of

[2] R. Löbl, *Demokrits Atomphysik*, 62.

[3] W. Nestle, *Vom Mythos zum Logos*, 205.

ethical idealism".[4] The goal of Leucippus and Democritus is to explain all phenomena in a strictly causal manner, excluding any influenc of an exterior (attraction, repulsion) or teleological (Mind) force, or the intervention of random events. Nor do they accept the Parmenidean position that tangible experiences are but "the opinions of mortals in which is no true belief at all".[28B1(30)] On the contrary, they believe that sense experience, however limited, constitutes objective knowledge of the physical world through which one can approach the truth.

Nevertheless, they are forced to accept the Eleatic ontology. Thus, they would attempt in a brilliant fashion to combine the 'theory' of nature as we perceive it through the senses, with Parmenidean metaphysics: "Through superb methodology", Aristotle admits, "based on a *one* logical cause only, they provided an explanation of *all things*, assuming one principle in agreement with nature as it actually is".[67A7] In this sense, "Atomism," Guthrie observes, "is the final and most successful, attempt to rescue the reality of the physical world from the fatal effects of Eleatic logic by means of a pluralistic theory".[5]

Fascinating as the atomic theory is -as we shall see immediately below- even more so is the study of the *historic causes* that led up to it. According to Burnet, "This is really the most important point in the history of early Greek philosophy and as such, rightly understood, it furnishes the key to the whole development".[6] We will firs turn, then, to an attempt to clarify the historical causes of atomistic thought.

The Roots of Atomic Theory

"Atomic theory", writes R. Löbl, expressing a view that is shared by virtually all commentators, "arose within a confrontation with, and as a reaction to, *Eleatic* thought, as a correction as well as further development of ideas formed by the Eleatic school".[7] These ideas, which were exhaustively developed by Parmenides and his pupils Melissus and Zeno, centered upon four fundamental natural phenomena: the *void, motion, plurality* and *divisibility*. According to the Eleatics, the 'void' is 'not being' and as such *does not exist*. If the *void* does not exist, *motion* cannot exist (since a movement from one place to another is possible only if the latter is void; nor is *plurality* or *divisibility* possible (since no void exists between beings to separate them). The Parmenidean metaphysical conception thus leads to the 'logical' conclusion that being is necessarily "one and immobile".[67A7]

---

[4] W. Nestle, *Die Vorsokratiker*, 54.

[5] W.K.C. Guthrie, *A History of Greek Philosophy. The Presocratic tradition from Parmenides to Democritus*, II, 389.

[6] J. Burnet, *Early Greek Philosophy*, 334.

[7] R. Löbl, *Demokrits Atomphysik*, 65.

All this, however, is in blatant contradiction to tangible *experience*, in accordance with which *motion, generation, destruction*, and *plurality* prevail in the natural world. Aristotle would comment in an extremely lively and pithy manner on this contradiction: Aristotle would not hesitate to say that "It amounts to madness for one to entertain such an opinion".[28A25] The issue is, however, how to refute this opinion. Empedocles and Anaxagoras, as we saw, would attempt refutation by introducing *exterior forces* (attraction-repulsion) or *teleological causes* (Mind) These responses to the Parmenidean challenge, although original, cannot be deemed either complete or fl wless.

The great achievement of Leucippus and Democritus was their succeeding to combine in a single theory the *empirical view* of the natural world and Parmenidean *ontology*, without resorting to transcendent forces or causes, or violating Eleatic concepts of being. They achieved this by (a) breaking the Gordian knot called *void*, and (b) transferring the Parmenidean attributes of 'One being' to every single one of the infinitel numerous microscopic '*beings*' -the *atoms*- that make up the world.

(a) As we have seen, the unrealistic exclusion of any possibility of motion, plurality and divisibility derives from the Parmenidean conception that the *void* as *not being, does not exist.* Leucippus and Democritus would agree that the *void* indeed is *not being*, but they would maintain that *it does exist*: "for *there is* void".[67A7]

On the basis of this fundamental assumption, the deadlock is at once broken, and motion, plurality, and divisibility immediately become 'logically' feasible.[67A7] Leucippus and Democritus perceive therefore -in contrast to Parmenides- that 'not being' should not necessarily be considered also not-existent. To 'being' exclusively they would attribute material, corporeal existence, while they would consider 'not being', the immaterial, the void, but without the implication that it does not exist. Geometric figure are also immaterial, yet they do *exist*, and the Pythagoreans include them in their natural *cosmo*-theory. In common-sense logic we also say, for example, "There 'is'*empty space* in the storeroom". Thus, according to Leucippus and Democritus, like 'being', also 'not being' -the void- does *exist.*

Furthermore, they would place 'not being' in a dialectical relationship with 'being'. "They would simply oppose"[8] the 'void' to the 'full', thus forming the *opposing pairs* 'full – void', 'being – not being', 'thing – nothing'. This relation of opposites, as R. Löbl observes, "constitutes the expression of a basic conception: when two equal possibilities exist, on principle there does not exist any binding reason to give the one absolute priority over the other".[9] Democritus would express this position of his in the phrase 'not more than . . . '. He would state: " 'Being' does not exist *to a greater extent* than 'not being', and both are equally the cause of the generation of things. Furthermore, assuming that the essence of things is compactness and fullness, he said that this is the 'being', able to move in the 'void', which constitutes 'not being', maintaining that [the latter] exists *no less than* 'being' ".[67A8] 'Being' is *matter* and 'not being' *void*, and they *both exist*; these

---

[8] I. Philoponus, *In Aristotelis Libros, De Generatione et Corraptione Commentaria*, 156, 25.

[9] R. Löbl, *Demokrits Atomphysik*, 95.

constitute the two components of the world, the two principles, "the full and the void".[67A8] J. Burnet would comment: "It is a curious fact that the Atomists, who are commonly regarded as the great materialists, were actually the firs to say distinctly that a thing might be real without being a *body*".[10]

(b) The two above principles mutually defin and set boundaries for one another; 'not being' (the void) intersects 'being' (the full), creating beings infinit in number and invisible due to their smallness. To these beings, Leucippus and Democritus would lend the Eleatic attributes of 'being'. 'Beings' are full, compact, uncreated, imperishable. They differ from the Parmenidean 'Being' only because it was One and immobile, while these are *infinit in number* and *mobile*, due to the natural consequence of the *existence* of void. In turn, the Parmenidean attribute of *indivisibility* would be lent to the infinitel numerous 'beings', which are named *a-toma* -in other words, un-dividable.

A fundamental issue now arises: Why *a-tomo* – in-divisible? The answer to this question, which undoubtedly constitutes the foundation of *atomic theory*, is provided in a superb fashion in an Aristotelian passage considered to reflec Democritus's arguments faithfully and directly. Democritus "would refute infinit divisibility through *reductio ad absurdum*".[11] To this end he would develop a series of syllogisms based on Zeno's arguments against the plurality and divisibility of being: *Infinit divisibility* could result either (i) in geometrical forms (points), or (ii) in nothing, or (iii) in some other body, or (iv) in an immaterial form. In the third instance (iii) we again fin ourselves facing the original dilemma concerning the divisibility of a body, while in instances (i), (ii) and (iv) we would have the absurd case, an initial body, extended in space, consisting of parts without size: infinit 'partition' would lead to total disintegration, which would make reshaping of matter impossible in the reverse process, and would thus constitute an irreversible destruction of matter. This reasoning, however, is invalid, for we assume that matter, as 'being', is indestructible.

Thus, as inferred from the above, atomic theory did not come into being, as the result of some special observation, or a particular experiment (which, of course, would have been impossible, given the scanty means available), or a sudden inspiration. It was the result of an unparalleled intuitive power and fruitful evaluation of *theoretical* data, combined with *empirical* knowledge.

Democritus uses both the inductive and deductive method in developing his theory. He begins with *observation*. Among the necessary criteria of knowledge are "phenomena that facilitate the understanding of unseen things".[68A111] "That which is useful," he would say, "is what leads to conclusions concerning the future, always inspired by what is seen and what exists".[75B2(26,1)][12] The deductive development

---

[10] Cited in B. Russell, *History of Western Philosophy*, 86–87.

[11] I. Philoponus, *In Aristotelis Libros, De Generatione et Corraptione Commentaria*, 35, 11.

[12] This method of investigation is not limited to the natural sciences, but is also extended to social thought, since "this way of thinking is also used by the most capable democratic or monarchial leaders, or leaders of any political type whatsoever".[75B2(26,1)]

of his atomic theory, which constitutes purely trustworthy knowledge -the 'pure knowledge' of ultimate reality- assumes as a presupposition (in contrast to the Eleatic school) *agreement* between the atomic theory and the perceived phenomena of the universe, even though the latter may be an unclear, obscure image ('dark knowledge'[68B11]) of the truth. Democritus believes that one can pursue 'truth', as an objective reality through tangible experience. This 'truth', however, is not unsupported, but on the contrary is embedded in the Eleatic ontological framework, with which Democritus agrees in most respects. Because of this, as Löbl rightfully observes, "Although Democritus's philosophy was primary 'physics' in the Aristotelian sense, it was also indeed ontological".[13]

Parmenides, as we have seen, was the creator of the firs *deductive* theory. This is a theory, however, which is subject to *empirical validation*. On this basis it appears that Parmenides's conclusion that motion and change do not exist, is incorrect. Because of the incorrectness of the conclusion, Leucippus and Democritus were led to view Parmenides's syllogism as faulty. Since motion and change exist, the void must also exist. "Atomism", Popper observes, "became the firs physical hypothesis that was a direct result of a falsifying deductive argument".[14] Leucippus and Democritus would transform the *axiomatic* deductive method of Parmenides into the *hypothetic* deductive method. "In this way, the greatest physical theory ever was born from a critically inspired discussion of Parmenides' thought that led to the refutation of his theory".[15]

Atoms and Atomic Compounds

Void and *atoms* are equal in rank. Accepting the *existence of the void*, Leucippus and Democritus provide an unforced natural explanation of what is confirme by sense perception, namely *plurality* and *movement*. Space *divides* portions of matter from one another and provides the possibility of their motion within it. The objective existence of *atoms* and *motion* arises as a direct result of the existence of the *void*. As *existing* 'not being', the void constitutes a negation of 'being', a principle that includes the *negative essence of Being*. 'Being' and 'Void' constitute an ontologically antithetical pair.

In the '*a-tomon*'[68A58] – in-divisible Leucippus and Democritus would fin all of the attributes of Parmenidean 'Being', except *oneness* and *immobility*: The 'atom' is: 'indivisible';[67A14] 'compact';[67A14] 'lacking void within';[67A6] 'solid, full',[67A6] 'partless'[67A13] (homogeneous); 'eternal',[68A37] uncreated and immortal, because time itself also is unending;[68A71] 'unchangeable',[68A57] indestructible and immutable. Atoms are substances 'unlimited in number',[68A57] 'qualitiless',[68A57] without

---

[13] R. Löbl, *Demokrits Atomphysik*, 91.

[14] K. Popper, *The World of Parmenides*, 91.

[15] K. Popper, *The World of Parmenides*, 76.

qualitative differences, 'their nature being one'[68A57] –that is, of the same material constitution. These innumerable Democritean atoms constitute in this sense a *unity*, being of the same essence.

If, however, atoms are 'qualitiless', without qualitative differences, and of the same nature, which 'lacks any sensible qualities',[68A59] how can the vast qualitative variety of tangible things arise? The natural differentiation of macroscopic bodies is due, not to an essence, but to *spatial* and *quantitative* differences of the innumerable atoms.

The *individual* atoms differ among themselves as to:

(a) 'shape'[68A6; 68A37] and
(b) 'size'.[68A47]

"Democritus gave [atoms] two [properties], size and shape".[68A47]

The *compounds*, 'combination of atoms'[67A14] (i.e., today's molecules) differ - beyond the shape and size- as regards:

(c) the number of the atoms in the compound,
(d) the 'arrangement'.[67A6; 67A125] (For example, "'AN' [differs from] 'NA' in arrangement"[67A6]),[16] and
(e) the 'position' of the atoms among themselves ("as regards position [there is] above and below, forward and back".[68A45] For example, "'H' [differs] from '工' as regards position"[67A6]).

"The atoms move in the void, and overtaking each other they collide, and some deflec in random directions, while others become entangled, depending on the symmetry of their shapes, sizes, positions and arrangements, remaining together and thus creating the synthesis of compound bodies".[67A14] In this simple and ingenious way, all the *qualitative* differentiations between the bodies derive from *quantitative* differences (number of atoms, size), and from *spatial* differences (shape, arrangement, position) of the atoms in relation to one another in the various compounds. The *weight* of an atom, for example, is proportional to its size. The weight of a compound, again, is dependent [a] on the weight of the atoms of which it is composed, and [b] on the void space between them; that is, the number of atoms per volume.[68A135(61)] In a similar way, the *hardness* of a compound depends on the symmetry of its structure. The more asymmetrical the arrangement of atoms in a compound, the greater the hardness of the compound.[68A135] Worthy of note is the

[16] This use of letters (*stoichion:* 'letter' of the alphabet) for the effective expression of these concepts could have been the reason for the atoms later to be called *stoichion* (the component into which matter is ultimately divisible > 'element'): "From these small substances, then, *as from letters* [elements], the visible and tangible bodies are formed".[68A37]

In English, today: '*Stoichiometry*': The quantitative measurement of the composition of chemical compounds and of the proportions in chemical reactions.

fact that Democritus explains the concepts of weight and hardness in terms that are used in today's physical chemistry and crystallography.

Motion

The existence of *void*, as we have seen, constitutes the condition for movement of atoms within it: atoms "move due to the void",[68A58] Democritus would maintain. The void constitutes the necessary cause; but does it also constitute the sufficien cause? Even if void exists, why should there be motion? On firs sight Democritus does not seem to treat this question. He simply considers it self-evident that: "the atoms are always in motion in the infinit void[67A16] ... separated from one another".[67A14] The atoms' basic motion is not caused from without, but is an inherent attribute of matter, and -in contrast to the universe- has no beginning but is eternal: "In the infinit void the infinit ... atoms have been moving through infinit time before cosmogony".[17] Later Aristotle wonders why Democritus provides no explanation for this conclusion; "but the question of motion, whence or how it belongs to things, they too, like the others thoughtlessly ignored"[18] ... It is not correct to assume as a sufficien principle that things exist or became so eternally. Democritus refers physical causes to this principle, "maintaining that things were always this way in the past. He disdains, however to seek the beginning of 'always' ".[68A65] This reproach is, of course, invalid, since it is based on Aristotle's own syllogism, which Democritus does not accept -quite rightly, according to today's scientifi perspective: "A healthy scientifi instinct protected the founders of atomic theory from getting lost in the woods of theoretical argumentation, as did Aristotle. They avoided making the problem of the cause of motion the starting point of their considerations; instead they perceived motion as given, like the existence of matter", Sambursky notes. "A scientifi sense is shown when at the outset of theoretical inquiry the questions posed are limited so that one accepts a portion of phenomena as given prime facts, and from those deduces other facts".[19] For Democritus, that which has no 'temporal beginning', such as motion, needs no 'logical principle' to explain it. If the atoms were at firs immobile and suddenly began to move, then there would be a reason to provide an explanation for the cause of motion which suddenly arose in the midst of inertness. However, given that he posits *eternal* motion, it is meaningless to seek a cause.[20] Perpetual motion is "by natural necessity"[68A66] given and inherent in atoms, since they move "continuously and very rapidly"[67A24] "on their own power".[68A43] In this way, Democritus also liberates movement from the last trace of animism, viewing it as a purely physical

---

17 Simplicius, *Aristotelis De Caelo Commentaria*, 591, 14.

18 Aristotle, *The Metaphysics*, 985b19.

19 S. Sambursky, *Das physikalische Weltbild der Antike*, 157.

20 Aristotle, *Generation of Animals*, 742b20.

phenomenon. At the same time, amazingly, he touches on the contemporary kinetic theory of gases, according to which the atoms or molecules of a gas move ceaselessly and rapidly in all directions, continually colliding with one another.

What, however, is the *nature* of the atoms' motion? The question is difficult to answer, since the texts of the later commentators introduce diverse terms like 'momentum',[68A43] 'downward momentum',[21] 'thrusting impulse',[22] 'heaviness',[68A58] 'blow'[68A66] (stroke; impact), 'impetus',[68A66] 'repulse',[68A66] 'mutual impact'.[68A47] The firs two terms are of misleading Peripatetic origin and it is doubtful that they render Democritus's thought.

The central issue is which motion is considered *initial*, and which motions are its *derivatives*. 'Heaviness' led at firs to faulty conclusions, due to conflictin commentaries, but today most scholars accept the view that it represented an inherent secondary property of atoms, and as we saw above, it is proportionate to their size. It does not constitute an original kinetic property of atoms which move freely in all directions, "in infinit void, [where] there is neither top nor bottom nor center nor extremity".[68A56] This Democritean *isotropy* in space constitutes today a fundamental universal law of nature. For Democritus atoms acquire weight *only* during the creation of the universe; that is, under the influenc of the cosmic whirl, when a 'center' and the direction 'up-down' f rst appear.

Democritus viewed 'oscillation' as the sole basic motion of atoms: "There is only one kind of motion, that due to oscillation".[68A47] Within the 'combinations of atoms' (today's molecules), the various atoms exhibit a sustained oscillation, a sort of *tuning*[23] between one another.[24] This image renders with amazing fidelit today's theory of the harmonious oscillations of atoms in a molecule. As Sambursky observes, "Once more the scientifi intuition and visionary power evoke astonishing, as expressed here in the elaboration of an essential feature of the molecule, the sum of its possible oscillations".[25]

This *fundamental*, perpetual *oscillation* of atoms leads to the *derivative* forms of motion brought about by the collision of atoms amongst themselves. Democritus was undoubtedly influence in this conception of motion also by observation of particles of dust that, caught a certain way in sunbeams, are seen to oscillate in all directions.[67A28] This perpetual motion and collision between atoms leads to a sudden change in course, direction, and speed, leading to the joining of compound bodies, as well as the breaking up of the latter. Thus is explained the *change* which we observe in the natural world.

According to the record, Democritus frequently resorted to experiments in support of his theories. He observed the floatin of thin fla es of metal on the surface of

---

21 Simplicius, *Aristotelis Physicorum Commentaria*, 679, 26.

22 Theophrastus, *Fragment 1, Theophrasti Eresii (De sensibus)*, 71.

23 Lucretius, *De rerum natura*, II, 109–111.

24 Epicurus, *The Extant Remains (Letter to Herodotus)*, 43.

25 S. Sambursky, *Das physikalische Weltbild der Antike*, 173.

water, as opposed to the sinking of lighter objects spherical in shape.[26] As an experiment proving the existence of void, he used a container full of ashes, which "receives the same quantity of water as the empty container".[67A19] It is suggested that he was concerned with the phenomenon of osmosis as well as with the extraction of salt from saltwater through the use of a semi-permeable ceramic pot immersed in the sea,[27] thus recovering drinkable water from saltwater for the firs time.[28]

'Necessity' and 'Chance'

"As far back as we go in infinit time, all things that were and are and are to come are foreordained by natural necessity".[68A39] Democritus is unequivocal: all things are controlled by 'necessity', by natural law, which stems from the very existence of beings. Introducing *absolute causality* in his natural system, he totally excludes a spectrum of ancient and modern assumptions, such as the intervention of external forces, a teleological principle, the intervention of the divine or the supernatural in the becoming of the universe, randomness. "Nothing comes about by chance, but everything that we say happened by chance or by itself, has some specifi cause".[68A68] 'Necessity', conceived as 'natural law', was to be the foundation-stone of his system", C. Bailey observes. "The consequences of this decision were momentous. In the sphere of physical speculation it introduced for the firs time the possibility of a strictly scientifi conception of the world, and therefore immensely strengthened the Atomic Theory as a system".[29]

This natural law, 'necessity', is inherent in beings and derives from their kinetic state: "For Democritus the nature of necessity consists in the repercussion, motion, and impact of matter"[68A66] ... "The universe ... moves by necessity and because of the vortex".[68A83] A mechanical causal consistency regulates the universe: "Just as a *cosmos* is born, so also it grows, declines and perishes in accordance with some concrete natural necessity".[67A1(33)] At this point, however, we encounter what seems at firs an odd 'contradiction', since both Aristotle and Simplicius maintain that the atomists attributed the creation of the entire universe not to 'necessity' but to 'chance': The atomists "allege chance as the cause of this heaven and all the worlds; because by chance the vortex arose, and the movement which separated and brought the whole universe to its present order".[68A69] While, on the one hand, we have the categorical statement, "nothing occurs at random, but everything for a reason and by necessity",[67B2] on the other, "Democritus ... appears to believe

---

[26] Aristotle, *On the Heavens*, 313a14.

[27] Aristotle, *Meteorologica*, 358b5.

[28] For an exhaustive description of the experimental work of Democritus, see A. Stückelberger, *Empirische Ansätze in der antiken Atomphysik*, 123–140 and A. Stückelberger, *Einführung in die antiken Naturwissenschaften.*

[29] C. Bailey, *The Greek Atomists and Epicurus*, 122.

that the vortex had a chance, spontaneous genesis".[68A67] This seeming dissonance troubled the commentators. It is resolved, however, when we examine the meanings which the word 'chance' (*tyche*) and the word 'spontaneous' (*automaton*) had for Aristotle: Whatever does not come about on the basis of a define teleological plan, a 'fina cause', but takes place only accidentally, either 'by virtue of a concomitant', or as an 'action-for-a purpose which involves rational choice', we say happens by chance or spontaneously.[30] "Necessarily, then, the causes resulting in chance events are indefinite As such, chance seems to belong to the sphere of the indeterminate and to be inscrutable to man",[31] chance is 'inscrutable to human intelligence'.[32]

In this sense, Democritus's concept of an all-pervading necessity remains unaffected. *All things* do depend on physical causation, come about 'by necessity', have a determinate natural cause. Yet the *subjective* factor is also involved: in certain cases we know the cause, while in others it is as yet unknown, hidden, 'inscrutable'. The second case involving chance concerns an event, which is the result of necessity, is causal, but the cause evades human comprehension. The following statement should be understood in this light: "Indeed, even if [Democritus] appears to employ chance as a cause in his cosmogony, as regards details he does not consider chance to be the cause, but ascribes all events to other causes".[68A68] In the second case, the cause is known, while in the first though determinate causation still applies, Democritus refers to 'chance' because its cause is unknown.

Cosmogony

Democritus posits a universe spatially and temporally infinit that contains an infi nite number of atoms. "Everything comes about by natural necessity; the cause of creation of all beings is the vortex, which he calls 'necessity'".[68A1(44–45)] "There are innumerable worlds of different sizes. In some there is neither sun nor moon, in others they are larger than in ours and others have more than one. These worlds are at irregular distances, more in one direction and less in another, and some are flourishing others declining. Here they come into being, there they die, and they are destroyed by collision with one another. Some of the worlds have no animal or vegetable life nor any water[68A40(2–4)] ... The disintegration and decline of the world does not end in its material ingredients, which possessed the power to become a world, but in another world. Since the worlds are innumerable and succeed one another there is no need of return to the same world".[68A82] As Guthrie remarks, "One cannot but admire a man whose scientifi imagination reached so far beyond the

---

30 Aristotle, *The Physics*, 197a32.

31 Aristotle, *The Physics*, 197a8.

32 Aristotle, *The Physics*, 196b5.

limited experience of his time as to paint this picture of an infinit variety of cosmic systems, in some ways so suggestive of modern cosmological knowledge".[33]

Democritean time remains -in contrast to the cyclical concept of Anaximander, of the Pythagoreans, of Empedocles- *linear*, with no periodicity, which could suggest the existence of a cosmic intelligence. While atoms are eternal, the innumerable worlds have a beginning just as they have an end. The precondition of the genesis of a world is the creation for unknown reasons and, therefore, 'by chance', of a swirl of atoms in some region of the universe: "When Democritus says that 'a whirling of a variety of atoms is manifested [somewhere] in the universe' (but does not tell us how and by what cause), it appears that it is produced accidentally by chance".[34] From this moment atoms assume a 'weight', which is proportionate to their size, the heavier (and larger) move toward the center of the 'vortex', while the lighter (and smaller), pressed by the larger, make way toward the outer surface: "as the atoms concentrate in the same place, the larger and heavier settle downward; and those that are small, round, smooth and slippery are pushed upward by the amassed atoms".[67A24(2)]

In addition to *weight* resulting from the whirl, the creation of the world also follows another fundamental Democritean law: the *attraction of like to* like.[68A63] Mutual interaction occurs only between like or related beings and is not the result of some exterior force. The likeness between atoms is mainly of shape.[68A131]

Cosmology

In general terms Democritus's cosmology lacks inventiveness. Although it partly surpasses the views of Leucippus, as regards consistency and coherence, it nevertheless presents concepts similar to those of the earlier Ionian Presocratics (Anaximander, Anaximenes, Anaxagoras) and Empedocles. As one would expect, he completely ignores the cosmology of the Pythagoreans, to the degree that it contains the element of the divine and of life, conceptions which Democritus rejects. Very clearly, he insistently strives to provide a mechanical explanation of the heavenly bodies and natural phenomena, based on his atomic theory. Although this line of thinking frequently leads to paradoxes, it nevertheless establishes a fertile, strictly scientifi foundation upon which later theories of the natural creation and evolution of the cosmos will build.

Democritus's cosmology is the logical extension of his cosmogony.[68B5.1] As for the heavenly bodies, Democritus assumes their proper positioning, while he distinguishes between the planets and the fi ed stars.[68A40(4)] Theorizing that the closer a body is to the center of the vortex, the slower its motion, he attempts an interpretation of the orbits of the sun and the moon and the deviation of the

[33] W.K.C. Guthrie, *A History of Greek Philosophy. The Presocratic tradition from Parmenides to Democritus*, II, 405.

[34] Simplicius, *Aristotelis Physicorum Commentaria*, 327, 24.

equinoxes.[68A86] In regard to the nature of the heavenly bodies, "the sun and the moon have come about through the combining of such smooth and round masses [compounds of atoms]".[68A1(44)] According to Democritus, the sun is a fier iron or stone mass.[68A87] ... The moon appears to be like the earth ... This is shown by the shadows cast by its high areas; indeed, the moon has valleys and ravines.[68A90] The Milky Way is the light of certain stars ... [it is] consisting of very small stars and in such dense concentration that they appear to us joined because of the distance between the sky and the earth, as if one had spread many small grains of salt.[68A91] "The planets are Saturn, Jupiter, Venus, Mars, and Mercury ... Democritus suspects that there are more planets".[68A92]

Democritus sought a strictly mechanical cause for every phenomenon.[68A95; 68A94; 68B15; 13A20; 68A96] The organic world is created by fermentation processes within membrane bubbles under the proper conditions of heat and moisture. Such a creation of life is not possible today because "the earth will not again be found mixed in the same way with water, nor will the stars again be in the same position".[68B5.2(6)] As for mankind, "they were created in the beginning from water and mud .... They emerged from the ground like worms, without the existence of a creator or of any sort of reason".[68A139]

Natural Phenomena

In the same manner Democritus would attempt to provide purely *mechanical* explanations for a variety of natural phenomena.[68A99a; 68A99a; 68A93; 68A98; 68A98; 68A97; 68A165] As Guthrie observes, many of these statements "afford a good example of the extraordinary contortions to which Democritus was prepared to resort in applying his general principles to minor phenomena".[35]

Biology – Medicine

Democritus was intensively involved with scientifi observation, aiming at investigating matters of biology, physiology, and medicine. His medical research, which included both diagnosis and therapy, was recorded in three non-extant works – *Prognosis, On diet* or *Diets*, and *Medical Instruction.* According to him, "medicine treats bodily illness, while wisdom alleviates passions of the soul".[68B31]

Among biological subjects, Democritus's main interests must have been reproduction and embryology.[68B32; 68A143; 68B148] A variety of passages give us the impression not only of the aptness of particular observations, but also of Democritus's ability "to probe the causes of things inexplicable and unparalleled".[68A150a]

---

[35] W.K.C. Guthrie, *A History of Greek Philosophy. The Presocratic tradition from Parmenides to Democritus*, II, 426.

In his shaping of biological observations, the core of his approach remains always his theory of atoms and void.[68A145]

Soul – Life – Death

Soul, life, and death are examined through purely *mechanical atomic* causes. The soul has a fier atomic consistency and engages in exchange with the atoms of the body.[68A104a; 68A104; 68A108] Constituting the moving force of the body, the soul - through the breath- sustains life in living things.[68A101; 68A104; 67A28; 68A108]

Upon death the soul disappears,[68A109] but it is lost gradually.[68A117; 68A160] The loss of the soul after death consequently renders meaningless the idea of post-mortem judgment and punishment.[68B297] This strictly materialist view is softened somewhat by the fact that Democritus makes the soul ethically superior to the body: "Happiness and unhappiness are conditions of the soul".[68B170] "It is fittin for men to pay attention to the soul rather than the body; for perfection in the soul corrects the faults of the body, but strength of body without reasoning makes the soul no whit better;[68B18] he who chooses the goods of the soul, chooses the more divine; while he who chooses those of the body, chooses human goods".[68B37]

Sensation

"Democritus considers senses as conventional beliefs and teaches that none of these appears as they are in reality, but only in accordance with our subjective judgment. What is true about beings is only 'atoms' and 'void'. For he says that sweetness, bitterness, heat, cold, color are merely conventional beliefs, while the true and the real are solely atoms and void. (That is, we imagine and believe that the perceived exists, while these things do not truly exist, but only the atoms and the void[68B9]). He believes that all sensible properties for us who perceive them are caused by contact and joining of atoms, and that in reality there is no white or black or yellow or red or bitter or sweet . . . and all such properties. What exists is only 'being' (atoms) and 'not-being' (void)".[68A49]

Democritus would base his theory of sensation on three fundamental principles: (a) Each sense impression is created exclusively through physical *contact.*[68A119] (b) This physical contact coming from the striking of atomic 'effluences or 'images' (*idola*) of exterior objects, causes alteration in the sense organs of the subject's body, and thus produces the corresponding sensation.[67A30; 68A135(63)] (c) Thus, sensation is purely *subjective* and does not inhere as a corresponding property in the object presented to our perception: "Perceptions are relative – that is, they come from our impressions and disposition. Nothing true and sound exists aside from the primary elements, atom and void".[67A32]

Democritus would -though frequently with obvious difficulty exhaustively elaborate and explain his atomic theory on the subjectivity of sensations. He would attribute sensations both (i) to the atoms, which as effluent emanate from external *objects*, and (ii) to the very *subject* who experiences and perceives:

(i) Each object consists of a mixture of atoms of different types (i.e., in size and shape), and the atoms that are dominant in number determine its characteristics. [68A135(65–67)]

(ii) The relative quality of the sensation does not, however, derive only from the composition of the object. At the same time "it depends on the disposition and condition [of the organism] into which [the atoms dominant in number] enter".[68A135(67)] Here the subjective factor figures since "the constitution [of the organism] is affected according to its passive state and age. From this it follows that the disposition and condition [of the subject] are the cause of impressions".[68A135(64)] Furthermore, Democritus also in this case provided a mechanical interpretation of the process involved.[68A77; 68A135(67)]

In sum, "in reality, we perceive nothing that is true and certain, but only what affects us according to the disposition of our body and [the atoms] which enter it as well as those that resist".[68B9] Thus, "nothing prevents what seems sweet to us from seeming bitter to other animals, and the same being true for the other tastes,[68A130] but even the same person does not always experience the same sensation and perception from the same object".[68A11]

On the basis of the above principles, Democritus engages in an enormous effort to interpret all of the senses. In this effort, he would not be able to avoid omissions, contradictions, and paradoxes, which would later evoke the fertile criticism of Aristotle and Theophrastus. Nevertheless, this interpretation remains the firs comprehensive natural theory of sensations:

Intuition: It is significan that Democritus includes intuitive sensory powers possessed by certain beings: "There are more [than f ve senses] in irrational animals, in wise men, and in gods.".[68A116]

Touch: Cold and hot are not properties of the object itself, but are a strictly subjective sensation derived from the atoms that come into contact with our bodies.[68A120]

Taste: All tastes are relative and are due, as stated above, to the *shape* and *size* of the various atoms: "Democritus, assigns a shape to each fl vor. *Sweetness* is round and large; *astringency* large, rough, polygonal, not rounded; *sour* taste as its name indicates has a sharp body, angular, bent, small, not rounded; *pungency* is round, small, with angles and bends; *saltiness* angular, large, crooked, with equal sides; *bitterness* round and smooth but with irregularities and small in size; ... *oily* fl vor is fine round, and small".[68A129] In addition to the atoms of the object, a decisive role is played by the condition or 'disposition' of the sense organ – that is, the tongue or the roof of the mouth with which the atoms come into contact: "We must know not merely the active body but also the passive body acted upon, especially as 'the same taste does not appear the same to all' as Democritus says, because nothing prevents what seems sweet to us from seeming bitter to other animals, and the same being true for the other senses".[68A130]

Smell: "In regard to smelling, Democritus only said that the sense of smell is produced by something subtle emanating from heavy substances",[68A135(82)] which, as in the case of taste, is caused by contact with the sense organ of smell.

Hearing: "Sound is corporeal",[68A127] that is, it is composed of material atoms emitted from the sound source. Air is the medium which spreads a sound in all directions.[68A128; 68A135]

Sight: As C. Bailey remarks, "It is evident that the sense of sight is by no means so easy to reduce to terms of touch: the object that you see does not touch your eye and a less immediate form of contact must be found".[36] To this end, Leucippus and Democritus would resort to 'images' or 'imprints' (that is, icons, representations), which, "according to Democritus, are 'effluences similar in form to the thing [from which are emitted]".[68B123] Sight, then, is nothing else but the reception of an image that comes from the object we are looking at ... "The cause of sight is the 'images' that, continually emanating from visible objects, retain the object's form and fall upon the eye".[67A29; 68A135(51); 68A135(50); 68A135(54)] Colors do not constitute an inherent property of atoms.[68A135(73,74,75,76,78); 67A29] In reality, "no color exists in nature, since the elements – that is, the 'solid atoms' and the 'void' – lack properties. The combinations of these elements take on color according to their arrangement, shape and position of the atoms; that is, on their placement, form and orientation images in fact depend. And the colors of optical impressions are divided into four: white, black, red, and greenish yellow".[68A125]

Both Democritus and contemporary atomic physics and chemistry attempt to explain the properties of the macrocosm through processes, which take place in the microcosm. In this regard, "the direct sense perception of a phenomenon tells us nothing concerning its objective essence and from the start must be excluded as a source of information. Nevertheless, in spite of this, the theoretical image that we finall arrive at is based exclusively on a variety of information all of which was gained through direct sense perception", physicist E. Schrödinger notes. And he admits: "I was amazed by the revelation that this situation had already been fully understood in the 5th century B.C. by the great Democritus, who had no knowledge of any sort of measuring equipment".[37] It was precisely this realization that caused Democritus to state in despair and self-derision: "Wretched mind, though you gain your evidence from us (the senses), are you trying to overthrow us? Our overthrow will be your downfall"![68B125]

[36] C. Bailey, *The Greek Atomists and Epicurus*, 165.

[37] E. Schrödinger, *Geist und Materie*, 77.

Thought

Democritus maintains that "thoughts are sensations and sensations are alterations of the body [of the subject].[68A112] Sensation and thought take place by the impact of images from outside. Neither occurs to anyone without the impact of an image."[67A30] Democritus would identify thought with the soul: "Soul and mind are the same".[68A101] Soul and mind are composed of the *same* small, spherical atoms, and the only difference is that the atoms of the soul are spread though the entire mixture of atoms in the body, while those of thought are concentrated in the chest, or in another version, in the brain. Indeed, the assembled atoms of thought develop a special mental force, since "what is massed together has a powerful effect, but what is widely diffused is imperceptible".[68A135(63)] Thought "derives its reliability from the sensation".[68B125; 68A135(58)]

For the firs time in the history of western thought, soul and thought assume *corporeal*, atomic state of existence, since thought and sensation are conceived exclusively as the product of movement and touch. It is to be expected that there will be gaps and weaknesses in Democritus's theory, due in part to loss of the original texts. Nevertheless, as Bailey observes, "it is impossible not to admire the consistency of the whole and the economy with which all the different experiences of sensation are reduced to the one central notion of touch".[38]

Knowledge

"Truth is in the depths".[68B117] This terse statement of Democritus's -condensed into f ve words- hardly suggests the enormous effort it took for him to fin a path beyond both from the Eleatic metaphysics of a 'reality' that contradicts human experience and the unrestrained subjectivity and relativity of his contemporaries, the Sophists. He would have to struggle: with the 'logical' conclusion of Parmenides regarding the illusory phenomenal world of motion and plurality;[28B1(30)] with the idea of Protagoras (480–411 B.C.) that "man is the measure of all things",[80B1] which rejects objective truth; and with the thought of Gorgias (483–385 B.C.) that "firs of all, nothing exists, but if even if it does exist, it is incomprehensible to man, and even if it were comprehensible to someone, it is not communicable to anyone else",[82B3] which supports a nihilistic view regarding knowledge.

According to Democritus, "man must [begin to] learn on this principle – that he is deprived of reality.[68B6] In reality, we know nothing, for truth is in the depths.[68B117] The truth in things that exist is that there are 'atoms' and 'void' ".[68B9] Ultimate reality for Democritus is in essence, as we have seen, strictly *physical*, like the perceived world surrounding us. In this sense, "truth resides is what is shown.[67A9] The phenomenon which is presented through the senses is necessarily true".[68A112]

---

[38] C. Bailey, *The Greek Atomists and Epicurus*, 174.

Nevertheless, the phenomenal world cannot reveal the ultimate truth of atoms, which because of their smallness are imperceptible to the senses. In this respect, "nothing perceptible by sense exists in nature, given that the atoms whose combinations form everything, by nature lack any sense quality".[68A59] "The truth in things that exist is that there are *atoms* and *void*".[68B9] Democritus's not equating the perceived world with ultimate reality unquestionably leads to true skepticism: "The real character of things is inaccessible;[68B8] we do not comprehend what is or is not the true character of each thing;[68B10] we know nothing truly about anything, but each man's opinion is a reshaping by influx" [68B7]

However, these realizations would not lead Democritus to agnosticism. The means to knowledge are *sensations* and *thought.* The senses provide us with a subjective, faulty perception of the macrocosm: "What appears simply permits the understanding of unrevealed things".[68A111] When, however, we wish to penetrate into the microcosm where the ultimate truth of the atoms is to be found, then it is the mind which undertakes further investigation: "There are two forms of cognition, one through the senses and the other through the intellect... one uncelar, one genuine. To the unclear belong all these: sight, hearing, smell, taste, touch. The other is genuine and does not belong to these .... When the unclear cognition cannot see any further in the direction of smallness, or hear or smell or taste or perceive by touch, and exact <investigation is required, then genuine knowledge appears which provides fine organs of perception>".[68B11]

Thought is the only path leading to the truth hidden from the senses. The working mechanisms of thought are not specifie in the extant fragments. Perhaps thought achieves its ends through a purely *spiritual* process of indirect intuition in which empirical experience functions as a rough prototype for the deduction of more advanced mental conclusions. Given, however, that sensation and thought are -as discussed above- corporeal, it is more likely that the intellect's approach to truth is based on a strictly physical process in which the atoms detached from the external world, which cannot be perceived by the senses because of their smallness, penetrate and activate the *hyper-receptive* atoms of thought *amassed* in the brain or chest, rendering knowledge of absolute truth possible in a mental level, yet on a physical, mechanical basis.

Other Sciences

Portions of Democritus's concepts about knowledge must have been contained in his non-extant treatises *On Logic* or *Canon* (three volumes) and *Supporting Arguments*, a critical reinforcement of the foregoing works. According to Aristotle, Democritus would be the firs to be concerned with the definitio of the essence of things and with the criteria of human knowledge, laying the groundwork for the -still inseparable- disciplines of *logic* and *gnoseology*.

- His *Mathematical* works (twelve volumes) bear witness to his serious involvement in this field The extant fragments are, unfortunately, very few. It appears that he arrived at the conclusion, without proving it,[39] that "a cone is a third part of a cylinder and a pyramid of a prism with the same base and equal height".[40] It seems that for Democritus *geometry* was not an abstract discipline but was indissolubly tied to the *physical* reality of material bodies.[68B155] A cone is not generated by a straight line, but under magnification by a broken line.[68B155.68B155] Thus, it is believed that Democritus also brought his atomic theory into geometry, positing a *limit* of atomic size in the division of geometrical bodies.

- Democritus is the avid supporter of the Xenophanean theory of *progress*. He believes in a gradual cultural *evolution* motivated mainly by 'necessity': "Generally necessity has been the instructor of mankind in all things".[68B5.1,8,1(7)] Man develops gradually from a primitive state to today's elevated way of life.[68B5.1,8[1–7]; 68B154]
- *Language* would be the product of the need for communication.[68B5.1,8] In his book *On Names* Democritus would oppose the Pythagorean view that names are formed through *physis* (by nature) -that though man-made, they possess a 'natural' relation to the objects they signify. Such a correspondence of *names* with the very essence of the signifie object derives from older, magical notions, which Democritus rejects. He would ally himself with the view of Parmenides and Empedocles that names of things are determined by *thesei* (by arbitrary determination); in other words, by convention and by agreement, according to the exigencies of communication.[68B26] As Nestle notes, "While in Parmenides and Empedocles the theory of language is even more deeply interwoven with the metaphysical fundamentals of their system, in Democritus it is completely independent of this, and the foundation of his view is based on the observation of unshakable linguistic facts. Consequently it is no longer speculative but is scientifi in character".[41]

Theology

Given that Democritus perceives the evolution of civilization as the result of purely human factors – of *necessity* and of common *interest*, he excludes from the outset any mediation or influenc from divine agencies in the life of man. Nonetheless, he cannot deny that a sense of the divine exists in man. Consequently, "we arrive at the concept of god through the paradoxical events that take place in the universe . . . Thus, as earlier peoples viewed meteorological phenomena, e.g., thunder, lightning, thunderbolts, syzygy of stars, and eclipses of the sun and the moon, their fear caused them to regard gods as responsible for these [events]".[68A75] Predictably, Democritus would also integrate theology into his physical system. He would accept the existence of 'images', which travel in the air, enter the body, and stimulate the

[39] The proof was later provided by Eudoxus.

[40] Cited in: H., Diels – W. Kranz, *Die Fragmente der Vorsokratiker*, Vol. II, 174.

[41] W. Nestle, *Vom Mythos zum Logos*, 197.

mind or feelings, creating the impression of the divine.[68B166] Our relation with the divine thus does not constitute a metaphysical event, but is enrolled in the material world where we live.

These presentations of the divine, the 'images', "show men the future in advance, by their appearance and by uttering sounds".[68B166] Through this mechanical approach, Democritus interprets not only the oracular powers of certain individuals, but also provides a general interpretation of dreams, "attributing them to the appearance of 'images' ".[68A136] He would extend this theory even to telepathic phenomena: the images.[68A77] He would also provide a rational explanation of 'spells' and of the 'evil eye'. When images "emanate from malicious individuals . . . they are full of the malice and evil of those from which they come. Bearing this malice and evil they enter, adhere and permanently entrench themselves within those given the evil eye, violating and troubling their body and spirit".[68A77] Indeed, one never ceases to be astounded by Democritus's continual effort to integrate every possible expression of human life within his *universal* mechanical, physical system.

Concerning the question, as to whether the 'materialist' Democritus was an atheist, opinion is divided. While on the one hand he attributes religiousness to demonstrable causes -for example, to human *awe* before the wonders of nature, to 'images' derived from dreams, and to the sayings of great visionaries- on the other hand he speaks of *divinity* in a positive way, considering it, in one conception, as "the mind of god in a spherical fire" [68A74] as well as the source of all human good.[68B34] V. Vitsaxis, having carefully examined the range of opinions, thus summarizes the atomists' position on the divine: "Democritus's intense interest concerning the origin of the idea of god and his various attempts at explanation show. . . that he had a deep awareness that his investigation of the nature of the universe also constituted a portion of the ancient and universal human effort to approach the divine nature of reality"[42] Einstein would agree: "indeed, among the heretics of the ages there are those people who though frequently seeming to their contemporaries to be atheists, were swept up in the loftiest [cosmic] *religiousness* . . . as was Democritus."[43]

Ethical Thought

Democritus, the last of the Presocratics, is known as the supreme *physicos*, the 'natural philosopher', the founder of the most complete physical theory. Yet few are aware that he was also the firs Greek thinker to be exhaustively concerned with *man* – his destiny, passions, and above all his *responsibility as an individual.* He who provided the most advanced interpretation of the 'being' and 'becoming' of the *macrocosm* would also focus his attention on man, who for him is a 'little world', a *microcosm.* "As in the observed universe (macrocosm) some (beings), like

[42] V. Vitsaxis, *O Stochasmos kai i Pisti (Thought and Faith)*, II, 285–286.

[43] A. Einstein, *Mein Weltbild*, 16.

the divine, only dominate, others, like man, dominate and are dominated, and still others, like the animals, are only dominated, the same applies to man, according to Democritus, since some of his parts only domi+nate, like logic, others dominate and are dominated, like the thymic, and yet others are only dominated, like desire".[68B175] Nestle does not hesitate to express his amazement: "It is most remarkable that the founder of materialism would at the same time become the founder of Greek philosophical *ethics*".[44]

Among the works of Democritus, eight are classifie as *ethical*; for example, *On felicity*, *On the character of a good man* or *On virtue, Commentary on ethics, Tritogenia, On the idiosyncracy of the wise*. It is worth noting that while our knowledge of the *physical theory* of Democritus is based mainly on the ancient commentators, since few fragments of his works remained extant, the contrary is true of his *ethical teachings* of which a large number of maxims survive.

The foundation on which Democritus would build his ethical teachings is his conviction that a human being is the *only party responsible* for his actions and his life. Chance, fate, supernatural powers, constitute for him merely a remote mythical echo of man's historical development. For the firs time the word 'consciousness'[68B297] appears, and for the firs time the influenc of chance is rejected in such a decisive and disdainful manner: "Men have fashioned the image of chance as cloak for their own ill-council .... Chance does not intervene greatly in the life of a wise man. The greatest and most significan things are regulated by reason, which governs them and will govern as long as life endures".[68B119]

Just as in Democritus's *physical* deterministic world, 'necessity', takes the place of 'chance', so in the *ethical* sphere man is called upon to command 'chance' with 'prudence': "Chance rarely fight against prudence, and most things in life keen sight with understanding guides aright".[68B119] For Democritus, man is *free* and capable of choosing right-mindedly. This freedom does not entail the suspension of the strict determinism, which pervades the natural world. Our actions are also subject to causative determinism, where, however, the cause often remains unknown to us. It is in such cases that Democritus refers to 'chance'; that is, something which is beyond our understanding of the causality, the relation between cause and effect. In this sense, the meaning of his saying becomes clear: "courage is the beginning of an action, but chance is the master of the end".[68B269]

From the moment that Democritus affirm that man is free to choose, the question naturally arises: *Free* to choose *what*? André Gide wrote that "Happiness is not merely a natural need, but even more, it is an *ethical obligation*."[45] What is the ideal ethical goal toward which man should strive? Democritus's answer is just this: *Euthemia* (felicity; cheerfulness; contentment; fin spirits; a good disposition; a 'kind heart'). This constitutes the 'end', the supreme purpose, of life: "The aim in life is the good disposition of the soul";[68A45] he also calls this 'well being', 'harmony', 'symmetry' (internal balance), 'tranquility' (calmness; peacefulness),[68A167]

---

[44] W. Nestle, *Die Vorsokratiker*, 58.

[45] A. Gide, *Les nourritures terrestres*, 201.

'fearlessnrss'.[68B4] The 'felicity' of Democritus does not mean apathy, insouciance, frivolity. On the contrary, it presupposes ceaseless *alertness*,[68B87] a constant spiritual struggle for *self-control* based on *moderation* and *refinemen* .[68B149]

'Felicity' constitutes a spiritual state which is not related to bodily pleasure.[68B170] "[Felicity] is not the same thing as pleasure, as some have erroneously taken it to be".[68A1(45)] Democritus is categorical[68A167]: "He who chooses the goods of the soul, chooses the more divine: he who chooses those of the body, chooses human goods".[68B37] Uncontrolled pleasures can only bring troubles.[68B71; 68B235; 68B234; 68B195; 68B105; 68B189] "Love cleanses the erotic act from anything objectionable".[68B271] According to Democritus, the soul is responsible for the state of the body: "If the body were to bring charges against the soul for the pains and ill-treatment it has suffered all its life, and he [Democritus] were appointed judge in the case, he would readily condemn the soul because on the one side, it has ruined the body by carelessness and weakened it by drunkenness,[46] and on the other side it has destroyed and dissipated it by voluptuousness; and he would condemn it, just as in the case of some defective instrument or tool he would hold responsible the one who used it recklessly".[68B159]

Furthermore, 'felicity' is not linked to material things: "felicity dwells not in herds nor in gold";[68B171] men are happy not through strength of body or through possessions, but through righteousness and wide thoughts.[68B40; 68B146; 68B50; 68B285; 68B95; 68B92; 68B96]

'Due measure' constitutes the means of achieving felicity and internal harmony.[68B102; 68B233; 68B70; 68B198; 68B72; 68B286] Greed leads to discontent: "what the body needs we can provide easily, without effort and worry. That which leads to anxiety and worry, making life unbearable is not what the body is hungering for, but preoccupations that lack a purpose;[68B223] a craving for possessions, when it knows not satiation, is far worse than the most extreme poverty; for the greater the craving, the greater the need;[68B219] the excessive accumulation of wealth for the sake of children is a pretext for avarice, which thus reveals its true character.[68B222] Profitee - ing is the worst thing that exists;[68B221] excess profi is the ruination of virtue;[68B220] wealth gained by ill practices bears the stamp of the most blatant shame".[68B218] Due measure is also recommended in learning: "seek not to learn everything, lest you end up lame in everything",[68B169] just as it is in regard to time: "the foolish wish to live many years but do not enjoy their many years".[68B201] Frugality is good, but recreation in moderation is also prescribed: "Thrift and tightening the belt are of value... But sometimes expenditure is also good. And the worthy man knows when this is appropriate;[68B229] life without feasting is like a long journey without an inn".[68B230]

In addition to 'measure', aesthetic enjoyment of the "*kalon*" (of the good and beautiful) can contribute to inner serenity and felicity;[68B207] the great pleasures are derived from the contemplation of beautiful works.[68B194; 68B73]

---

[46] Smoking and narcotics were yet to be used.

The free man uses but one means to fin the path leading to felicity: *prudence.* "Three things come from prudence – good counsel, unerring speech, and right action".[68B2] Notwithstanding the effect of civil law,[68B47] 'right action' is not imposed on the individual, who for Democritus is morally independent and totally responsible for his actions – those duties which spring only from his inner 'persuasion toward duty'[68B181]: "We should avoid committing wrongs, not out of fear [of punishment], but because that is our duty".[68B41] This sense of 'duty' – the *inner ethical imperative* – springs from the prudence, self-respect, decency, propriety, and dignity of the individual, that which the ancient Greeks termed *aedos*, He who commits shameful deeds should firs feel shame in his own eyes;[68B84] even when you are by yourself, say and do nothing ugly; learn to feel more shame before your self than before others.[68B244] This calls for continual self-control, and self-awareness in thought and deed: "it is better to examine your own faults than those of others".[68B60]

For a person, there is no higher judge of his actions than his own conscience – the '*law of his soul*'.[68B264] This is what determines what is just, or unjust – what is duty: "It is good not merely to refrain from injustice but not even to wish to do it".[68B62] The sense of justice in itself constitutes a source of happiness: "He who commits an offense is more unhappy than the one against whom it is committed".[68B45] Finally, the committing of a wrong leads to remorse.[68B174]

For Democritus, 'good' and 'bad' are not fi ed objective entities but are determined by the way in which a person pursues his objectives: "Things that bring us good also bring bad";[68B172] "for human beings bad comes from good, when they do not know how to handle and control their good purposes. Also, it is possible, if we wish to do so, to use good things for bad ends".[68B173] This prophetic observation by Democritus is true today more than ever in relation to the dizzying development of technology (atomic energy, biotechnology) and its possible applications and implications. Democritus counsels that we can prevent the bad in two ways – through *learning* and *wisdom.* For example, deep water is indeed useful; yet it is also bad because of the danger of drowning. But a solution can be found: learning to swim.[68B172] It is wisdom that will guide us to the proper choices[68B175]: "The doctor's art heals the diseases of the body, but wisdom releases the soul from suffering".[68B31] In the fina analysis, "self-control increases enjoyment and makes pleasure greater".[68B211] Profound spirituality enables one, finall , to embrace the entire world: "for the wise man the whole earth is open: for the entire world is the native land of an honorable soul".[68B247]

Democritus believes unshakably not only in the development of community life, of science, of technical knowledge, and civilization generally, but also in the progress and fulfillmen of the individual himself. He holds the highest good to be wisdom.[68B216] True wisdom is called 'fearless', for it liberates the soul from all feelings of fear and awe that are caused by ignorance. Wisdom is not only innate; it is also acquired. It can be won through devoted *study*, hard *work*, and *determination.* Education *transforms* a man, giving him a *second nature.*[68B33] Thus, Democritus powerfully stresses the value of education;[68B183; 68B182; 68B179; 68B180; 68B242; 68B241; 68B243] "no one can possess craft or wisdom without study";[68B59] The best method of teaching is by example: "the self-control of the father is the greatest exhortation to the children".[68B208] Conversely, "The frequent company of bad people increases one's bad propensities".[68B184]

On the basis of these concepts one can outline the proper *way of life* professed by Democritus: "Cheerfulness comes to man through moderation in enjoyment and harmony of life: excess and deficien y tend to produce change and great movements (disturbances) in the soul; and souls in great movement are neither stable nor well disposed;[47] therefore you must fi your mind on things within your power and be content with what you have, taking little thought of those who are envied and admired and not associating with them in your mind: rather you must contemplate the lives of those in hard times and mark their sore suffering, so that what you have and possess already may appear great and enviable to you, and you may not desire more and so come to suffer in the mind . . . If you cling to this state of mind, you will live more cheerfully and will be rid of many misfortunes in your life – envy, jealousy, and ill-will".[68B191] Democritus strives for an optimistic attitude toward old age: "strength and beauty are the blessing of youth, prudence is the fl wer of old age;[68B294] the old man was once young, but it is uncertain whether the young man will attain to old age; therefore the completed good is superior to that which is still to come and is uncertain".[68B295] Yet, at bottom, he acknowledges that "old age is a wholesale crippling: it possesses everything, but lacks all".[68B296] Death remains the sole recourse, as Democritus demonstrated by his own action: "it is the foolish who fearing death linger into old age".[68B206] Democritus does not aim at a *universal* ethical concept but at a *subjective* inward balance and well-being: "the man who intends to be cheerful must not be overactive either in private or in public, and whatever he does must not choose what is beyond his capacity and nature. But he must so be on his guard that even when chance falls in his way and seems to lead him to advancement, he can lay it aside and not engage in what is more than he can do. For proper moderation is safer than excess".[68B3]

Without doubt, the ethical teachings of Democritus are self-centered. What, then, would be his conception of a person engaged in the life of the *family* and the *city*? In domestic life, Democritus subordinates the woman,[68B273; 68B110; 68B274; 68B122a] views offspring cynically,[68B275; 68B280; 68B276; 68B277; 68B272] and places great value on friendship.[68B186; 68B98; 68B99; 68B103; 68B106; 68B109]

It is obvious that Democritus preferred abstention rather than involvement in *public life*.[68B3] But he is not naïve. He recognizes that every individual's personal welfare directly depends on the political condition of his country. Thus, reluctantly but conscientiously, every citizen must be responsibly involved in political concerns, so as to regulate communal life.[68B252] Thus, Democritus advises us to "learn the statesman's art as the greatest of all and pursue those toils from which great and brilliant results accrue to men".[68B157] Involvement in community life is not rewarding in itself, but a necessary means to an end. For, "it is unbearable to be ruled by an inferior".[68B49] This is especially true, given that "when bad men come into places of

---

[47] This statement shows that even in the sphere of ethics Democritus thinks in terms derived from his physical theory, such as the *motion* of the soul, which is generated by atoms.

honor, the more unworthy they are, the more careless they show themselves and the more are they swollen with imprudence and impudence".[68B254]

As a consequence of this position, Democritus is a champion of *democratic principles*[68B251] and a strong supporter of obedience to *law*, favoring strict *penalties* and opposing *strife*.[68B250; 68B249] Democritus chooses democracy as the political system that most effectively secures *equality* between citizens. The prototype of this equality is found in *nature*.[68A81; 68A166; 68B267]

Laws are made for the protection of the state from enemies both *domestic* and *foreign*. Domestic dangers -and corresponding laws- would not exist if people were mature and conscientious, with understanding and respect for their fellow man.[68B181; 68B245; 68B248] Strict punishment is necessary and just.[68B261] In turn, the threat to society from *foreign* enemies is considered a grave danger, calling for the severest punishment.[68B260; 68B259; 68B262]Democritus agrees that "it is proper to be obedient to the law, the authorities, and those who are wiser".[68B47]

Self-control, self-respect, courage, cheerfulness, inner serenity and balance, constitute the optimistic personal ethical message of Democritus, who was "not only the most scientifi of the ancients but in industry also second to none of whom we know from history".[68B144]

Overview

"If in some cataclysm, all of scientifi knowledge were to be destroyed, and only one sentence passed on to the next generations of creatures, what statement would contain the most information in the fewest words? I believe it is the atomic hypothesis . . . that all things are made of atoms. . . . In that one sentence there is an enormous amount of information about the world, if just a little imagination and thinking are applied".[48] Those words of R.P. Feynman, one of the most prominent theoretical physicists of the 20th century, only confir what is generally accepted: The atomism of Leucippus and Democritus was unquestionably a brilliant achievement. As Russell observes, "atomism is the outcome not of fanciful speculation, but a serious answer, one hundred and fift years in the making, to the Milesian question".[49] It was a response, as we have seen, with a strong ontological foundation, which promised to be decisive in the development of scientifi thought. Beyond the fact that it offered a consistent theoretical framework for the interpretation of most physical -and also mental- phenomena, perhaps one of its major strengths is that for the firs time it introduced and established as an indubitable methodological principle the condition that a deductive theory must be in agreement with the given empirical data.

---

48 T. Ferris (Ed.), *The World Treasure of Physics, Astronomy and Mathematics*, 3.

49 B. Russell, *Wisdom of the West*, 45.

Democritus's contribution to the advancement of Western thought is so multidimensional that each scientist or thinker casts light on a different facet of the whole. Thus, to refer to but a few approaches: For C.F. von Weizsäcker, "atomic theory had a decisive significanc for the comprehension of the world and thus established the factor of the *infinit divisibility* of matter – one of the fundamental issues of philosophy".[50] According to Sambursky, "The greatest achievement of the Greek atomists lies in the introduction in scientifi argumentation of conclusions concerning the *invisible* derived from the *visible*".[51] For W. Heisenberg, "The power of atomic theory resides in *abstraction*". Given, however, that this abstraction shows that the 'qualitative' complexity of the universe is based on the variety of atomic geometrical forms and combinations, it would have a profound consequence: "The abandonment of the 'direct' comprehension of qualities".[52] For physicist P. Jordan, the recognition of inviolable laws on which all phenomena depend constitutes the essential discovery of atomic theory: "This magnificen conception of *natural law*, which remained decisive throughout subsequent developments in scientifi research, was for the firs time in the history of human thought conceived and clearly presented in the atomic philosophy of Democritus".[53] K. Popper maintains that "the main point about the [atomic] theory is that it gives a rational account of *change*, ... [which] still remains the fundamental problem of Natural Philosophy. Democritus' theory of change was of tremendous importance for the development of physical science".[54] For D. Bohm, Democritus's concept of atoms was "an important mode of realization of *wholeness*, for it enabled men to understand the enormous variety of the whole world in terms of the movements of one single set of basic constituents, through a single void that permeates the whole of existence".[55]

The history of the development of atomic theory -an epic and continuous struggle in western thought- unquestionably constitutes one of the most spectacular adventures of the human mind. It is worthwhile to trace the main phases of this struggle, for thus one grasps the great intellectual leap achieved by Democritus – a leap which scientists from that time up to today have attempted to follow.

Immediate follower in the work of Democritus was, among others, Epicurus (341–271 B.C.); he would attempt the simplificatio of the atomic theory, making it more materialistic. He would ignore Democritus's reservations and skepticism, and would support the absolute trustworthiness of the senses in a simplistic way, which would lead contemporary physicist E. Schrödinger to observe somewhat cynically that "the difference between the two men Democritus and Epicurus is that with humility the former had realized that he knew nothing, while the latter was

---

50 C.F. von Weizacker, *Zum Weltbild der Physik*, 33.

51 S. Sambursky, *Das Physikalische Weltbild der Antike*, 162.

52 W. Heisenberg, *Wandlungen in den Grunlagen der Naturwissenschaft*, 12.

53 J. Jordan, *Die weltanschauliche Beduutung der modernen Physik*, 208.

54 K. Popper, *Conjectures and Refutations*, 81, 79.

55 D. Bohm, *Wholeness and the implicate Order*, 8.

convinced that he knew more or less everything".[56] The doctoral thesis of the young Karl Marx refers precisely to the difference between the Democritean and Epicurean physical philosophy.[57]

With Plato we have the firs clash between the *materialistic*, *mechanistic* system of Democritus, in which "intelligence, direction, and purpose [are] phenomena emerging at a late stage from nothing but the undersigned clash and recoil of individually inanimate particles",[58] and the *idealistic*, *theological* natural philosophy of Plato, as developed in one of his last great works, *Timaeus*, in which the central point of the dialogue is the action of the Divine creator – that cosmic 'sympathy' which unites the microcosm of man with the macrocosm of the universe, the mathematisation of nature. While Plato knows the work of Democritus and is clearly influence by it, he never -even once- refers to his name, and his antipathy to his teachings was so great that he allegedly stated somewhere his desire to toss all Democritus's works into the fire In spite of all this, many elements in the *Timaeus* dialogue amount to indirect allusions to the work of Democritus. He would, however, maintain that the *teleology* holds sway over *mechanical* causation. He would oppose any sort of experimental proof;[59] he would reject the existence of void and introduce the divine *Creator* of a universe spatially and temporally limited, as a second cosmic power besides Democritus's 'necessity'.

Functioning as a Pythagorean supporter of *atomism*, Plato would construct on *mathematical* foundations a grandiose mechanical system of vast extent and completeness: he would assign the four elemental roots of Empedocles to the four 'most perfect bodies',[60] the regular tetrahedron, (fire) the regular octahedron (air), the regular icosahedron (water), and the cube (earth). All of the *sides* of these four regular polyhedrons are made up of *right triangles*, the firs three of right scalene triangles (with sides $1,\sqrt{3},2$), and the latter of right isosceles triangles (with sides $1,1,\sqrt{2}$). These Platonic regular polyhedrons can, to a degree, be compared with Democritus's 'atoms', but there is a *significan difference*. They are not indivisible, but can be broken down into their triangular sides, which can once more be combined, giving rise to other regular polyhedrons and, respectively, elements. [Thus, for example, from the division of a regular octahedron-*air* we get two regular tetrahedrons-*f re*]. This bold innovation of Plato's in opposition to Democritus contains the possibility of mutual *transformation of the elements* and is of major significance since the ultimate structural constituents of the universe cease now to be *material bodies* (i.e., three-dimensional regular polyhedrons), but are *immaterial geometrical forms* (i.e., triangles). This is the reason that contemporary physicists like Heisenberg maintain that the physical theory of Plato is closer to the spirit of contemporary

---

56 E. Schrödinger, *Die Natur und die Griechen*, 138.

57 K. Marx, *Differenz der demokritischen und epikureischen Naturphilosophie.*

58 W.K.C. Guthrie, *A History of Greek Philosophy. The Presocratic tradition from Parmenides to Democritus*, II, 501.

59 Plato, *Timaeus*, 68d2.

60 Plato, *Timaeus*, 53e.

physics than that of Democritus, since -as we shall see below- "in today's quantum theory there can be hardly any doubt that the elementary particles are also in the fina analysis *mathematical forms*, though of a much more complicated and abstract type".[61] This is the *idealistic* response of Plato to Democritus's *materialistic* theory. "Democritus and Plato", Guthrie writes significantl , "fought the firs round in a contest which still continues and can never be decided by reason and observation alone".[62]

Aristotle, in contrast to Plato, would speak of Democritus's teachings with great admiration and respect. Nevertheless, he would not accept atomic theory in relation to the issue of the genesis and decline of beings. He posits a *continuous*, indivisible structure of matter in which both discrete atoms and the void have no place whatsoever. Movement and change for him are the transition of an essentially unalterable matter from the condition of simple *potentiality* to the condition of *actuality*. The Aristotelian *teleological*, *organic* view of nature is the polar opposite of the mechanical conception of the atomists. Here we fin the grounds of another conflic – between Democritus's *discrete* and Aristotle's *continuum*, which continues with unslacking intensity to our day.

The last advocate of atomism prior to the dark period of the Middle Ages would be Titus Lucretius Carus (98–55 B.C.) in his poem *De rerum natura* (*On the nature of things*), supporter and warm advocate of the Epicurean philosophy. During the following fiftee centuries -with their prevalence of Christianity and Aristotelianism- Democritus's atomic theory would fall into oblivion and disrepute.

During the Renaissance, atomism gained a new impetus as both a philosophic and scientifi concept, though significan divergences began to emerge. The French philosopher, R. Gassendi (1592–1655) would be the firs supporter of the two Democritean entities, 'atoms' and the 'void', opposing the view of his contemporary, Descartes (1596–1655) that *body* is, in its essence, *extension*, and consequently matter is identical with extension, infinitel divisible, space filled and void non-existent. Nevertheless, Descartes would agree with Democritus regarding absolute *causation* prevailing in nature, as well as the *mechanical* interpretation of the latter; that is, that action is not induced from a distance, but only *by contact*. Leibniz (1646–1716) would not accept the latter conception. For him, matter is space fille by repulsive forces acting at a distance and resisting penetration of one body by another. Instead of 'atoms' and 'void', he introduces the doctrine of *point-atomism*: the 'monads' are unextended points in space that emanate central forces. Kant and Boscovitch would later expand this purely metaphysical theory, attempting a synthesis of the ideas of Democritus, Leibniz, and Newton. They would accept Democritus's concept of 'void' but not of 'atoms'. According to these thinkers, "the presence of [extended] matter in a certain region of space is a phenomenon consisting of the presence of repulsive forces in that region, forces capable of stopping

---

[61] W. Heisenberg, *Physik und Philosophie*, 52.

[62] W.K.C. Guthrie, *A History of Greek Philosophy. The Presocratic tradition from Parmenides to Democritus*, II, 502.

penetration".[63] They accept, then, the dynamic 'continuity' of matter, considering immobility as a Heraclitean *balance* of equal opposing forces. According to Popper, this theory constitutes the forerunner of all modern theories of the structure of matter, and also of the 'dualism of matter and field' In regard to the *properties* of matter, British empiricist, John Locke (1632–1704), would restate the Democritean concept of 'primary' (size, shape, impenetrability) and 'secondary' (color, smell, taste, etc.) qualities of bodies – considering, however, secondary qualities to be purely subjective, as opposed to Democritus who considered them to be a combination of subjective and objective factors. G. Berkeley (1684–1753) would take this theory to an extreme, maintaining that even 'primary' qualities constitute purely subjective ideas.

The enlightened scientifi spirits of the Renaissance were favorably disposed toward atomism. After centuries of oblivion, the republication of the works of Lucretius and Diogenes Laertius (at the end of the 15th and beginning of the 16th century), together with the tendency of the time to move away from Aristotle's natural theories, brought Democritus's atomism once more to the fore. The firs to refer to Democritus -in the beginning elliptically- were, among others, Leonardo da Vinci (1548–1600), Giordano Bruno (1548–1600), and Galileo Galilei (1564–1642). The latter reexamined Democritus's views in relation to the influenc that the differences between atoms (number, shape, velocity and weight) might have on sense experiences of taste, smell, sound etc. It is believed, indeed, that it was precisely Galilei's ideas supporting atomism that were one of the causes of his persecution.[64] Francis Bacon (1561–1626), like Democritus, would consider the motion of atoms to be the cause of heat.[65] Newton (1642–1727) would write: "I believe it possible that God in the Beginning formed matter in solid, compact, hard, impenetrable, moving particles".[66] On this point Democritus appears to be more scientifi than Newton, since he excludes the intervention of a God-Creator in cosmic becoming. Newton deviates from Democritus's theory by introducing forces of attraction and repulsion acting on bodies not by contact but at a distance. Thus, for the firs time in physics emerges the co-existence between the *Aristotelian* concept of 'continuum' regarding the fiel of forces – and the Democritean concept of 'descrete' particles, regarding the structure of matter.

The true renaissance of Democritus's atomic theory would, however, firs appear in *chemistry.* "The laurels belong", Stückelberger notes, "to German physician and chemist, Daniel Sennert (1572–1637), for once more bringing atomic theory to the epicenter of attention, and for further developing it through his own observations

---

63 K. Popper, *The Myth of the Framework*, 116.

64 R. Redondi, *Galilée hèrètique.*

65 "For neither cold nor hot exist by nature, but the movement of atoms induces these different impressions".[68A135(63)]

66 I. Newton, *Mathematical Principles of Natural Philosophy*, Query 31; cited by J. Powers, *Philosophy and the New Physics*, 42.

and experiments beyond the bounds of the atomists of antiquity".[67] During the same period (1636) the French physician and chemist Sébastien Basso would compare atoms with the letters of the alphabet – a comparison which we have seen was made two millennia before by Democritus: "Thus, from these atoms -as from letters- are created and constructed the visible and tangible badies".[68A37]

At the beginning of the Enlightenment period, this conception of the atom would constitute yet another step toward its acceptance. "As suggested by the alphabet metaphor", B. Bensaude-Vincent and I. Stengers observe, "atomism appears to promise the building of a science upon stable foundations – the atom comprises a constructural principle both of reality and of knowledge"[68]. In 1661, J. Chrysostomus Magnier, in his study significantl entitled *Democritus Reviviscens* (*Democritus Reborn*), would be the firs to attempt the *quantitative* determination of atoms, estimating that a granule of incense contains around 7.7×10^{17} atoms. French chemist N. Lémery (1645–1715) would hold that the acidity of a liquid depends on it shape, that is, upon "the thinness of the edges of its particles",[69] a view that Democritus had already supported, stating that "sour taste has sharp shape, many pleated, small and thin".[68A135(65)] R. Boyle (1627–1691) would be the one of the firs proponents of the Democritean atomic doctrine, maintaining that all phenomena could be explained in terms of the motion, shape and position of simple, imperceptible corpuscles. He tried to assign a modern term for 'elements'. In his book *Chymista Scepticus* (*The Skeptical Chemist*, 1667), he would refer explicitly to those 'excellent atomists of antiquity, Leucippus and Democritus'.[70] His atomic-mechanical view laid the groundwork for the *theory of mechanism* that governed the course of science for the next two centuries.

At the beginning of the 19th century, J. Dalton (1766–1844), in his work *New Systems of Chemical Philosophy*, founded modern chemistry. Based on his *experiments* with gases and on related *experiments* by Lavoisier, J. Proust, and J. Richter, he would state that the 'atom' is the smallest indivisible unit of a substance that as yet retains its chemical properties. A chemical combination can be define as the union of discrete particles of definit weights characteristic of each element. Thus, the central principles of Democritus (De) reappear, and it is fascinating to set them side by side with those of Dalton (Da),[71] never forgetting that the two scientists are separated by a chronological interval of more than two millennia:

De: "*Nothing is created from not being, or is destroyed ending in not being*".[68A1(44)]
Da: "No new creation or destruction of matter is within the reach of chemical agency".

[67] A Stückelberger, *Antike Atomphysik*, 58.

[68] B. Bensaude-Vincent – I. Stengers, *Histoire de la chimie*, 45.

[69] Cited in: B. Bensaude-Vincent – I. Stengers, *Histoire de la chimie*, 48.

[70] A. Stückelberger, *Antike Atomphysik*, 60.

[71] J. Dalton, *New Systems of Chimical Philosophy*; rpt. *Cambridge Readings in Science*, 93.

De: *"These [atoms] move within the void, and their joining creates the genesis, their dissolution the perishing, of things.*[67A7] *Since genesis is the joining of atoms and perishing the dissociation of atoms, then ... genesis must be change"*.[6BA37]

Da: "All the changes we can produce consist in separating particles that are in a state of cohesion or combination, and joining those that were previously at a distance".

De: *"The larger [of the atoms] is heavier*[68A69] *... All things are created by the inter-linking [of the atoms] and their combining ... Because in some sense they [the atomists] maintain that all beings are numbers and are made up of numbers"*.[67A15]

Da: "[We must ascertain] the relative weights of the ultimate particles, both of simple and compound bodies, the *number* of simple elementary particles, which constitute one compound particle".

Here the amazing visionary capacity of Democritus appears closer to the chemistry of our day, since the founder of the atomic theory does not refer merely -as does Dalton- to the stoichiometric proportions of the atoms (today's *molecular* formula), but also foresees the 'order' (see p. 173), that is, the *arrangement* of atoms in the molecule (*structutral* formula) and the 'position', that is, the *orientation* of the combined atoms in space (*stereochemical* formula).[72]

This conception of the chemical atom would be further developed, in turn, by J. Berzelius (1779–1848), J. L. Gay Lussac (1778–1895), A. Avogadro (1776–1856), S. Cannizaro (1826–1910), L. Meyer (1830–1895), and D. Mendelejew (1834–1907), so that by the end of the 19th century the *atomic* chemical composition of matter would be considered more or less established. This was 'more or less' so, because reservations were still held by some distinguished chemists like M. Berthelot (1827–1907), W. Ostwald (1853–1932), Nobel laureate in chemistry, J.B. Dumas (1800–1884), who announced that "if I were master of the situation, I would efface the word atom from science".[73] In spite of these disputes,

---

[72] This prediction of Democritus would be verifie in 1828 in F. Wöhler's historic synthesis of *urine* from *ammonium cyanide*: $H_4NOC{\equiv}N \rightarrow H_2NCONH_2$. It was then found for the firs time [Cited in: B. Bensaude-Vincent – I. Stengers, *Histoire de la chimie*, 189] that two compounds of precisely the same constituents and the same proportions (CH_4ON_2) differ in properties because of differing 'arrangement' of the atoms. This phenomenon was named *structural isomerism*. The same is true of *stereo isomerism*. For example, *lactic acid*, depending on the 'orientation' of atoms in space, appears in two stereochemical arrangements, D and L:

```
     COOH     and         COOH, each with reciprocal optical activity.
      |                    |
  H– C–OH             HO– C–H
      |                    |
     CH3                  CH3
```

[73] Cited in: N. Herbert, *Quantum Reality: Beyond the New Physics*, 9.

accumulating evidence established atomic theory as an indispensable part of chemistry, a fact that would be corroborated by the great Chemistry Conference of Karlsruhe in 1860. There the generally accepted definitio of the atom would be set as 'the smallest mass that is capable of existing within a combination, and that of a molecule as "the smallest quantity that is capable of existing in a free state".[74]

While Chemistry had brought back Democritus's 'atoms', most physicists continued to be skeptical. English physicist Lord Kelvin (1824–1907) seems to have voiced the opinion of the overwhelming majority of physicists, when he still maintained in 1870 that "the notion of the atom has always been tied to fantastic assumptions of infinit durability, absolute cohesion, mystical action at a distance and indivisibility, so that chemists and other conscientious contemporary scientists have lost their patience and displaced the atom to the realm of metaphysics".[75] Physics indeed had to face specifi difficultie regarding the acceptance of the atom's existence. The electromagnetic theory of J.C. Maxwell (1831–1879), in its introduction of the concept of 'field' had already relegated simplistic mechanical models with ether to the sidelines. The only physical theory of the time that is based on the *atomic* hypothesis is the kinetic theory of gases of Maxwell and Boltzmann (1844–1906), who attempted a statistical mechanical interpretation of heat by attributing it to the motion of the innumerable atoms comprised in a gas. Democritus's atomic theory can be considered the forerunner of this *statistical* conception of the natural world, as we are dealing with processes of a large number of single atoms whose macroscopic effects only are perceptible to us. Thus, for example, the ancient atomists maintained that "while *all* atoms are in motion, the *sum* of them *appears* to be *still* . . . for the nature of atoms lies below the threshold of our perception".[76]

This theory, however, is in conflic with the *concept of entropy*, which anticipates a single, *non-reversible* direction for all thermal processes, a direction, which is determined mathematically by the increase of entropy. Conversely, the motion of atoms occurs as much in one direction as in the other, and thus the atomic kinetic theory of gases, like any mechanical theory, entails the *reversibility* of all thermal processes, in direct contradiction of everyday experience and the principle of entropy. This fact would place the whole atomic theory in extreme danger, and outstanding physicists like Planck (1858–1947), Poincarè (1854–1912) and others would oppose it up to the last decade of the 19th century.

The *positivists* would play a major role in this insistent attack. According to these 'philosophers of experience' we can only know what presents itself to tangible (experimental) perception. Recognized scientists like physicist-chemist W. Ostwald, R. Avenarius, and others, would reject atomic theory as an unfounded metaphysical conjecture of Democritus's; and physicist E. Mach (1838–1916) would inquire

---

[74] Cited in: B. Bensaude-Vincent – I. Stengers, *Histoire de la chimie*, 179.

[75] Cited in: R.U. Sexl, *Was die Welt zusammenhält*, 85–86.

[76] Lucretius, *De rerum natura*, II, 309–314.

sarcastically, "to this day, have you ever set eyes on an atom"?[77] The upshot of the battle over the atom in the fiel of physics appears at this juncture to favor the positivists. A bold philosophical response would appear, mainly from the Marxist position, expressed, above all, in the book, *Materialism and Empirical Criticism* (1909), where V.I. Lenin would attempt to respond to the positivist argument; the basis of physical philosophy must not be sense perception, but matter.

The ultimate outcome, however, would not be determined on philosophical grounds, but by the *experimental* physicists. The crucial issues were: Were Democritean atoms simply a useful hypothesis for the interpretation of natural phenomena, or did they really exist? And if the latter were true, did they indeed constitute the ultimate, indivisible unit of matter? A series of decisive experiments at the turn of the 19th century would provide the answer to these two fundamental questions: 'Yes' to the firs question; atoms do exist. 'No' to the second; atoms are not indivisible or compact, but are composed of sub-atomic particles. In 1895, W.C. Röntgen discovered Röntgen-rays; in 1896, H.C. Beckerel, radioactivity; in 1897, J.J. Thomson the electron; in 1898 M. Curie isolated radium; and in 1903, Thompson offered the firs interpretation of the atom's structure. In 1905, Einstein would proceed to the statistic calculation of the irregular movement of dust particles, which Democritus had noted. He would demonstrate that this so-called 'Brownian motion' is due to the impact of invisible *atoms* upon particles of ash, a phenomenon, which would soon be experimentally confirme by J.B. Perrin. Today, this is considered the firs convincing proof of the existence of atoms.

At the end of 1910, E. Rutherford would announce his atomic model, which resembles the solar system – a positively charged nucleus in the position of the sun; negatively charged electrons revolving around it like the planets, driven by electromagnetic, instead of gravitational, forces. More than 99.9% of the atom's mass is concentrated in the nucleus, which is, however, around ten thousand times smaller than the atom, whose size is determined by the external orbits of the electrons. Between the microscopic nucleus and the external electron orbits there is a vast void space. At the *Solvay Conference* of leading physicists in 1911, it was triumphantly confirme that atoms exist in the natural world. Two and a half thousand years were needed for the experimental verificatio of Democritus's bold hypothesis. Democritus was vindicated in his battle with Aristotle on the issue of matter being "discrete" and not continuous. As for Democritus's second hypothesis concerning the 'indivisibility' and 'compactness' of atoms, the question was now transferred from the atom to the three sub-atomic particles: electron, proton, neutron (discovered later, in 1932). Do these constitute the ultimate, indivisible, compact building blocks of physical reality? The answer would come a few decades later, and would be negative, as we shall see below. In the meantime, however, a dramatic revelation would

[77] Cited in: R.U. Sexl, *Was die Welt zusammenhält*, 118. In 1905, W. Ostwald was finall swayed to believe in the existence of atoms in 1905, following the experimental proof of Einstein and J.B. Perrin. E. Mach appears to have remained unconvinced to the end of his life.

occur: the Democritean concept of 'discrete' is not limited only to *matter*, but also applies to *energy*!

It is of great interest that the cause of two of the most decisive discoveries in the history of physical science corresponded to two profound deadlocks. The firs stalemate arose, as we have seen, in the 5th century B.C. with the Parmenidean *ontology* of one single being, immobile and unchanging, in total contradiction to *experience*. Two and a half thousand years later, the second stalemate would arise with the *theoretical* conclusions of Lord Rayleigh and J.H. Jeans, which also were in direct opposition to *experience* and the *experiments*: the total energy of black-body thermal radiation at all frequencies cannot tend to *infinit* , as predicted by *classical theory*, leading to an 'ultraviolet catastrophe', but is *bounded.* The firs deadlock would be overcome by Democritus through his ingenious theory of 'discrete' atoms. The overcoming of the second deadlock would be achieved by Planck in 1900 through the revolutionary 'discrete' theory of 'quanta'. The former introduces the 'discrete' property of *matter*; the latter, of *energy* ($E = h \cdot \nu$).[78] The gigantic effort of both men was to develop -in opposition to up-to-date ontological or theoretical assumptions- a new theory that would be in accord with empirical reality. Both 'discrete' discoveries would be decisive in the subsequent development of natural philosophy and science. Aristotle would admit that "no one besides Democritus could probe beneath the surface of these problems",[68A35] and Einstein would confir concerning Planck's quantum theory that "it was as if the ground had been pulled out from under one with no fir foundation to be seen anywhere, upon which one could have built".[79]

While Democritus's atomic theory had to wait two and a half thousand years for experimental verification Planck's quantum theory waited only f ve years. In 1905 Einstein -in his interpretation of the 'photoelectric effect', for which he received the Nobel Prize in Physics- would offer the firs solid experimental proof for the new theory. Beyond the prevalent wave theory of light and the assumption that electromagnetic energy is 'continuous' – Einstein's experimental conclusion, which would be confirme by A.H. Compton's experiments in 1923, would be that light behaves in a manner as if it were composed of 'atoms' or 'quantums' of light, which would later (1926) be named 'photons'. A 'photon' has a dualistic nature, possessing both undulatory and corpuscular characteristics. L. de Broglie (1892–1987) would demonstrate, in turn, that matter has also a dualistic nature, it too behaving as particles or waves. With the simple mathematical formula $\lambda = h/p$ he links for the firs time an *undulatory* property (wavelength, λ) with a *corpuscular* property (impulse, $p = m.v.$). "His way of thinking was rather akin to that of the ancient Greeks", observe F. Richthmeyer and E. Kennard. "The reasoning used might

[78] The great but misunderstood L. Boltzmann had already predicted in 1891 the possibility of a 'quantified energy. Unfortunately, he did not live to see his guess verified Four years before, he committed suicide sick and embittered by the non-recognition of his work on statistical mechanics as the basis of thermodynamics, which accepted the existence of atoms.

[79] P.A. Schilpp (Ed.), *Albert Einstein, Philosopher-Scientist*, 45.

almost be paraphrased as follows: (1) Nature loves symmetry; (2) therefore the two great entities, matter and energy, must be mutually symmetrical; (3) if [radiant] *energy* is undulatory and/or corpuscular, therefore *matter must be corpuscular* and/or undulatory."[80] Three years later, Davisson and Germer would offer the experimental proof, showing that during the reflectio of a beam of *particles* (electrons), the *undulatory* phenomenon of diffraction occurs. The moment it seems that the Democritean *atomic* conception of matter finall prevails, its *wave-like* nature appears. Although these two concepts are contradictory, all evidence supports the *dual* wave-particle nature of matter and of energy, according to the experimental technique, which is applied. How correct was Heraclitus in stating: "that which is in conflic with itself is simultaneously in agreement with itself",[22B51] and how right was Democritus in affirmin that "in reality we know nothing about anything, but the opinion of each is created by what fl ws to him"![68B7]

In the meantime, investigation of atomic structure would continue. The atomic model proposed in 1910 by Rutherford presented serious theoretical disadvantages. Bohr (1885–1962) revised it in 1913, including the new quantum theory in his calculations, assigning fi ed orbits in which electrons do not radiate energy. A. Sommerfeld would advance additional improvements (elliptical electron orbits), applying the theory of relativity. However, though the Bohr-Sommerfeld model explained the electromagnetic spectrum of atomic hydrogen, it presented serious difficultie when extended to atoms of more than one electron, like atomic helium. In 1925, Bohr himself had already realized that "the required generalization of the classical electrodynamical theory demands a profound revolution in the concepts on which the description of nature has until now been founded".[81] As it turned out, this revolution began that very year.

In 1925, the young German physicist W. Heisenberg (1901–1976) would conceive a new theory based no longer on what the inside of the invisible atom *might be*, but exclusively on what the atom *does* so that we can *measure* it – that is, on its *atomic* energy, emission and absorption. This new 'quantum matrix mechanics' theory would be independently developed on the one hand by W. Heisenberg, M. Born, and P. Jordan, and on the other by P.A.M. Dirac. During the same period Schrödinger (1887–1961) would approach the problem from another angle; extending de Broglie's theory mathematically, he created his 'wave mechanics' theory, according to which "material points consist of, or are nothing but, wave systems"[82] described through Schrödinger's 'wave equation' ψ. P.A. M. Dirac would immediately show the *physical equivalence* of wave and matrix mechanics, which are simply different mathematical representations of the same theory. Many would fin here a revival of the thought of Pythagoras, who had envisioned in arithmetical harmonies of vibrating strings the deep secret of nature. A. Sommerfeld would write enthusiastically in the introduction of his book, *Atombau und Spektrallinien*: "What

---

[80] F.K. Richtmeyer and E.H. Kennard, *Introduction to Modern Physics*, 236.

[81] Cited in: H.R. Pagels, *The Cosmic Code*, 55.

[82] Cited in: W.C. Dampier, *A History of Science*, 396.

we are listening to today in the language of the spectra is a true music of the spheres of the atom, an accord of whole numbers relations, an increasing order and harmony within a variety of forms . . . All the whole number laws of the spectral lines and of atomism derive, finall , from the quantum theory. It is the mysterious instrument on which nature plays spectral music, and on the basis of its rhythm tunes the structure of atoms and nuclei".[83]

One question, nonetheless, continued to persist: What is the *physical* meaning of Schrödinger's 'wave-function' ψ? M. Born (1882–1970) would provide the answer: The square of the wave-function $|\psi|^2$ specifie the *probability* of findin an electron at a specifi point in space. This interpretation, which relates the *undulatory* and the *corpuscular* nature of a wave-particle, met at firs strong objections, since it introduces *probability*, replacing the strict classical *determinism* and *causality* firs taught by Democritus. Today it is widely accepted, "although in truth it poses as many questions as it answers".[84] It is known as the 'Cophenhagen interpretation', and besides Born, architects in its development were Bohr and Heisenberg. In 1927, the firs would introduce the 'complementarity principle', and the same year, the second 'the uncertainty principle'. According to the first two concepts are complementary when one sets limits on the other; or two properties of the same object are complementary, when one, if known, excludes the simultaneous precise knowledge of the other. Thus, both the *corpuscular* and the *undulatory* concepts constitute *complementary* descriptions of physical reality, although the physical reality itself may not be either the one or the other. According to the second principle of 'uncertainty', the observer is an inseparable part of the experiment (observer – experimental method – object of experiment) and his observation *always* influence the result of measurement. He is not, for example, capable of simultaneously establishing with the same accuracy both the *position* and the *momentum* of an electron. The greater the accuracy of determining the position, the lesser known the momentum, and vice-versa. The 'uncertainty principle' is not the result, as some wrongly believe, of experimental shortcomings, but constitutes a fundamental principle inherent in nature. "The orbit of an electron is created only when we observe it",[85] Heisenberg notes, and at the same time it is destroyed by this very observation. Specifi phenomena are created only through observation. A particular electron, for example, is *neither* wave *nor* particle as long as we are not observing it, but within the framework of the complementarity principle, depending on our choice of experimental method, it behaves at times like a wave and at times like a particle. The characteristic experiment showing this 'complementary' behavior is that of 'the two holes' – a phenomenon which is *absolutely impossible* to explain by 'classical' physics, constituting, according to the distinguished physicist R. Feynman, the 'only mystery' of quantum mechanics. In this experiment, electrons seem not only to know whether only one or both holes are open, but also whether we are observing them or not, and

---

[83] Cited in: R.U. Sexl, *Was die Welt zusammenhält*, 149.

[84] P. Coveney and R. Highfield *The Arrow of Time*, 121.

[85] Cited in: K. Baumann and R.U. Sexl, *Die Deutungen der Quantentheorie*, 29.

accordingly they exhibit their wave or corpuscular behavior. The quantum world is *holistic*, since all of its part -in some inexplicable manner- are interrelated. Nature appears to keep all of its options and choices open, up to the moment that it is placed *under observation*. In the end, theoretical physicist Sexl notes, "the answer one receives, depends on the question one puts to nature".[86] "In reality we know nothing about anything", as Democritus puts it, "but the opinion of everybody is created by what fl ws to him".[68B7]

Some distinguished scientists would never agree with the introduction of the quantum concept of probability. Einstein would make his famous pronouncement, "God does not play dice with the world",[87] and Schrödinger would confess in despair, "if we are to remain with these cursed quantum leaps, then I am sorry I was involved in any way with quantum theory".[88] In 1935, both would contrive 'thought experiments' (Gedankenexperimente) like the so-called 'Schrödinger's cat' and the 'EPR experiment',[89] intended to prove the non-validity of the 'Copenhagen interpretation'. While the 'paradox of Schrödinger's cat' continues to be challenging, the "EPR paradox" was shown *experimentally* by A. Aspect in 1982 to *confir* the 'Cophenhagen interpetation'!

At the same time, the Democritean principle of the 'discrete' and 'indivisible' of the ultimate structure of matter would continue to be the object of intensive scrutiny. After the discovery of the neutron by J. Chadwick in 1932, it was believed at firs that the ultimate constituents of matter are three: *Protons* and *neutrons* (which constitute the atom's nucleus), and *electrons* (which revolve in specifi orbits around the nucleus). Soon, however, this simplistic conception would prove incorrect: with the discovery of anti-matter, the positron (1932), the anti-proton (1955), the anti-neutron (1955), and other particles from cosmic rays, like the mesons, the number of 'elementary particles' began to increase, and with the use of ever more powerful accelerators, the sum soon exceeded two hundred! There was an attempt by theoretical physicists to introduce order into this chaos. "Once more there were resources in antiquity", notes Sexl. "In his theory of matter, Plato's starting point was the grand concept of *symmetry*... Could one not [in our day] accept the idea of symmetry"?[90] Indeed, based on some 'symmetries', the years saw the development of the so-called 'Standard Model', which is in agreement with current experimental data and limits the total number of all the hypothetically 'indivisible' fundamental particles to 24 fermions: 18 quarks and 6 leptons, and their antiparticles. Besides the above fermions, there are four types of *force carriers*, the bosons W^+, W^-, Z^o, γ, eight gluons, the graviton, and the mysterious Higgs particle, which has not as yet been detected experimentally. Does the Standard Model, which still contains some debatable points, indeed represent the ultimate, indivisible structural constituents of

86 R.U. Sexl, *Was die Welt zusammenhält*, 165.

87 A. Pais, *'Subtle is the Lord'. The Science and Life of Albert Einstein*, 443.

88 W. Heisenberg, *Die Entwickung der Deutung der Quantentheorie*, 292.

89 That is, **E**instein-**P**odolsky-**R**osen.

90 R.U. Sexl, *Was die Welt zusammenhält*, 192.

the universe -the 'atoms' of Democritus- or is there a deeper inner structure? This will be revealed in the future. Along with the development of the concept of the 'atom', the original sense of Democritean 'void' has also been transformed. The 'vacuum' has ceased to be empty. Indeed, it is fille up with short-living particles, waves, forces, and fields as these arise from the 'quantum fiel theory', that is, from *relativistic* quantum mechanics. Heisenberg's 'uncertainty principle' permits within the framework of the relativistic quantum fiel a variety of energy *fluctuation* in a space which lead to momentary *materialization* and, again, disappearance of particle pairs. Each fiel of forces arises from an interchange of particles, and each particle in space represents a 'materialization' at a specifi point in the field On account of this, 'void' ceases any longer to be a vacuum. "Nowadays", comments physicist C.L. Smith, "we don't even understand the vacuum".[91] And physicists Davies and Gribbin add: "Solid matter dissolves away, to be replaced by weird excitations and vibrations of invisible fiel energy. In this theory, little distinction remains between material substance and apparently empty space, which itself seethes with ephemeral quantum activity. The culmination of these ideas is the so-called 'superstring theory', which seeks to unite space, time, matter, and to build all of them from the *vibrations* of sub-microscopic loops of invisible string inhabiting a ten-dimensional imaginary universe".[92] How visionary were Democritus's words when he declared, "there is only one kind of motion, the *vibrational*"![68A47]

Considering this state of affairs, briefl reviewed above, what is left today of Democritus's vanguard atomic theory? Theoretical physicist and thinker B. d'Espagnat would give the following answer: "Finally one ends up with universal concepts in which solid *matter* seems to have been dissolved away into *equations*. This is a concept where *materialism* is more and more compelled to develop toward *mathematization*, and where -if one can put it thus- Democritus must seek refuge in Pythagoras".[93] Where, then, did Democritus go wrong? His error did not derive from faulty reasoning, but from something unknown which remained so up to the beginning of the 20th century. As we saw (p. 172), Democritus theoretically rejected *infinit divisibility* through reductio ad adsurdum, maintaining, among other things, that it is impossible for a *material* being, extensible in space, to be subject to division ad infinitum because in that case it would end up in spatially non-extensible parts. However, from parts without spatial extension, a material extended body cannot be resynthesized. We would have, then, an irreversible disintegration of matter, which is in opposition to Democritus's principle that "nothing can be reduced to 'not being' ".[68A1(44)] This Democritean principle would be revised only in 1905 by Einstein, who for the firs time correlates energy and mass in the famous formula, $e = mc^2$, indicating that the elementary particles -which readily correspond to Democritus's atoms- are neither indestructible nor stable in number, and that, on the contrary, it is possible for them to 'end up' as immaterial energy. During the

---

91 Cited in: P. Coveney and R. Highfield *The Arrow of Time*, 142.

92 P. Davies and J. Gribbin, *The Matter Myth*, 8.

93 B. D'Espagnat, *Auf der Suche nach dem Wirklichen*, 11.

same period, this phenomenon was verifie experimentally. According to today's 'quantum fiel theory' – which combines the 'quantum theory' with the 'special theory of relativity' – each *transformation* of elementary particles may occur with a certain probability, as define by the theory, provided that it conforms to the fundamental *laws of conservation* of (i) the total energy, (ii) the total momentum and angular momentum, (iii) the total electric charge, (iv) the total color charge (the charge of the strong interactions), (v) the total lepton charge (which appears in three variations), and (vi) the total baryonic charge. How would it have been possible for Democritus to have imagined that it is not *matter* itself and the number of atoms that are conserved, but certain *physical perameters* that characterize matter? While his spirit was superbly rational and visionary, what was not available to him in his time was the scientifi achievement of the last three hundred years: the *quantifie physical magnitude.*

As we have seen, Democritus refers not only to the senses, but also to thought itself involving *physical* atomic processes. On the other hand, he considers the individual to be free to arrive at decisions with reason as his guide. This Democritean view, which combines in a seemingly contradictory manner an atomic concept of *causality* with human *free will*, is now being seriously reexamined under the lens of contemporary neurophysiology. Today it is generally accepted that the functioning of mind is based on the interactions of atoms and molecules, which obey the causal laws of physics and chemistry. In the fina decades of the twentieth century, continuous research by both physicists and neurophysiologists indicates that within the neurophysiological processes of the mind and on a 'deeper level' than that of atomic interactions, *quantum* phenomena occur, which introduce a quantum *indeterminacy* or *probability* field [94] These non-causal regions of the mind may perhaps explain the existence of *free* will, which may influenc possible quantum options. If and how such a process may work remains unknown. According to the *dualistic* hypothesis, "the mind-brain interaction is analogous to a probability fiel of quantum mechanics, which has neither mass nor energy yet can cause effective action at microsites"[95] of the brain. Some scientists do not agree with this interpretation: How could an 'external' 'immaterial' mental intention be the cause of neurological events? According to the *monistic* view, the answer will be found by probing "far more deeply within the existing physical 'material' structures that constitute the portions of the brain – and, in turn, much more deeply within the very question, what precisely *is* a 'material structure' on the quantum level"?[96]

All of the above developments make more than ever timely Democritus's insight that "man must acknowledge as a fundamental principle that he is very distant from the truth;[68B6] in reality, he knows nothing, for truth lies in the depths".[68B117] "Truth resides in the abyss"[97] – these are the words of Schiller, often repeated by

---

94 H. Margenau, *The Miracle of Existence.*

95 H.J. Eccles, *Evolution of the Brain, Creation of the Self*, 189.

96 R. Penrose, *Shadows of the Mind*, 350–351.

97 W. Heisenberg, *Positivismus, Metaphysik und Religion*, 313.

Heisenberg. "[Here] it is implied", D. Bohm admits, "that the ultimate source is immeasurable and cannot be captured within our knowledge".[98] This ignorance and doubt, however -as taught by the personal example of the Presocratics- must not constitute a reason for withdrawal from the path of scientifi pursuit. Their spirit is reflecte today in the words of R. Feynman: "I feel a responsibility as a scientist who knows the great value of a satisfactory philosophy of ignorance, and the progress made possible by such a philosophy, progress which is the fruit of freedom of thought. I feel a responsibility to proclaim the value of this freedom and to teach that doubt is not to be feared, but it is to be welcomed as the possibility of a new potential for human beings. If you know you are not sure, you have a chance to improve the situation".[99]

The turn taken by natural sciences toward the end of the 20th century does not in the least diminish Democritus's inspired achievements. As K. Reinhardt remarks, "The carry over of Leucippus's concepts and his way of thinking to human relationships, the idea that here too the end can be interpreted by the stages of becoming, and that the great and all – encompassing is derived from the motion of microscopic parts, constitute the enormous and unique contribution of Democritus, which -though unrecognized- influence all ages most intensely and through the longest stretch of time".[100] At the same time, he would lead the way to the culmination of Greek philosophy: "We know nothing", Democritus proclaims, "for truth is in the depths".[68B117] "With thoughts like this", Popper remarks, "the critical stance of the Presocratics prefigure and prepared for the ethical rationalism of a Socrates: his conviction that the *pursuit of truth* through *critical discussion* engenders a *mode of living* – the fines he could conceive".[101]

[98] Cited in: R. Weber, *Dialogues with Scientists and Sages*, 103.

[99] R. Feynman, *The Meaning of it All*, 28.

[100] K. Reinhardt, *Hekataios von Abdera und Demokrit*, 513.

[101] K. Popper, *Die Anfänge des Rationalismus*, 11.

Epilogue

> *We cannot read the Presocratics save through our own theoretical restlessness From their point of departure we too set off, following different and often diverging paths of thought, all of which, however, take root in a beginning inexhaustibly rich, precisely because it contains everything.*
>
> Jean Brun

Democritus rounds out the great age of the Presocratics. By the mid-fift century, B.C., there had already been a turn in Greek thought from *nature* to *man*. It appears that man's common sense was troubled by the conflictin theories of the Presocratics. It was not easy to deal with their conclusions involving a Parmenidean being, 'single, immobile, and unchanging', or a Heraclitean cosmos in which 'all things are continually changing and nothing remains stable', or a Democritean universe constituted exclusively by 'the void and the invisible atoms'. For the common man all this is unrelated to everyday life and tangible experience. The fact that the Presocratics themselves concede that ultimate truth is inaccessible would discourage people even more from dealing with them. Furthermore, during this period *philosophy* and the various *sciences* began to split off from one another. Democritus would be the last Greek thinker who as yet offered an all-encompassing theory of the *cosmos*.

The core of Presocratic inquiry was, as we have seen, the search for an ultimate reality hidden beneath the phenomenal world of sense experience. The question of the existence of such an absolute truth would mark the two basic directions of post-Presocratic thought: the path of *empiricism*, *phenomenalism*, and *positivism*, which would be pursued by the Sophists (Protagoras, Hippias, Gorgias, et al.), and the diametrically opposed path of *idealism* and *transcendentalism*, which would be laid down by Socrates, Plato, and Aristotle. The former do not seen absolute truth, are not concerned with cosmological problems, and do not fin theoretical concepts important, but focus their attention on practical matters, on man and his relationship with the community. The law is not the expression of some cosmic truth, but is exclusively the result of a 'position' taken – a wholly human product. The *realativism* and *individualism*, which pervade their theories conduct to *skepticism*, *agnosticism*, and even full-fledge *nihilism*. On the other hand, the *anthropological* view of the *cosmos* introduced by Socrates also centers on the human being – without, however, detaching it from his social, natural, and ontological foundation, which is exhaustively explored in the work of Plato and Aristotle. Attic philosophy is not only a new creation; it is also a recapitulation. All the themes that were treated during the Presocratic age are examined anew, evaluated, and ordered within a wider and fuller unity.

Thus is created the unparalleled edific of Platonic and Aristotelian *philosophy*. As to the development of science after the Presocratics, however, there is a long-lasting containment. "Democritus", Russell observes, "is the last of the Greek philosophers to be free from a certain fault that vitiated all later ancient and medieval thought. All the [Presocratic] philosophers we have been considering so far were engaged in a disinterested effort to understand the world.... Their attitude, in the main, was genuinely scientifi whenever it did not merely embody the prejudices of their age. But it was not *only* scientific it was imaginative and vigorous and fille with the delight of adventure... From this point onwards, there are firs certain seeds of decay, in spite of previously unmatched achievement, and then a gradual decadence. What is amiss, even in the best philosophy after Democritus, is an undue emphasis on man as compared to the universe. First comes skepticism, with Sophists, leading to a study of *how* we know rather then to the attempt to acquire fresh knowledge. Then comes, with Socrates, the emphasis on ethics; with Plato, the rejection of the world of sense in favor of the self-created world of pure thought; with Aristotle, the belief in purpose as the fundamental concept in science. In spite of the genius of Plato and Aristotle, their thought has vices, which proved infinitel harmful. After their time, there was a decay of vigour, and a gradual recrudescence of popular superstition. A partially new outlook arose as a result of the victory of Catholic orthodoxy; but it was not until the Renaissance that philosophy regained the vigor and independence that characterize the predecessors of Socrates".[1]

What role does Presocratic thinking play in our time? First of all, science as a whole is based on the tradition of rational, *critical* -as opposed to dogmatic- research and theory, a tradition, which the Presocratics firs established. Without this, no science could exist. "There can be little doubt", Popper notes, "that the Greek tradition of philosophical criticism had its main source in Ionia. It was a momentous innovation. It meant a break with the dogmatic tradition which permits only *one* school or doctrine, and the introduction in its place of a tradition that admits a *plurality* of doctrines which all try to approach the truth by means of critical discussion... It thus leads to the tradition of bold conjectures and of free discussion, the tradition, which created the rational or scientifi attitude, and with it our Western civilization, the only civilization which is based on science (though, of course, not upon science alone)".[2]

Beyond this general assessment, there are a number of fundamental Presocratic principles and concepts which prevail in today's natural sciences,[3] such as

---

[1] B. Russell, *History of Western Philosophy*, 89–90.

[2] K. Popper, *The World of Parmenides*, 23.

[3] Anticipating possible reservations based on the fact that there are many *opposing* views among the Presocratics, the answer would be that today also there are many *conflictin* physical theories as to the problem of cosmogony, the ultimate structure of reality, and the underlying unity in nature. It is precisely this multiplicity of views that constitutes the decisive *critical* element marking both the thought of the Presocratics and science in general.

- Cosmic order.
- Universality of natural laws.
- Unity underlying seeming plurality.
- Interrelation and interconnection of all beings.
- Man-observer as an inseparable part of the natural world.
- Merging of opposite concepts.
- Shift in emphasis from matter to process (Heraclitus).
- Inherent dynamic balance (Heraclitus).
- Mathematization of nature (Pythagoras).
- Atomic Theory (Democritus).
- Attribution of quality to quantity.
- Evolutionary process
- Intelligent-design theory (Anaxagoras).
- Natural selection.
- Hidden ultimate reality underlying phenomena.
- Symmetry, proportion.
- Homogeneity, isotropy.
- Rhythm, measure.

In turn, Greek words/concepts, through which the Presocratics articulated ideas, were established as philosophic and scientifi terms, many of them being also words of the international vocabulary.

The world of the Presocratics, as we have seen, is above all *holistic*. Nature and man, macrocosm and microcosm, universal and human law, cosmic and ethical order, inorganic matter and living beings co-exist, totally interrelated and interdependent, in a singular *union*. This unique union is marked by *harmony*, *logos* (proportion), and *measure*, which endow the universe with order and beauty. Precisely this feature of 'due measure' is one of the decisive Greek elements which marks the dividing line between Western thought and Eastern meditation, in which the im-measurable and in-expressible prevail. As Bohm notes, "this is an extraordinarily significan point. Whereas in Western society, as it derives from the Greeks, *measure*, with all that this word implies, is the very essence of reality, or at least the key to this essence, in the East measure has now come to be regarded commonly as being in some way false and deceitful".[4]

Back to the Presocratics was the title of the presidential address delivered by K. Popper to the Aristotelian Society on October 13, 1958. Is such a return to the world of the Presocratics, which is marked, as noted above, by *wholeness* and *measure*, possible? Before approaching an answer, let us examine the development Presocratic thought after the Renaissance – when it began, to a degree, once more to revive. Heidegger would summarize the situation in one sentence: "What the word *cosmos* expresses for us today is the unregulated confusion of a technical mechanism of information which arose in the face of unfragmented *nature* and usurped her place, while its functioning remained accessible and controllable only through

[4] D. Bohm, *Wholeness and the implicate order*, 23.

calculation".[5] In such a world, *wholeness* disintegrates. The 'fragmentation' into separate sciences and specializations would become the exclusive context of thinking and investigation of things -a phenomenon which leads, finall , to the perception that this detached state of affairs corresponds to ultimate reality itself. Presocratic *due measure*, which marks a cosmic structure embodying harmony and beauty, would cease to constitute a visionary concept of *logos*, which pervades *wholeness*-material, personal, social, ethical, spiritual, aesthetic – and would be reduced to a mere formal process of *measuring*. Finally, the mechanistic, materialistic, deterministic attitude, which prevails, leaves no room for a universal Mind or the mind of God.

The *splitting* of the Whole and the introduction of *measurement* constitute, to be sure, a fundamental achievement of science for approaching truth. Among other things, quantitative data have unquestionably contributed to the training of man in a more precise, critical thought. And this critical, accurate thinking must be maintained, lest our culture be marred by neo-medieval tendencies. However, *quantificatio* alone will not suffice While it may be accompanied by spectacular technological developments, it has tragic impacts – social, psychological, environmental, aesthetic, to name but a few. *Measurement*, the achievement of contemporary civilization, must be accompanied by *measure*, the achievement of Presocratic thought.

Without underestimating the power of already established technology, we can perceive the beginning of a return to the *holistic* thought of the Presocratics. There are signs that "philosophy and the natural sciences, and religion and the natural sciences as well, are once more *converging*, in opposition to the impression that prevailed at the end of the 19th and the beginning of the 20th century, that religious experience, philosophic thought, and scientifi knowledge were increasingly departing from one another. The theory of relativity and quantum physics gave rise to the firs sign of a change in direction".[6] With these words on May 21, 1988, E. Albrecht greeted over sixty distinguished scientists, philosophers, and theologians in his opening address to the International Conference on 'Spirit and Nature' in Hannover. The *whole* is not simply the *sum of its parts*. The noted physicist and philosopher, C.F. von Weizsäcker remarks that "the quantum theory is, as it turns out, totally holistic. It refuses to accept any separation... The world is not composed of objects. It is merely limited human knowledge that fragments into object the Whole -to which man himself belongs- so as to comprehend it".[7] D. Bohm agrees: "Wholeness is the entity that is real. Fragmentation is only the effect of human actions on wholeness... The process of splitting is a means of thinking about things".[8] For Heisenberg, "The task is renewed . . . to penetrate into a knowledge of reality in

---

[5] M. Heidegger, *Aufenthalte*, 20.

[6] E. Albrecht, *Zur Eröffnung des Kongresses, Geist und Natur*, 13.

[7] C.F. von Weizsäcker, *Welt, Wissenschaft, Wirklichkeit*, 25.

[8] D. Bohm, *Fragmentierung und Ganzheit*, 270, 264.

such a manner that the various interdependencies can be perceived as parts of a single, meaningfully ordered world".[9]

This fresh holistic approach is not confine to the narrow limits of contemporary physics. Recent research on the temporal development of open systems by I. Prigogine and his collaborators, as well as that of many other scientists, is narrowing the gap between *physics* and *biology*. Today, many scientists and thinkers are promoting the new 'biopsychosocial' approach,[10] according to which the human body and soul are parts of a larger whole. For the firs time since the Presocratic age, there is an effort in the Western world toward a 'reconciliation' and 'reunification of the natural sciences, as also described by the noted authority on evolutionary biology and founder of sociobiology, E.O. Wilson, in his book, *Consilience: The Unity of Knowledge*: "It means a belief in the unity of the sciences –a conviction, far deeper than a mere working proposition, that the world is orderly and can be explained by a small number of natural laws. Its roots go back to Thales of Miletus., in Ionia, in the sixth century B.C."[11] Harvard University physicist and historian of Science, Gerald Holton, coined the term "Ionian Enchantment",[12] an expression that links the idea back in the 6th century B.C. to the ancient Ionian physical philosophers. Physicist H. P. Dürr would state the same position, thus: "The *natural* and the *transcendental* have a place in the thought of today's physicist, no longer in antagonism, but in a complementary relationship, marking the different sides of a single reality".[13]

This transcendental factor, however, is perceived differently by each scientist: Some grasp it as *cosmic mind*: "The mental and the physical", Bohm observes, "are two aspects, like the form and content of something that is separable in thought, not in reality".[14] Others live this insight as *religious experience*: Einstein would maintain that "the cosmic religious feeling is the strongest and noblest motive for scientifi research".[15] Still others, finall , identify transcendence with *consciousness*: Eddington would claim that "We can never achieve a complete conception [of the *cosmos*] as long as we withdraw our consciousness from the world of which it is a part".[16] Schrödinger would be even more categorical: "Multiplicity is mere appearance. In essence, there is simply *one* consciousness".[17] This epistemological position can be summarized in the words of Holton: "The *synthesis* of rationality and intuition -rather than their opposition- is the key to answering all questions of science, as we now understand the term".[18]

---

[9] W. Heisenberg. Cited in: H.P. Dürr, *Das Netz des Physikers*, 138.

[10] E. Sarafino *Health Psychology: Biopsychosocial Interactions*.

[11] E.O. Wilson, *Consilience*, 4.

[12] Cited in: E.O. Wilson, *Consilience*, 4.

[13] H.P. Dürr, *Vorwort, Physik und Transzendenz*, 11, 16–19.

[14] D. Bohm, cited in R. Weber, *Dialogues with Scientists and Sages*, 106.

[15] A. Einstein, cited in: R. Weber, *Dialogues with Scientists and Sages*, xv.

[16] A. Eddington, *Wissenschaft und Mystizismus*, 109.

[17] E. Schrödinger, *Das arithmetische Paradoxon – Die Einheit des Bewusstseins*, 160.

[18] G. Holton, *The Controversy over the End of Science*, 168.

All the above views of contemporary scientists reflec anew the spirit of the Presocratics. To be sure, a 'return' to them is impossible. What is feasible, however, is to *assimilate their thought* in an attempt to *transcend* today's situation: "What is needed is to learn afresh, to observe, and to discover for ourselves the meaning of wholeness ... Insight [is required] ... Such insight implies an original and creative act of perception into all aspects of life, mental, and physical, both through the senses and through the mind, and this is perhaps the true meaning of meditation. . .Then the whole fiel of *measure* will come into *harmony*, as fragmentation within it comes to an end. . . .When such a harmony prevails, man can then not only have insight into the meaning of *wholeness* but, what is much more significant he can realize the *truth* of this insight in every phase and aspect of his life".[19] These words of D. Bohm, a distinguished 20th century physicist and philosopher, best reflec the spirit of the Presocratics.

In his book, *The End of Certitude*, Prigogine, Nobelprize winner (1977) writes: "Truly, the ancient Greeks bequeathed us two ideals which played a guiding role in our history: The firs is the understanding of nature; ... the second is democracy, which presupposes human freedom, creativity, and responsibility. Certainly we are very far from fulfillin these two ideals; at least we are capable of concluding that they are not contradictory".[20] Indeed, in the thought of the Presocratics such a contradiction does not exist, since for them the laws of nature are not rigid; they are not exclusively dictated by 'measurable' magnitudes, as they are today, but are determined by 'measure', 'proportion', and 'reason' (*logos*), which conduct to *harmony*. "Harmony, order, and measure constitute the steadfast ideals of the Greek",[21] Snell notes. Responsibility for definin *due measure* resides within the individual, in accord with *cosmic* order, of which man considers himself an inseparable part. In a unique moment, Presocratic thought provided a *rational* explanation of the universe, elevating at the same time man to the condition of a *free* being, exclusively *responsible* for his actions.

Approaching Athens during his visit to Greece in the spring of 1962, Heidegger would note: "Only one question confronts the mind: If man can as yet be granted a familiar dwelling place, as in ancient time the Greeks were granted a fundamental, great and rich, yet with measure. But that too had its time and suddenly collapsed ... This dwelling place remains inimitable. Nonetheless, it has not disappeared. It lives on as the point of departure, which after various transformations determines the technical and scientificall industrialized world of the present age ... An age whose historical course will be determined by its point of departure being forgotten or reaffirme as a Reminder".[22]

---

[19] D. Bohm, *Wholeness and the implicate Order*, 24–25.

[20] I. Prigogine, *La fi des certitudes*, 25.

[21] B. Snell, *Die Entdeckung des Geistes*, 159.

[22] M. Heidegger, *Aufenthalte*, 24.

Appendix: A Brief Historical Retrospect: On the Origins of the Greek People

With the possible exception of the Chinese, it is doubtful that there has been another people with such persistence and duration. In its long history, it undertook at least three times the cultural leadership of the world: in the high Mycaenean period, the classical age, and in the period of the peak of the Byzantine Empire.

Carl Blegen

In geographical terms, the territory of Greece constitutes the link between Europe, Asia, and Africa. Its varied landscape, with its continual alternation between mountainous land and sea, its sprinkling of islands, its mild climate, and its limited natural resources, undoubtedly influence the historical development of the Greek spirit.

The evidence is unclear as to when man firs appeared in Greece. Finds of stone tools make the existence of settlements likely from the *Early Paleolithic period* (ca. 600,000–100,000 B.C) and certainly (widespread remains of tools, the skull of the Neanderthal of Petralona) during the *Middle Paleolithic period* (100,000–33,000 B.C.). The beginning of the *Late Paleolithic period* (33,000–8,000 B.C.) is marked by the total disappearance of 'Neanderthal Man' and the appearance of Homo Sapiens, whose way of life was characterized by both a radical technical change and the production of cutting tools, which had fundamental consequences for the subsequent intellectual and social development of mankind. Most scholars maintain that human settlements spread to Europe by way of the Near East and that Greece was an area of transit and development.

In the Near East, during the 12th to 10th millennia, B.C., the 'Neolithic Revolution' – considered one of the most crucial milestones in the history of mankind – took place. Man gradually abandoned the life of the hunter and began to engage in farming and animal breeding, establishing permanent settlements. Due to its geographical position, Greece was the firs European region to see this basic change, around the 6th millennium, B.C. The firs Neolithic settlements were created (Argissa, Sesklo, Soufli-Magoul in Thessaly, et al.), where cultural advances (the 'Sesklo Culture') into the *Middle Neolithic period* (5,000–4,000 B.C.) took place and was extended to other island areas, such as Crete and Cyprus, which took on increased significanc in the *Late Neolithic period* (4,000–2,800/2,700 B.C.). The 'Saliagos Culture' (a small island in the Cyclades), the 'Dimini Culture' in Thessaly, as well as other find in Epirus, northwest Madedonia, and Thrace during this age present dissimilarities compared to features of their immediate predecessors, leading to conjectures about 'invasions' and influence from neighboring territories, for example Asia Minor. The small figurine of Neolithic Greece provide us with evidence of the social and intellectual life of man during this period. It is significan that the element of the occult, which appears in other geographical areas, is entirely lacking here.

The 'Urban Revolution' took place around 3,000 B.C. in the Near East, accompanied by the appearance of the firs great Cultures with writing. This second landmark in the history of mankind, which corresponds to the spread of the use of metal (*Early Bronze Age*, 2,800/2,700–1,900 B.C.), marks the beginning of a new epoch. The new culture would spread through Greece to Europe by way of the islands and the shores of the Aegean, thus falling under the Aegean ilsand influence During this age the 'Prepalatial Minoan Civilization' developed in Crete, the 'Protocycladic Culture', in the islands of the Aegean, and the 'Protohelladic Culture' on mainland Greece, from Thrace to the Peloponnese, under the influenc Eastern and local elements. It is believed that during the *Early Bronze age* the earliest signs of the unificatio of the Greek nation also appear, and possibly the Greek language, independent of influence from eastern monarchies and theocracies – an event which would prove decisive in the subsequent development of Greek thought.

During the *Middle Bronze Age* (1,900–1,600 B.C.) Crete developed the 'Protopalatial Culture' (Knossos, Phaistos, Malia), which was suddenly cut short by a catastrophic earthquake around 1,700 B.C. The 'Middle Cycladic' Culture appears to be the link between the 'Middle Minoan' and the 'Middle Helladic' Culture. The latter -in contrast to the 'Minoan'- possessed a retrograde character, the cause of which has given rise to many theories. According to the most recent and prevailing one, there was a migration of agrarian Greek-speaking tribes of Thessaly -the 'Protohellenes'- under the pressure of invasions by other, more northern, possibly Indoeuropean races, bringing once more to Southern Greece a closed, agrarian economy, with many settlements but no cultural center. This hypothesis is supported by ancient tradition, which places the Greek theogony and cosmogony in the Thessalian region.

The total destruction of all the important Cretan centers, around 1,700 B.C. was followed by the 'Neopalatial Culture' through three glowing centuries (1,700–1,450 B.C.), the firs high culture in Europe, which was marked by the development of large settlements around the three new palaces of Knossos, Phaistos, and Zakros. During this period, when Crete became the ruler of the seas, commerce flourishe and the dynamism of the 'Neopalatial Culture' led to its acme and spread to surrounding areas, an event, which would contribute three centuries later to the climax of the 'Mycenaean Culture'. During the same period, the Cretan hieroglyphic script was gradually replaced by the Minoan Linear Script (Linear System A and B). The end of this high Cretan Civilization was as sudden as it was tragic: the culture was wiped out around 1,500 B.C. by a gigantic tidal wave caused by the great eruption of the Thera volcano.

In southern mainland Greece during the *Late Helladic period* (1,600–1,100 B.C.), the Achaeans developed the 'Mycenaean Civilization', which would constitute the greatest cultural upsurge during the *Helladic Bronze Age*. The 'Mycenaean culture' was characterized by a competitive, organizational spirit, and was deeply influ enced by the advanced 'Minoan' and, in part, 'Cycladic' cultures, to such a degree that many scholars (e.g. A.J. Evans and others) maintained that its creation was the result of Minoan colonization – a theory that does not appear to stand today. Besides, the recent decoding of the Cretan-Mycenaean script of Linear B, proves that the language of the Mycenaean cultural centers was Greek, even that of the Knossos archives, a fact that underlines the parallel influenc exercized by the 'Mycenaean' culture on the 'Minoan' during the *Late Mycenaean period* (1,450–1,350 B.C.).

The *Early Mycenaean Phase* (1,500–1,425 B.C.) is marked by escape from isolation. Various cities, like Tiryns, Pylos, Athens, Thebes, Orchomenos, Iolkos, but especially Mycenae, developed commerce and shipping. During the *Middle Mycenaean period* (1,500–1,425 B.C.), the Mycenaeans took the lead, extending their influenc to the Aegean isles and the Asia Minor shore. The abrupt destruction of the palace at Knossos by fire around 1,400 B.C., also marked the end of the dominant Cretan rule and competitiveness, leaving the fiel free during the 14th and 13th century B.C. for the development and establishment of Achaean hegemony. It spread both toward the North in Hellenic territory (Macedonia, Epirus) and overseas, where Mycenaean settlements and commercial centers multiplied, in Crete, the Dodecanese, Cyprus, the Asia Minor seaboard (Miletus, Ephesus, et al.), and on the shores of the Eastern Mediterranean, all the way to southern Italy (Akragas, et al.). All these cultural centers were closely tied to commercial transactions, and were marked by a unifie advanced civilization with all its expressions in life and the arts. During the same period, the expedition against Troy also took place, a mobilization, which for the firs time would unite the Greeks of the Mycenaean world in a common cause. "Before Troy, it seems that Hellas never undertook a common effort".[1]

The 12th century B.C. marked the decline of the 'Mycenaean' Civilization. The displacement of the 'people of the sea' southward in the Eastern Mediterranean led to the violent dismemberment of the Hititie empire (around 1,175 B.C.) and a massive intermixing of populations from Asia Minor to Egypt, thus disorganizing the entire Achaean commercial network in the Near East. The inevitable decline of productivity and economic activity, with a corresponding drop in living standards, unavoidably led to a weakening of Mycenaean influenc and to a great wave of Achaean expeditions to Crete, the Dodecanese, Cyprus, and Southwestern Asia Minor. During the *Subminoan* and *Submycenaean period* (1,125–1,050 B.C.), which also forms the transitional period between the end of *prehistoric* and the beginning of *historic* times, the fina decay and disintegration take place in Mycenaean culture, which – following the Minoan civilization – was the next peak civilization in Europe. Common tradition and ethnic heritage, however, endure, and they

---

[1] Thucidides, *Peloponnesian Wars*, I, 3.

contributed the connective tissue enabling the crucial transition from the *Bronze Age* to the *Iron Age* – the beginning of *historic times.*

Into the vacuum created by the Achaean's desertion of central Greece, Greek tribes descended from the northern border, backward mountain people, who became the link between Helladic territories and the more northerly peoples of Europe. The firs three centuries of historic times (1,125–800 B.C.) were characterized by continual movements of Greek tribes – at times peaceful, at times violent – from less developed areas, economically, culturally, and socially, into the ruined sites of Mycenaean lands where, in cooperation with a scattering of natives, they initiated a new creative effort, which is on the one hand clearly inferior to the previous Mycenaean epoch (thus the unfortunate label, 'Greek Middle Ages' or 'Dark Ages'), yet on the other provides the base for the upcoming peak of Hellenic 'Classical Civilization' of the 5th century B.C.

The most significan regroupings took place in successive phases: During the *Submycenaean period* (1,125–1,050 B.C.), Dorians and Aetolians descended from Mainland Greece and settled in the Peloponnese, causing Achaeans and Arcadeans from Argolida and Laconia to migrate toward Lesvos, Tenedos, and Crete, as well as toward the northern shore of the Peloponnese, which the Ionians, in turn, abandoned to move to Attica and the shore of the Saronic Gulf. At the same time, Thessalians and Magnesians came down from Pindus to today's Thessaly and Pelio, respectively, expelling the Boeotians from Arni to today's Boeotia.

In turn, another branch of Dorians settled in Corinth, while Ionians from Attica and the Northern Peloponnese crossed over to Euboea and the Cyclades, to continue from there, at the beginning of the *Protogeometric period* (1,050–900 B.C.) the colonization of the central and southern Asia Minor seaboard (Miletus, et al.). At the same time, Aeolians, under pressure from Thessalians and Magnesians, took refuge in Tenedos, Lesvos, and the opposite areas of Asia Minor. The continual spread of the Dorians toward Northeastern and Southwestern Peloponnese created a new wave of colonies in the direction of Asia Minor (Kolophon, et al.), while a bit later, Dorians from the Peloponnese crossed over to Crete, Milos, Rhodes, Ko and the Asia Minor shore opposite (Alikarnassos, et al.).[2]

These overseas migrations continue to the middle of the *Geometric period* (900–700 B.C.). During the 8th century, B.C., colonization extended to the West, to the Eastern shores of Sicily (Syracuse, Leontinoi, Catania, et al.) and South Italy (Rhegium, Croton, Sybaris, Taras, et al.).

These significant widespread resettlings led to the creation of many small states, which in most cases took the form of cities, a phenomenon unique in human history. During the initial phase, the 'city-states', under monarchical rule, developed

[2] Two and a half millennia later, at the beginning of the 20th century, A.D., this flourishin Greek civilization in Asia Minor was violently uprooted by the Turks. The impossibility of imposing an inferior culture on a superior one, leads finall to failure and genocide. More than two and a half million Greeks were killed or purged from Asia Minor, Pontus, and Eastern Thrace in the decade, 1913–1923. "It was not only people but also the civilization that left", Hélène Ahrweiler recently stated; "but the memory has not fle ...".

handicrafts and trade. Although they were at firs more or less equal in power, toward the end of the 8th century, B.C., some (e.g., Corinth, Athens, Miletus) achieved greater power and wealth. The common belief in the Olympian gods, the panhellenic shaping of myth, the Olympic games, the ingenious addition of vowels to the norther Semitic consonant script and its transformation into a modern Greek alphabet, led to a closer bond among the 'city-states', and thus established for the firs time a panhellenic consciousness transcending its separate tribes, and marked by a common religion, common mores and traditions, common language, writing and art.

Now, for the firs time the name *Hellenes*[3] appears to designate this shared consciousness. By the 8th century, B.C. – the century of Homer – all these positive and fertile elements, which resulted from three centuries of continual movement and resettling, have matured. 'Geometric' art and epic poetry have reached perfection. This is the historical moment of the dawn of the European spirit.

[3] Hesiod, *Gynaikon Katalogos (List of Women)*, 5.9.

Bibliography

Superscript notes in small type refer to the Fragments of the Presocratics in the three volume edition by H. Diels and W. Kranz, (Hrsg.), Die Fragmente der Vorsokratiker, Weidmann, Zürich – Hildesheim, 1992.

[The designation '..A..' refers to entries of commentators, while '..B..' refers to original Presocratic statements].

Aeschylus, *Libation-Bearers*, The Loeb Classical Library, Cambridge, MA, London, 1926.

Aeschylus, *Prometheus Bound*, The Loeb Classical Library, London, New York, 1946.

Albrecht, E., Zur Eröffnung des Kongresses 'Geist und Natur'. In: H.-P. Dürr and W.C. Zimmerli (Hrsg.), *Geist und Natur*, Scherz Verlag, Bern-München-Wien, 1990.

Anton, J.P., The Pythagorean Way of Life: Morality and Religion. In: K. Boudouris (Ed.), *Pythagorean Philosophy*, Athens, 1992.

Archilochus, *Fragments*, M. Treu (Hrsg.), E. Heimeran, München, 1959.

Aristotle, *The Politics*, The Loeb Classical Library, London, New York, 1932.

Aristotle, *Generation of Animals*, The Loeb Classical Library, Cambridge, MA, London, 1953.

Aristotle, *Meteorologica*, The Loeb Classical Library, Cambridge, MA, London, 1952.

Aristotle, *On Sense and Sensible Objects*, The Loeb Classical Library, Cambridge, MA, London, 1935.

Aristotle, *On the Heavens*, The Loeb Classical Library, Cambridge, MA, London, 1939.

Aristotle, *On the Soul*, The Loeb Classical Library, Cambridge, MA, London, 1935.

Aristotle, *The Metaphysics*, The Loeb Classical Library, London, New York, 1933.

Aristotle, *The Nicomachean Ethics*, The Loeb Classical Library, Cambridge, MA, London, 1945.

Aristotle, *The Organon*, The Loeb Classical Library, Cambridge, MA, London, 1938.

Aristotle, *The Physics*, The Loeb Classical Library, London, New York, 1929.

Axelos, K., *Héraclite et la Philosophie*, Les éditions de Minuit, 1962.

Axelos, K., *Vers la Pensée Planétaire*, Les éditions de Minuit, 1970.

Bacon, F., *Advancement of Learning and Novum Organum*, The Colonial Press, New York, 1899.

Badash, L., *The Completeness of Nineteenth-Century Science*, Iris, 1973.

Bailey, C., *The Greek Atomists and Epicurus*, Russell & Russell Inc., New York, 1964.

Barrow, J., *The Origin of the Universe*, BasicBooks, New York, 1994.

Barrow, J. and Tripler, F.J., *The Anthropic Cosmological Principle*, Oxford University Press, Oxford, 1996.

Baumann, K. and Sexl, R.U., *Die Deutungen der Quantentheorie*, F. Vieweg & Sohn, Braunschweig-Wiesbaden, 1987.

Benfey, O.T., *Journal of Chemical Education*, 35, 1958.

Bensaude-Vincent, B. and Stengers, I., *Histoire de la chimie*, La Découverte & Syros, Paris, 2001.

Bernal, M., *Black Athena: The Afroasiatic Roots of Classical Civilization.* Vol. I The Fabrication of Ancient Greece 1785–1985. Rutgers University Press, New Brunswick, N.J., 1987.

Bernal, M., *Black Athena: The Afroasiatic Roots of Classical Civilization.* Vol. II The Archaeological and Documantary Evidence, Rutgers University Press, New Brunswick, N.J., 1991.
Bohm, D., Fragmentierung und Ganzheit. In: H.-P. Dürr (Hrsg.), *Physik und Transzendenz,* Scherz Verlag, Bern-München-Wien, 1986.
Bohm, D., *Wholeness and the Implicate Order*, ARK Paperbacks, London, New York, 1988.
Boman, T. *Das hebräische Denken im Vergleich mit dem Griechischen*, Vandenhoeck & Ruprecht, Göttingen, 1959.
Boudouris, K., The Pythagorean Community. Creation, Development and Downfall. In: K. Boudouris (Ed.), *Pythagorean Philosophy*, Athens, 1992.
Brun, J., *Les Présocratiques*, Presses Universitaires de France, Paris, 1998.
Bruno, J., De la causa, principio et Uno. In: G. Gentile and G. Aquilecchia (Hrsg.), *Dialoghi Italiani*, Sansoni, Firenze, 1958.
Bruno, J., De minimo. In: F. Fiorentino et al. (Hrsg.), *Jordani Bruni Nolani Opera latine conscripta*, Nachdruck, Frommann/Holzboog, Stuttgart-Bad Cannstatt, 1961.
Brush, S.G., *Die Temperatur der Geschichte*, Vieweg, Wiesbaden, 1982.
Bultmann, R., 'Zur Geschichte der Lichtsymbolik im Altertum', *Philologus*, Bd. 97, 1948.
Burkert, W., 'Weisheit und Wissenschaft: Studien zu Pythagoras, Philolaos und Platon', *Erlanger Beiträge zur Sprache und Kunstwissenschaft, Bd.X*, Verlag Hans Carl, Nürnberg 1962.
Burnet, J., *Early Greek Philosophy*, Adam & C.Black, London, 1930.
Calvin, M., *Journal of Chemical Education*, 35, 1958.
Capra, F., *The Tao of Physics*, Bantam Books, Toronto, New York, 1984.
Capra, F., *The Turning Point*, Simon and Schuster, New York, 1982.
Chargaff, E., Erforschung der Natur und Denaturierung des Menschen. In: H.-P. Dürr (Hrsg.), *Physik und Transzendenz*, Scherz Verlag, Bern-München-Wien, 1986.
Cornford, F.M., *Before and after Socrates*, Cambridge University Press, Cambridge, MA, 1990.
Cornford, F.M., Mysticism and Science in the Pythagorean Tradition. In: A.P.D. Mourelatos (Ed.), *The Pre-Socratics. Modern Studies in Philosophy*, Doubleday, Garden City, NY, 1974.
Coveney, P. and Highfield R., *The Arrow of Time*, Flamingo, London, 1991.
Cudworth, R., *The True Intellectual System of the Universe*, London, 1978.
Dalton, J., *New Systems of Chemical Philosophy*, Manchester, 1808 and 1810. Reprinted in the 'Cambridge Reading in Science'.
Dampier, W.C., *A History of Science*, Cambridge University Press, Cambridge, MA, 1968.
Darnell, J., Lodish, H., and Baltimore, D., *Molecular Cell Biology*, Scientifi American Books, New York, 1986.
Darwin, F., *The Life and Letters of Charles Darwin*, Vol. I, Appleton, New York, 1897.
Davies, P.,'That Mysterius Flow', *Scientifi American*, September 2002, 27.
Davies, P., *The Last Three Minutes*, Phoenix, London, 1997.
Davies, P. and Gribbin, J., *The Matter Myth*, Penguin Books, London, 1992.
Davies, P. and Park, D., *No Way. The Nature of the Impossible*, W.H. Freeman & Co., New York, 1987.
Dede, M., *Herakleitos (Heraklitus)*, Iolkos, Athens, 1970.
DeWitt, B.S. and Graham, N., *The Many-Worlds Intepretation of Quantum Mechanics*, Princeton University Press, Princeton, 1973.
Diels H.-Kranz W. (Hrsg.), *Die Fragmente der Vorsokratiker*, Weidmann, Zürich, Hildesheim, 1992.
Dillingham, S.G., 'Closing in on Zeno', *Scientifi American*, May, 1995.
Dirac, P.A.M., *The Principles of Quantum Mechanics*, Oxford University Press, Oxford, 1958.
Dodds E.R., *The Greeks and the Irrational*, University of California Press, Berkeley, Los Angeles, London, 1951.
Durant, W., *Kulturgeschichte der Menschheit*, Ed., Rencontre, Lausanne.
Dürr, H.-P., *Das Netz des Physikers*, DTV, München, 1988.

Dürr, H.-P., Vorwort. In: H.-P. Dürr (Hrsg.), *Physik und Transzendenz*, Scherz Verlag, Bern-München-Wien, 1986.
Dürr, H.-P., Wissenschaft und Wirklichkeit. Über die Beziehung zwischen dem Weltbild der Physik und der eigentlichen Wirklichkeit. In: H.-P. Dürr and W.C. Zimmerli (Hrsg.), *Geist und Natur*, Scherz Verlag, Bern-München-Wien, 1990.
Eccles, H.J., *Evolution of the Brain, Creation of the Self*, Routledge, London, New York, 1991.
Eddington, A., Die Naturwissenschaft auf neuen Bahnen. In: H.-P. Dürr, (Hrsg.), *Physik und Transzendenz*, Scherz Verlag, Bern-München-Wien, 1986.
Eddington, A., Wissenschaft und Mystizismus. In: H.-P. Dürr (Hrsg.), *Physik und Transzendenz*, Scherz Verlag, Bern-München-Wien, 1986.
Einstein, A., *Mein Weltbild*, Selig C. (Hrsg.), Ullstein, Frankfurt, M., Berlin, Wien 1972.
Eliade, M. and Couliano I.P., *Dictionaire des Religions*, Plon, Paris, 1990.
Ellis, G.F.R., 'Thinking Globally, Acting Univrsally', *Scientifi American*, October 1995.
Elytis, O., *Anichta Chartia (Open Papers)*, Ikaros, Athens, 1996.
Epicurus, *The Extant Remains*, Clarendon Press, Oxford, 1926.
Espagnat, d', B., *Auf der Suche nach dem Wirklichen*, Springer Verlag, Berlin, Heidelberg, New York, 1983.
Euripides, *Helen*, The Loeb Classical Library, London, New York, 1925.
Farrington, B., *Greek Science*, Penguin Books, London, 1949.
Feynman, R., Atoms in Motion. In: T. Ferris (Ed.), *The World Treasury of Physics, Astronomy and Mathematics*, Little, Brown & Co., Toronto, 1991.
Feynman, R., *The Character of Physical Law*, Penguin Books, London, 1992.
Feynman, R., *The Meaning of it All*, Addison-Wesley, California, 1998.
Forte, F., Vom Wettbewerb zur Kooperation - ein neuer Ansatz für die Weltwirtschaft. In: H. -P. Dürr and W.C. Zimmerli (Hrsg.) *Geist und Natur*, Scherz Verlag, Bern-München Wien, 1990.
Fränkel, H., Studies in Parmenides. In: D. Furley and R.E. Allen (Eds.), *Studies in Presocratic Philosophy*, Vol. II, Routledge and P. Kegan, London, 1970.
Fränkel, H., *Dichtung und Philosophie des frühen Griechentums*. American Philological Association, Lancaster, New York, 1951.
Fritz, K., von, Die Rolle des νoύς. In: H.-G. Gadamer (Hrsg.), *Um die Begriffswelt der Vorsokratiker*, Wissenschaftliche Buchgesellschaft, Darmstadt, 1968.
Fromm, E. *To have or to be*? Harper & Row, New York, 1976.
Genz, H., *Symmetrie, Bauplan der Natur*, Piper, München-Zürich, 1987.
Georgoulis, K. *Historia tis Hellinikis Philosophias (History of the Greek Philosophy)*, Papadimas, Athens, 1975.
Georgoulis, K. I pneumatiki parousia tou Herakleitou (The Spiritual Presence of Haraclitus). In: *Herakleitos*, Kaktos, Athens, 1995.
Gide, A., *Les nourritures terrestres*, Gallimard, Paris, 1969.
Gigon, O., *Der Ursprung der griechischen Philosophie*, Schwabe & Co., Basel, 1968.
Glucker, J., Who invented 'Hylozoism'? In: K. Boudouris (Ed.), *Ionian Philosophy*, Athens, 1989.
Gribbin, J., *In Search of Schrödinger's Cat*, Bantam Books, Toronto, New York, 1988.
Guthrie, W.K.C., *The Greek Philosophers from Thales to Aristotle*, Routledge, London, 1997.
Guthrie, W.K.C., *A History of Greek Philosophy. The earlier Presocratics and the Pythagoreans*, Vol. I, Cambridge University Press, Cambridge, MA, 1992.
Guthrie, W.K.C., *A History of Greek Philosophy. The Presocratic tradition from Parmenides to Democritus*, Vol. II, Cambridge University Press, Cambridge, MA, 1992.
Guthrie, W.K.C., *Orpheus and Greek Religion*, Princeton University Press, Princeton, NJ, 1993.
Hawking, S., *A brief History of Time*, Bantam Press, London, 1988.
Hawking, S., *Is the End in Sight for Theoretical Physics*? Cambridge University Press, Cambridge, MA, 1980.
Heath, T.L., *Aristarchus of Samos: The Ancient Copernicus*, Clarendon Press, Oxford, 1913.

Hegel, G.W.F., *18 Vorlesungen über die Geschichte der Philosophie*, Suhrkamp Verlag, Frankfurt/M., 1971.
Heidegger, M., Aletheia (Heraklit, Fr.16). In: *Vorträge und Aufsätze*, Teil III, Neske, Pfullinger, 1967.
Heidegger, M., *Aufenthalte*, Vittorio Klostermann, Frankfurt/M., 1989.
Heidegger, M., Der Spruch des Anaximander. In: *Holzwege*, V. Klostermann, Frankfurt/M., 1980.
Heidegger, M., *Introduction to Metaphysics*, Yale Univesrity Press, New Haven-London, 2000.
Heidegger, M., Logos (Heraklit, Fr.50). In: *Vorträge und Aufsätze*, Teil III, Neske, Pfullinger, 1967.
Heidegger, M., Moira (Parmenides, Fr.VIII,34-41). In: *Vorträge und Aufsätze*, Teil III, Neske, Pfullinger, 1967.
Heidegger, M., *Platons Lehre von der Wahrheit*, Francke Verlag, Bern, 1954.
Heidegger, M., *Was heisst Denken*? Max Niemeyer Verlag, Tübingen, 1984.
Heidegger, M. and Fink, E., *Heraklit*, Seminar Wintersemester 1966/1967, V. Klostermann, Frankfurt/M., 1970.
Heisenberg, W. 'Die Entwickung der Deutung der Quantentheorie', *Phys. Bl.*, 12, 1956.
Heisenberg, W. 'Über den anschaulichen Inhalt der quantentheoretischen Kinematik und Mechanik', *Z. Physik*, 43, 1927.
Heisenberg, W., *Das Naturbild der heutigen Physik*, Rowohlt, Hamburg, 1956.
Heisenberg, W., *Der Teil und das Ganze*, R. Piper & Co., München, 1969.
Heisenberg, W., *Physik und Philosophie*, Ullstein GmbH, Frankfurt/M., Berlin, 1990.
Heisenberg, W., Positivismus, Metaphysik und Religion. In: H.-P. Dürr (Hrsg.), *Physik und Transzendenz*, Scherz Verlag, Bern-München-Wien, 1986.
Heisenberg, W., *Schritte über Grenzen*, Piper, München-Zürich, 1989.
Heisenberg, W., *Wandlungen in den Grundlagen der Naturwissenschaft*, S. Hirzel Verlag, Stuttgart, 1980.
Hermann, A., *Max Planck*, Rowohlt, Reinbek bei Hamburg, 1970.
Hermann, A., *The Genesis of Quantum Theory*, MIT Press, Cambridge/MA, 1971.
Herodotus *Historiae*, The Loeb Classical Library, London-New York, 1926, 1928.
Hesiod, *Fragments*, Rzach, A. (Hrsg.), Hesiodus Carmina, B.G. Teubner, Leipzig, 1913.
Hesiod, *Theogony*, Rzach, A. (Hrsg.), Hesiodus Carmina, B.G.Teubner, Leipzig, 1913.
Hesiod, *Works and Days*, Rzach, A. (Hrsg.), Hesiodus Carmina, B.G. Teubner, Leipzig, 1913.
Hicks, L.E., *A critique of Design Arguments*, Scribner, New York, 1883.
Hierocles, *The Pythagorean Golden Verses*, Thom, J. (Ed.), E.J. Brill, Leiden, New York Köln, 1995.
Historia tou Hellinikou Ethnous (History of the Greek Nation), Ekdotiki Athinon, Athens, 1971.
Hoffmann, R., *The Same and not the Same*, Columbia University Press, New York, 1995.
Hölscher, U., *Anfängliches Fragen*, Vandenhoeck & Ruprecht, Göttingen, 1968.
Homer, *The Iliad*, The Loeb Classical Library, London, New York, 1930.
Homer, *The Odyssey*, The Loeb Classical Library, London, New York, 1930.
Hörisch, J. 'Die Gewalt des Waldes', *Neue Zürcher Zeitung*, July 14, 1998.
Huson, H., *Pythagoras and Christianity*, 1961.
Iamblichus, *De communi mathematica scientia*, Testa, B.G.(Hrsg.), Teubner, Leipzig, 1891.
Iamblichus, *On the Pythagorean Way of Life*, Dillon, J. and Hershbell, E. (Ed.), Scholars Press, Atlanta, Georgia, 1991.
Ioannidou, E. *Herakleitos, Glossa kai Skepsi (Heraclitus, Language and Thought)*, Diss. Phil., Athens, 1962.
Jaeger, W. *The Theology of the Early Greek Philosophers*, Oxford University Press, London, Oxford, New York, 1952.
James, G.G.M., *Stolen Legacy*, Africa World Press, Inc., Trendon, NJ, Asmara, Eritrea, 1998.
Jaspers, K., *Aus dem Ursprung denkende Metaphysiker*, R. Piper & Co., München, 1957.
Jaspers, K., *Vom Ursprung und Ziel der Geschichte*, Fischer, Frankfurt/M., Hamburg, 1955.
Jeans, J., In unerforschtes Gebiet. In: H.-P. Dürr (Hrsg), *Physik und Transzendenz*, Scherz Verlag, Bern-München-Wien, 1986.

Jordan, P., Die weltanschauliche Bedeutung der modernen Physik. In: H.-P. Dürr (Hrsg.), *Physik und Transzendenz*, Scherz Verlag, Bern-München-Wien, 1986.

Kahn, C., *Anaximander and the Origins of Greek Cosmology*, Hackett Publishing Co., Inc., Indianapolis, 1994.

Kahn, C., Pythagorean Philosophy before Plato. In: A.P.D. Mourelatos (Ed.), *The PreSocratics. Modern Studies in Philosophy*, Doubleday, Garden City, NY, 1974.

Kahn, C., Religion and natural philosophy in Empedocles' doctrine of the soul. In: A.P.D. Mourelatos (Ed.), *The Pre-Socratics. Modern Studies in Philosophy*, Anchor Press, Doubleday, Garden City, NY, 1974.

Kahn, C., The Historical Position of Anaxagoras. In: K. Boudouris (Ed.), *Ionian Philosophy*, Athens, 1989.

Kaku, M. and Trainer, J., *Beyond Einstein*, Bantam Books, Toronto, New York, 1987.

Kalfas, V., *Plato, Timaeus*, Polis, Athens, 1995.

Kearns, E., Kajita, T., and Totsuka, Y., 'Detecting Massive Neutrinos', *Scientifi American*, August 1999.

Kelessidou, A., Herakleitos kai Zinon o Kitieus (Heraclitus and Zeno of Citium). In: *12 Meletimata Prosokratikis Philosophias* (12 Studies of Presocratic Philosophy), Athens, 1992.

Kelessidou, A. *I Filosophia tou Xenophani (The Philosophy of Xenophanes)*, Athens Academy, KEEF, 1996.

Kelessidou, A. Le temps et l'espace chez Xénophane. In: *12 Meletimata Prosokratikis Philosophias* (12 Studies of Presocratic Philosophy), Athens, 1992.

Kelessidou, A. Legein-Eidenai stous Xenopani kai Parmenidi, (Xenophanes and Parmenides: 'Legein' and 'Eidenai'). In: *12 Meletimata Prosokratikis Philosophias* (12 Studies of Presocratic Philosophy), Athens, 1992.

Kerferd, G.B., Anaxagoras and the Concept of Matter before Aristotle. In: A.P.D. Mourelatos (Ed.), *The Pre-Socratics. Modern Studies in Philosophy*, Doubleday, Garden City, NY, 1974.

Kirk, G.S., Raven, J.E., and Schofield M., *The Presocratic Philosophers*, Cambridge University Press, Cambridge, MA, 1983.

Kitto, H., *Greek Tragedy*, Methuen & Co. Ltd., London, 1950.

Koestler, A., *The Sleepwalkers*, Arkana Penguin Books, London, 1989.

Kuhn, T., *The Structure of Scientifi Revolutions*, The University of Chicago Press, Chicago, 1970.

Lack, D., *Evolutionary Theory and Christian Belief. The Unresolved Conflic* , Methuen, London, 1961.

Leakey, R., *The Origin of Humankind*, Basicbooks, New York, 1994.

Lederman, L.M. and Schramm, D.N., *From Quarks to Cosmos*, Scientifi American Library, New York, 1995.

Lefkowitz M.R. and MacLean Rogers, G. (Ed.), *Black Athena Revisited*, University of North Carolina Press, Chapel Hill-London, 1996.

Lefkowitz, M.R., *Not out of Africa*, BasicBooks, 1996.

Liddell, H.G. and Scott, R., *A Greek-English Lexicon*, Clarendon Press, Oxford, 1996.

Litman, G.W. 'Sharks and the Origin of Vertebrate Immunity', *Scientifi American*, November 1996.

Lloyd, G.E.R., *Methods and Problems in Greek Science*, Cambridge University Press, Cambridge, MA, 1991.

Löbl, R., *Demokrits Atomphysik*, Wissen. Buchgesellschaft, Darmstadt, 1987.

Lossky, V., *The Mystical Theology of the Eastern Church*, St Vladimir's Seminary Press, Crestwood, NY, 1998.

Lucretius, C., *De rerum natura*, Brown M. (Ed.), Bristol Classical Press, Bristol, 1984.

Macarius of Egypt, St., Spiritual Homilies XXV. In: J.-P. Migne (Ed.), *Patrologiae*, PG34, 673D, 1903.

Malevitsis, C., *I Philosophia tou Heidegger (Heidegger's Philosophy)*, Dodoni, Athens, 1974.

Malevitsis, C., *M. Heidegger, Introduction to Metaphysics* (Greek version), Dodoni, Athens, 1990.

Malevitsis, C., O Marasmos tis Psychis (The Withering of the Soul). In: *Newspaper 'Kathimerini'*, p. 3, Athens, July 21, 1995.
Malevitsis, C., *Peri tou Tragikou (On the Tragic)*, Astrolavos/Euthyni, Athens, 1986.
Malevitsis, C., *Philosophia kai Thriskeia, (Philosophy and Religion)*, Idryma Goulandri Horn, Athens, 1985.
Malevitsis, C. *Ta mila ton Esperidon (The Apples of Western Europe)*, Imago, Athens, 1984.
Mansfeld, J., *Die Vorsokratiker*, Reclam, Stuttgart, 1987.
Margenau, H., *The Miracle of Existence*, Ox Bow Press, 1984.
Markos, A., *I idea tis proodou ston Xenophani (Xenophanes's Concept of Progress)*, Diss. Phil., Patras, 1983.
Markos, A., Paratiriseis pano ston stiho B1,24 tou Xenophani (Remarks on Xenopahnes's verse B1,24). In: *Elliniki Philosophiki Epitheorisi* (Greek Philosophical Review), 2, 1985.
Martin A., Primavesi O., *L' Empedocle de Strasbourg (P. Strasb. gr. Inv. 1665–1666)*. Bibliothèque Nationale et Universitaire de Strasbourg-Walter de Gruyter, Berlin, New York, 1999.
Marx K., *Differenz der demokritischen und epikureischen Naturphilosophie*, Diss. phil., Jena, 1841.
McLaughin, W.I. and Miller, S.L., 'An Epistemological Use of Nonstandard Analysis to Answer Zenos Objections against Motion', *Synthese*, 92, 1992.
Merchant, C., Entwurf einer ökologischen Ethik. In: H.-P. Dürr and W.C. Zimmerli (Hrsg.), *Geist und Natur*, Scherz Verlag, Bern-München-Wien, 1990.
Michaelides, K., *I Prokratiki Philosophi (The Presocratics)*, Imago, Athens, 1984.
Michaelides, K., *Mensch und Kosmos in ihrer Zusammengehörigkeit bei den frühen griechischen Denkern*, Dissertation LMU, München, The Cyprus Association of Greek Philologists: 'Stasinos', Nicosia, Cyprus, 1979.
Michaelides, K., To Ypervatiko tou Logou kata ton Herakleito (The Transcendence of Logos according to Heraclitus), In: K. Boudouris (Ed.), *Ionian Philosophy*, Athens. 1990.
Mikrogiannakis, E., *Pathologia ton politeumaton stin archaiotita (The Pathology of Political Systems in Antiquity)*, Kardamitsas, Athens, 1992.
Moore, W.J., *Schrödinger Life and Thought*, Cambridge University Press, Cambridge, MA, 1989.
Mourelatos, A., "X is realy Y": Ionian Origins of a Thought Pattern. In: Boudouris, K. (Ed.), *Ionian Philosophy*, Athens, 1989.
Nagel, E. and Newman, J.R., *Gödel's Proof*, Routledge and Kegan Paul, London, New York, 1958.
Nelson, E., 'Internal Set Theory: A New Approach to Nonstandard Analysis', *Bulletin of the Americal Mathematical Society*, 83, 1997.
Nestle, W., *Die Vorsokratiker*, VMA-Verlag, Wiesbaden, 1956.
Nestle, W., *Vom Mythos zum Logos*, A.Kröner Verlag, Stuttgart, 1975.
Neugebauer, O., *The exact Sciences in Antiquity*. Dover Publications, Inc., New York, 1969.
New Testament, *The Complete Bible*, The University of Chicago Press, Chicago, IL, 1944.
Newton, I., Mathematical Principles of Natural Philosophy, Query 31. In: J. Powers (Ed.), *Philosophy and New Physics*, Routledge, London, New York, 1991.
Nietzsche, F., *Die Geburt der Tragödie*, A. Kröner Verlag, Stuttgart, 1955.
Nietzsche, F., Die Philosophie im tragischen Zeitalter der Griechen. In: *Die Geburt der Tragödie*, A. Kröner Verlag, Stuttgart, 1955.
Nietzsche, F., Philologica, Bd.3. In: O. Crusius and W. Nestle (Hrsg), *Nietzsches Werke*, A. Kröner Verlag, Leipzig, 1913.
Nietzsche, F., Wissenschaft und Weisheit im Kampfe. In: *Die Geburt der Tragödie*, A. Kröner Verlag, Stuttgart, 1955.
Oeconomou, E., *I Physiki simera (Physics Today)*, University Press of Crete, Herakleion, 1986.
Old Testament, *The Complete Bible*, The University of Chicago Press, Chicago, IL, 1944.
Otto, W., *Die Götter Griechenlands*, F.Cohen, Bonn, 1929.
Pagels, H.R., *The Cosmic Code*, Bantam Books, Toronto, New York, 1984.
Pais, A., *'Subtle is the Lord', The Science and Life of Albert Einstein*, Oxford University Press, Oxford, 1982.

Papanikolaou, A.D., Aer: Anaximenes und Corpus Hippocraticum. In: K. Voudouris (Ed.), *Ionian Philosophy*, Athens, 1989.
Pauli, W. Naturwissenschaftliche und erkenntnistheoretische Aspekte der Ideen vom Unbewussten. In: *Aufsätze und Vorträge über Physik und Erkenntnistheorie*, F. Vieweg & Sohn, Braunschweig, 1961.
Pauli, W., Die Wissenschaft und das abendländische Denken. In: *Aufsätze und Vorträge über Physik und Erkenntnistheorie*, F. Vieweg & Sohn, Braunschweig, 1961.
Pavlos, G.P., *Nostos Asymmetros Prosopou (The Individual's Asymmetric Nostalgia)*, Domos, Athens, 1994.
Penrose, R., *Shadows of the Mind*, Vintage, London. 1995.
Philoponus, I., Vitelli, H. (Hrsg.), *In Aristotelis Libros 'De Generatione et Corruptione' Comentaria*, Berlin, 1897.
Pietschmann, H., *Phänomenologie der Naturwissenschaft*, Springer, Berlin, Heidelberg, 1996.
Pindar, Bowra, C.M., (Ed.), *Pindari Carmina*, Oxford, 1965.
Pindar, *Nemean Odes*, The Loeb Classical Library, London, Cambridge, MA, 1937.
Pindar, *Olympian Odes*, The Loeb Classical Library, London, Cambridge, MA, 1937.
Pindar, *Pythean Odes*, The Loeb Classical Library, London, Cambridge, MA, 1937.
Planck, M., *Religion und Naturwissenschaft*, J. Ambrosius Barth Verlag, Leipzig, 1952.
Plato, *Cratylus*, The Loeb Classical Library, London, New York, 1926.
Plato, *Epinomis*, The Loeb Classical Library, London, New York, 1927.
Plato, *Gorgias*, The Loeb Classical Library, London, New York, 1925.
Plato, *Laws*, The Loeb Classical Library, London, Cambridge, MA, 1952.
Plato, *Menexenus*, The Loeb Classical Library, London, New York, 1929.
Plato, *Meno*, The Loeb Classical Library, London, New York, 1924.
Plato, *Phaedo*, The Loeb Classical Library, London, Cambridge, MA, 1938.
Plato, *Protagoras*, The Loeb Classical Library, London, New York, 1924.
Plato, *Sophist*, The Loeb Classical Library, London, Cambridge, MA, 1952.
Plato, *Theaetetus*, The Loeb Classical Library, London, Cambridge, MA, 1952.
Plato, *Timaeus*, The Loeb Classical Library, London, New York, 1929.
Plato, *The Republic*, The Loeb Classical Library, London, Cambridge, MA, 1935.
Plotinus, *Enneads*, The Loeb Classical Library VII, London, Cambridge, MA, 1938.
Poe, R., *Black Spark, White Fire*, Prima Publishing, Rocklin, CA, 1997.
Popper, K., Back to the Presocratics. In: A.F. Petersen (Ed.), *The World of Parmenides*, Routledge, London, New York, 1998.
Popper, K., *Conjectures and Refutations*, Routledge, London, 1989.
Popper, K., *Die Anfänge des Rationalismus*, Lesebuch UTB, Tübingen-Mohr, 1995.
Popper, K., *The Myth of the Framework*, Routledge, London, New York, 1994.
Popper, K., *The World of Parmenides*, Petersen, A.F. (Ed.), Routledge, London, New York, 1998.
Powers, J., *Philosophy and the New Physics*, Routledge, London, New York, 1991.
Prigogine, I. and Stengers, I., *Order out of Chaos*, Collins Publ. Group, London, 1984.
Prigogine, I., *La fi des certitudes*, Editions Odile Jacob, 1996.
Reinhardt, K., 'Hekataios von Abdera und Demokrit', *Hermes*, 47, 1912.
Reinhardt, K., *Parmenides*, V. Klostermann, Frankfurt/M., 1985.
Richtmeyer, F.K. and Kennard, E.H., *Introduction to Modern Physics*, McGraw-Hill, London, 1950.
Roberts, R.M., *Serendipity. Accidental Discoveries in Science*, John Willey & Sons, Inc., New York, 1989.
Russell, B., *Wisdom of the West*, Doubleday, Garden City, New York, 1959.
Russell, B., *History of Western Philosophy*, W.W. Norton & Co., Inc., 1929.
Russell, B., *Our Knowledge of the External World*, G. Allen and Unwin Ltd, London, 1961.
Russell, B., *Why I am not a Christian*, G. Allen & Unwin, New York, 1957.
Sakellariou, G., *Pythagoras, o Didaskalos ton aionon (Pythagoras, Teacher of the Ages)*, Athens.
Sambursky, S., *Das physikalische Weltbild der Antike*, Artemis Verlag, Zürich, 1965.

Sappho, *Lyra Graeca*, The Loeb Classical Library I, London-New York, 1928.
Schelling, F., *Abh.* In: K.F.A. Schelling (Hrsg.), Schellings Werke, 1797.
Schelling, F., *Ideen zu einer Philosophie der Natur.* In: K.F.A. Schelling (Hrsg.), Schellings Werke, 1797.
Schilpp, A., (Ed.), *Albert Einstein, Philosopher-Scientist*, The Open Court, La Salle, IL, 1949.
Schoinas, F., O Pythagoeios tropos tou biou kai i Christianniki askesi (The Pythagorean Way of Life and Christian Practice). In: K. Boudouris (Ed.), *Pythagoreia Philosophia (Pythagorean Philosophy)*, Athens, 1992.
Schrödinger, E., Das arithmetische Paradoxon - Die Einheit des Bewusstseins. In: H. P. Dürr (Hrsg.), *Physik und Transzendenz*, Scherz Verlag, Bern-München-Wien, 1986.
Schrödinger, E., *Die Natur und die Griechen*, Diogenes Verlag AG, Zürich, 1989.
Schrödinger, E., *Geist und Materie*, F. Vieweg & Sohn, Braunschweig, 1965.
Seneca, *Naturales Questiones*, The Loeb Classical Library, London-Cambridge, MA, 1971.
Sexl, R.U., *Was die Welt zusammenhält*, DVA, Stuttgart, 1984.
Simonides, *Lyra Graeca*, The Loeb Classical Library II, London, New York, 1924.
Simplicius, Diels, H. (Hrsg.), *Aristotelis De Caelo Commentaria*, Berlin, 1882.
Simplicius, Diels, H. (Hrsg.), *Aristotelis Physicorum Commentaria*, Berlin, 1882.
Snell, B., *Die Entdeckung des Geistes*, Vandelhoek & Ruprecht, Göttingen, 1975.
Solon, Diehl, E. (Hrsg.), *Anthologia Lyrica Graeca*, Fasc. 1, B.G. Terbneri, Leibzig, 1940.
Solon, *Elegy und Iambus*, The Loeb Classical Library, London, New York, 1931.
Sophocles, *Antigone*, The Loeb Classical Library, London, New York, 1915.
Sophocles, *Ayax*, MacMillan & Co Ltd, London, New York, 1963.
Sophocles, Fragments, *Sophokles, Tragödien und Fragmente*, Willige, W. (Hrsg.), E. Heimeran, München, 1966.
Sophocles, *Oedipus the King*, The Loeb Classical Library, London, New York, 1915.
Sophrony, Archimandrite, *St Silouan the Athonite*, St. Vladimir's Seminary Press, Crestwood, New York, 1999.
Spencer, H., *First Principles of a New System of Philosophy*, D. Appleton & Co, New York, 1882.
Stegmüller, W., *Hauptströmungen der Gegenwarts-Philosophie*, 2 Bände, A. Kröner, Stuttgart, 1978.
Stillman, D., Cp., *G. Galileo. Dialogue Concerning the Two Chief World Systems*, 1953.
Stobaeus, Wachsmuth, C., (Hrsg.), *Eclogae Physicae et Ethicae*, Weidmann, Berlin, 1884.
Stückelberger, A., 'Empirische Ansätze in der antiken Atomphysik', *Archiv für Kulturgeschichte*, 56, 1974.
Stückelberger, A., *Antike Atomphysik*, Heimeran, München, 1979.
Sweeney, L.S.J., *Infinit in the Presocratics: A Bibliographical and Philosophical Study*, Martinus Nijhoff, The Hague, 1972.
Thanasas, P., *O protos 'defteros plous' (The frst 'second sailing')*, University Press of Crete, Herakleion, 1998.
Theophrastus, *Theophrasti* Eresii, Vol. 3, B.G. Teubner, Leipzig, 1862.
Thucydides, *History of the Peloponnesian War*, The Loeb Classical Library, London, Cambridge, MA, 1951.
Tsatsos, *I Zoi se Apostasi (Life at a Distance)*, Oi Ekdoseis ton filon Athens, 1996.
Tzermias, P., *Für eine Hellenistik mit Zukunft*, Universitätsverlag Freiburg, Freibunrg, CH, 1998.
Veikos, T., *I Prosokratiki (The Presocratics)*, I. Zacharopoulos, Athens, 1988.
Veikos, T., *Philosopohia kai Politiki (Philsophy and Politics)*, Iridanos, Athens.
Vitsaxis, V., *O Stochasmos kai i Pisti (Thought and Faith)*, Estia, Athens, 1991.
Vlastos, G., *Simeiosi gia to Velos tou Zinona (Note on Zino's arrow)*, Deukalion, 11, Athens, 1974.
Vlastos, G., Anaxagoras and the Atomists. The physical Theory of Anazagoras. In: A.P.D. Mourelatos (Ed.), *The Pre-Socratics. Modern Studies in Philosophy*, Doubleday, Garden City, NY, 1974.
Vlastos, G., Cornford's Principium Sapientiae. In: D. Furley and R.E. Allen (Eds.), *Studies in Presocratic Philosophy*, Routledge and P. Kegan, Vol. I, London, 1970.

Vrettakos, P. Christos Malevitsis, the Lyrical Philosopher. In: K. Tsiropoulos (Ed.), Testimony on Christos Malevitsis, *Tetradia 'Efthinis'* 36, Athens, 1998.
Waerden, B.L. van der, 'Die Arithmetik der Pythagoreer', *Mathematische Annalen*, 127 153, 1948.
Waerden, B.L. van der, *Erwachende Wissenschaft: Ägyptische, Babylonische und Griechische Mathematik*, 1.Band, Birkhäuser Verlag, Basel-Boston-Stuttgart, 1956.
Waerden, B.L. van der, *Erwachende Wissenschaft: Die Anfänge der Astronomie*, 2.Band, Birkhäuser Verlag, Basel-Boston-Stuttgart, 1980.
Weber, R., *Dialogues with Scientists and Sages*, Arkana, London, 1990.
Wehrli, F., *Die Schule des Aristoteles: Texte und Kommentar. Herakleides Pontikos*, Heft VII, Basel, 1969.
Weinberg S., *The First Three Minutes*, Bantam (Ed.), BasicBooks Inc., New York, 1979.
Weizsäcker, C.F. von, Welt, Wissenschaft, Wirklichkeit. In: H.-P. Dürr and W.C. Zimmerli (Hrsg.), *Geist und Natur*, Scherz Verlag, Bern-München-Wien, 1990.
Weizsäcker, C.F. von, *Zum Weltbild der Physik*, Hirzel Verlag, Stuttgart, 1963.
Weyl, H., *Philosophy of Mathematics and Natural Science*, Princeton University Press, Princeton, 1949.
Whitehead, A.N., *Science and the Modern World*, Cambridge University Press, Cambridge, MA, 1953.
Wightman, W.P.D., *The Growth of Scientifi Ideas*, Edinburgh, 1950.
Wilson, E.O., *Consilience*, Alfred A. Knopf, New York, 1998.
Wittgenstein, L., *Tractatus Logico Philosophicus*, Suhrkamp Verlag, Frankfurt a/Main, 1960.
Xanthopoulos, *Peri asteron kai sympanton (On Stars and Universes)*, University Press of Crete, Herakleion, 1985.
Yukawa, H., Intuition and Abstraction in Scientifi Thinking. In: *Frontiers of Modern Scientifi Philosophy and Humanism*, The Athens Meeting, 1964, Elsevier, Amsterdam-London, New York, 1966.
Zeller, E. and Nestle, W., *Grundriss der Geschichte der Griechischen Philosophie*, Scientia Verlag Aalen, Darmstadt, 1971.

Index of Names

A
Aeschylus, 85
Ahrweiler, H., 258n.2
Akousilaus, 11
Albrecht, E., 252
Alcaeus, 14
Alcman, 14
Alcmeon of Croton, 53n.1
Alexander Aphrodisiensis, 210
Ameinias, 137
Anaxagoras, 20, 26, 51, 151, 159, 170, 197–208, 212, 220
Anaximander, 16, 23, 24, 35–43, 45, 48, 63, 74, 164, 220
Anaximenes, 45–51, 63, 169, 220
Antisthenes, 159
Aquinas, T., 81, 204
Archelaus of Athens, 198
Archilochus, 14
Archimedes, 76n.68, 77, 156
Architas of Tarentum, 53n.1
Aristarchus of Samos, 76n.68
Aristippus, 159
Aristotle, 2, 3, 6n.9, 7, 18, 20, 29, 31, 39, 76, 77, 89, 103, 105, 117, 123, 125, 132, 153, 156, 159, 161, 164, 172, 173, 204, 210, 216, 218, 236, 237, 250
Arnold, M., 168, 182
Ashmole, A., 15
Aspect, A., 245
Augustine, St., 81
Avenarius, R., 240
Avogadro, A., 239
Axelos, K., 108n.15, 119, 122

B
Bacon, F., 20, 21, 82, 96, 183, 204, 237
Badash, L., 97n.31
Bailey, C., 218, 224, 225
Baltimore, D., 31n.3
Barrow, J.D., 190, 195
Basso, S., 238
Baumann, K., 244n.85
Beckerel, H.C., 241
Becker, O., 64n.35
Beethoven, L.van, 69, 135
Bensaude-Vincent, B., 238, 240n.74
Bentley, R., 145n.13
Bergson, H., 205
Berkeley, G., 156, 183, 237
Berthelot, M., 239
Berzelius, J., 239
Blegen, C., 255
Bloch, E., 82n.86
Bohm, D., 81n.85, 83, 101, 129, 234, 248, 251n.4, 253, 254
Böhme, J., 125
Bohr, N., 23, 243, 244
Boltzmann, L.E., 128, 163, 240, 242
Bondi, H., 189
Born, M., 243, 244
Boscovitch, R., 236
Bowra, 38n.6
Boyle, R., 187, 238
Bradley, F., 155
Brandis, C.A., 155n.32
Breton, A., 126
Brock, E., 182
Broglie, L. De, 242, 243
Brown, M.E., 74
Brun, J., 70, 82, 176, 249
Bruno, G., 125, 237
Buddha, 3, 126
Bultmann, R., 5
Burkert, W., 54
Burnet, J., 211, 213

C

Callinus, 14
Cannizaro, S., 239
Capra, F., 78n.75, 97n.28, 126n.45, 127n.46
Carnap, R., 97, 98n.35
Chadwick, J., 245
Char, R., 126
Cherniss, K., 126
Chuangtse, 13
Compte-Sponville, A., 62
Compton, A.H., 128, 242
Confucius, 3
Copernicus, 43, 76n.68
Cornford, F.M., 57
Couliano, I.P., 54n.9
Coveney, P., 244n.84, 246n.91
Cratylus, 108, 125, 159
Critias, 15
Curie, M., 241
Cusanus, N., 125

D

Dalton, J., 186, 187, 238, 239
Damon, 60
Dampier, W.C., 243n.82
Darius, 101
Darnell, J., 31n.3
Darwin, F., 206n.20
Davies, P., 163, 184, 194, 246
Da Vinci, L., 237
Davis, P., 94n.16, 194, 195n.55
Davisson, C.J., 128, 243
Dedekind, R., 79
Democritus, 20, 22, 23, 26, 51, 74, 91, 127, 133, 162, 167, 187, 188, 209–248, 249, 250
Descartes, R., 77, 81, 97, 133, 140, 162, 164, 183, 236
DeWitt, B.S., 42, 163
Diels, H., 1, 38n.5, 227n.40
Dillingham, S.G., 157
Diogenes of Apollonia, 49
Diogenes Laertius, 237
Dirac, P.A.M., 75, 128, 145, 243
Dodds, E.R., 54n.4, 176n.17, 179, 180, 181
Driesch, H., 205
Dumas, J.B., 239
Dürr, H.-P., 130

E

Eccles, H.J., 202n.9, 207n.24, 247n.95
Eckhart, M., 125
Ecphantus, 53
Eddington, A., 253
Einstein, A., 21, 23, 63, 70, 79, 83, 128, 131, 145, 162, 163, 228, 241, 245, 246, 253
Eliade, M., 54n.9
Eliot, T.S., 126
Elytis, O., 5
Empedocles, 17, 20, 26, 51, 54, 159, 167–196, 197, 201, 212, 220, 227, 235
Engels, F., 210
Epicharmus, 85, 89
Epicurus, 55, 217n.24, 234
Epimenides, 11, 17
Ermotimus, 17
Espagnat, B.D', 77, 79n.81, 83, 246
Euclid, 65, 67n.40, 76n.68
Euclid of Megara, 159
Eudemus, 210
Eudoxus, 156, 227n.39
Everett, H.III, 42
Ezekiel, 3

F

Ferris, T., 233n.48
Feyerabend, P., 98
Feynman, R.P., 21, 80, 83n.87, 233, 244, 248
Fichte, J.G., 164
Fink, E., 109n.16, 114n.25, 115n.26
Fränkel, H., 90n.5, 91, 104, 105n.7
Fresnel, A., 127
Friedmann, A., 189
Fromm, E., 126

G

Galilei, G., 20, 27, 43, 50, 77, 187, 237
Gamow, G., 189
Gassendi, P., 162, 236
Genz, H., 78, 79n.78
Germer, L.H., 128, 243
Gide, A., 229
Gigon, O., 36, 126, 136
Glashow, S., 43
Gödel, K., 98, 163
Goethe, J.W. von, 4n, 3, 96n.22, 125
Gold, T., 189
Gorgias, 125, 159, 182, 225, 249
Graham, N., 42n.14
Gribbin, J., 246
Guérin, R., 182
Guth, A., 192
Guthrie, W.K.C., 9, 15, 60, 73, 93, 169, 171, 173, 178, 196, 200, 201, 211, 219, 221, 236

H

Hartle, J., 190
Hawking, S.W., 97, 183, 184, 190, 191, 193

Heath, T., 17n.43, 64n.35
Hegel, G.W.F., 117, 125, 126, 135, 155, 164, 205
Heidegger, M., 1, 35, 109n.16, 111, 112, 115, 121, 126, 135, 136, 137, 146, 147, 148, 149, 157, 160, 164, 251, 254
Heisenberg, W., 21, 22, 41, 48, 49, 69, 77, 78, 80, 83, 129, 132, 234, 235, 243, 244, 246, 248, 252
Heraclitus, 15, 20, 23, 25, 27, 54, 95, 101–134, 136, 137, 141, 157–159, 164, 169, 170, 178, 209, 243, 251
Herbart, J., 155
Herbert, N., 239n.73
Hermann, A., 1, 97n.32
Herodotus, 5, 17, 19, 55
Hesiod, 10–12, 13, 17, 19, 36, 37, 89, 136, 137
Hicks, L.E., 203n.11
Hiero, 59n.18, 60n.22, 85, 256
Hierocles, 59n.18, 60n.22
Highfield R., 244n.84, 246n.91
Hippasus of Metapontum, 53n.1, 66
Hipponax, 14
Hoffmann, R., 131n.57
Hölderlin, J.C.F., 126, 168, 182
Hölscher, U., 18, 124
Holton, G., 45, 253
Homer, 5, 6, 10, 11, 14, 31, 56, 88, 105, 137, 259
Hoyle, F., 189
Hubble, E., 189
Hume, D., 183, 206
Huygens, C., 77, 127, 162

I

Iamblichus, 70n.49, 80n.82
Iliopoulos, I., 43
Ioannidou, E., 104

J

Jäger, W., 42, 126
Jaspers, K., 3n, 1, 35n.3, 42n.16
Jeans, J.H., 23, 197, 242
Jeremiah, 3
Jordan, P., 234, 243

K

Kahn, C., 42, 181
Kajita, T., 94n.53
Kant, I., 81, 155, 164, 169, 192, 205, 236
Kearns, E., 194n.53
Kelessidou, A., 90
Kelvin, Lord, 97, 240
Kennard, E.H., 242, 243n.80
Kepler, J., 43, 70, 75, 76n.68, 77, 78
Kierkegaard, S., 125
Kirk, G.S., 115n.28
Kitto, H.D.F., 208
Koestler, A., 51, 54n.5
Kranz, W., 227n.40
Kuhn, T., 98, 206

L

Lack, D., 205n.14
Lakatos, S., 98
Laotse, 3
Laplace, P.S., 192
Lassalle, F., 125
Lavoisier, A., 186, 187, 238
Leakey, R., 176n.16
Lederman, L.M., 184, 185n.32, 192n.48
Lefkowitz, M., 18n.47
Leibniz, G.W., 40, 70, 81, 156, 164, 183, 205, 236
Lémery, N., 238
Lenin, V.I., 125, 241
Leucippus, 20, 26, 40, 127, 133, 155, 162, 186, 197, 209, 210–211, 213, 220, 224, 233, 238, 248
Liddell, H.G., 47n.3
Litman, G.W., 40n.9
Lloyd, G.E.R., 16, 18
Löbl, R., 210n.2, 211, 212, 214n.13
Locke, J., 183, 237
Lodish, H., 31n.3
Lomonosoff, M.V., 187
Lucretius, 217n.23, 236, 237, 240n.76

M

Machavira, 3
Mach, E., 240, 241n.77
McLaughlin, W.I., 157n.35
MacLean, R.G., 18n.47
Malevitsis, C., 2, 16, 19, 85, 94, 95, 164
Malreaux, A., 1
Mansfeld, J., 203n.10
Margenau, H., 247n.94
Markos, A., 90n.5
Martin, A., 181n.27
Maxwell, J.C., 127, 128, 240
Meer, S. van der, 185
Melissus of Samos, 55, 137, 150–151, 159, 210, 211
Mendelejew, D., 239
Meyer, L., 239
Mikrogiannakis, E., 60n.24
Miller, S.L., 157n.35

Mimnermus, 14
Minkowski, H., 163
Monod, J., 205
Mourelatos, A., 87n.4
Musaius, 11

N
Nagel, E., 98n.34
Nearchus, 151
Nestle, W., 4, 9, 10, 58, 121, 206n.18, 209, 210, 227, 229
Neugebauer, O., 64
Newman, J.R., 98n.34
Newton, I., 48, 77, 78, 82, 127, 128, 145, 156, 162, 167, 187, 204, 205, 236, 237
Nietzsche, F., 19, 29, 38n.5, 68, 95, 102, 125, 145, 164, 167, 182
Novalis, G.F.P., 126

O
Oeconomou, E.N., 42n.15, 43n.19
Orpheus, 11, 17
Ostwald, W., 239, 240, 241n.77
Otto, W., 6

P
Packard, D., 98
Pagels, H.R., 243n.81
Pais, A., 245n.87
Papanikolaou, A.D., 49n.8
Park, P., 94n.16
Parmenides, 20, 25, 26, 50, 51, 85, 89, 95, 101, 103, 105, 116, 123, 124, 125, 133, 134, 135–165, 168, 169, 170, 177, 210, 211, 212, 214, 225, 227
Pascal, B., 81
Pauli, W., 23, 48, 53, 75, 127, 129
Pausanias, 168, 182
Penrose, R., 207, 208n.28, 247n.96
Penzias, A., 189
Periandrus, the Corinthian, 6
Pericles, 197, 208
Perrin, J.B., 241
Perse, S.-J., 126
Petron, 53n.1
Pherekydes, 11, 38n.6
Philolaus, 20, 53n.1
Philoponus, I., 210, 212n.8, 213n.11
Phintias, 60
Pietschmann, H., 131, 132n.64, 157
Pindar, 38n.6, 85
Planck, M., 23, 80, 97, 128, 131, 240, 242
Plato, 2, 3, 16, 37, 61, 65, 76, 78, 117, 125, 136, 137, 151, 159, 164, 182, 204, 210, 235, 236, 249, 250
Poe, E.A., 135
Poincaré, H., 240
Polycrates of Samos, 55
Popper, K.R., 23, 27, 42, 49, 83, 93, 95, 98, 99, 109, 133, 134, 145, 163, 214, 234, 237, 248, 250, 251
Prigogine, I., 131, 193, 253, 254
Primavesi, O., 181n.27
Proclus, 17
Protagoras, 125, 159, 225, 249
Proudhon, P.J., 125
Proust, J., 187, 238
Pythagoras, 16, 17, 20, 24, 27, 50, 51, 53–84, 85, 91, 94, 144, 151, 167, 169, 181, 243, 246, 251

R
Raleigh, J., 23
Rayleigh, Lord, 242
Redondi, R., 237n.64
Reichenbach, H., 97
Reinhardt, K., 54, 86, 117, 126, 133, 248
Renan, E., 167
Richter, J.J., 187, 238
Richtmeyer, F.K., 243n.80
Ritter, 38n.5
Rohde, E., 38n.5, 54n.3
Röntgen, W.C., 241
Roussos, E., 10, 12
Rreinhardt, K., 54, 86, 114n.24, 117, 122n.35, 126, 133, 248
Rubbia, C., 185
Russell, B., 13, 21, 39, 54, 63, 68, 76, 155, 205, 233, 250
Rutherford, E., 23, 241, 243

S
Salam, A., 43
Sambursky, S., 22, 23, 24n.15, 49, 67, 76, 185, 216, 217, 234
Sappho, 12, 14
Sarafino E., 253
Schelling, F., 8, 164, 205
Schiller, F. von, 247
Schleiermacher, F.D.E., 125
Schlick, M., 97
Schopenhauer, A., 82n.86, 182
Schramm, D.N., 184, 185n.32
Schrödinger, E., 2, 46, 83, 131, 145, 162, 163, 224, 234, 235n.56, 243, 244, 245, 253

Scott, R., 47n.3
Scotus, D., 204
Sennert, D., 237
Sexl, R.U., 240n.75, 241n.77, 244n.83, 245
Sextus, 210
Simmel, G., 126
Simonides, 14, 85
Simplicius, 46n.2, 108n.14, 173n.12, 203n.12, 210, 218
Smith, C.L., 246
Snell, B., 9, 14, 92, 254
Socrates, 3, 8, 19, 62, 63, 104, 136, 137, 159, 198, 248, 249, 250
Solon, 14, 38n.6
Sommerfeld, A., 243
Sophocles, 39n.6, 148n.23
Spencer, H., 41n.11
Spinosa, B., 81, 125
Stegmüller, W., 97n.29, 98n.35
Stenger, I., 131, 238, 239n.72
Stesichorus, 14
Stückelberger, A., 218n.28, 237, 238n.67

T
Tennant, E., 205
Terpandrus, 14
Thales, 16, 17, 24, 29–33, 36, 42, 43, 45, 49, 53, 90, 169, 253
Theaetetus, 65
Theodorakopoulos, J., 95
Theophrastus, 2, 210, 223
Thompson, G., 128, 241
Thomson, J.J., 241
Thucydides, 197, 208n.30
Totsuka, Y., 194n.53
Tripler, F.J., 206n.19, 207n.22
Tsatsos, K., 4
Tycho de Brache, 76n.68
Tyrtaeus, 14

V
Valery, P., 1
Vitsaxis, V., 228
Vlastos, G., 112n.22, 113n.23, 154

W
Waerden, B.L. van der, 16n.38, 17, 18n.45, 54n.8, 64n.35
Wallace, A., 205
Weber, M., 82n.86, 248n.98
Wehrli, F., 54n.2, 56n.12
Weierstrass, K., 156
Weinberg, S., 12, 35, 43, 190, 191n.41, 193
Weizsäcker, C.F. von, 192, 234, 252
Weyl, H., 163
Wheeler, J., 207
Whitehead, A.N., 76n.67, 77, 79, 129
Wien, W., 23
Wightman, W.P.D., 32
Wigner, E., 79
William of Occam, 204
Wilson, E.O., 253
Wilson, R., 189
Wittgenstein, L., 75, 102, 103
Wöhler, F., 239n.72
Wolf, A., 125

X
Xanthopoulos, V., 191
Xeniades, 159
Xenophanes, 17, 20, 23, 24, 25, 50, 85–99, 120, 137

Y
Young, T., 127
Yukawa, H., 12, 13

Z
Zeller, E., 209
Zeno of Elea, 137, 150, 151
Zoroaster, 3

Index of Subjects

Note: The letter 'n' introduces references to note numbers.

A

Abstraction, Greek, 13, 15
 in scientifi thinking, 12ff
Achsenzeit (pivotal period), 3
Acusmata (oral instructions), Pythagorean, 57ff
Acusmatici, Pythagorean, 57ff
Aer (the air)
 Anaximenes' primary element, 24, 45ff, 49
 Empedocles' concept of, 170ff, 175f, 183ff
Aether
 Plato's concept of, 182
 Pythagorean concept of, 74
Affinit , Pythagorean concept of, 56
Afro-centric theory, 18n.47
Agnosticism, Sophists', 249
Air, *see Aer*
Alchemy, Newton's, 167
A-lethia (un-concealment, the truth), 126
 Heraclitus' concept of, 111
 Parmenides' concept of, 147f
A-logos (absurd), Pythagorean concept of, 66
Alphabet
 Greek, 7
 Latin, 7
 Phoenician, 7
 phonetic, 7
 Semitic, 7
 Slavic, 7
Analogy
 Anaximenes' concept of, 48
 Milesians' concept of, 50
Anamnesis (reminiscence), Pythagorean concept of, 63
Anange (necessity)
 Anaxagoras' concept of, 203
 Empedocles' concept of, 171, 174f
Anthropic Cosmological Theory, 207
Anthropocentric theory, Xenophanes', 89f
Anthropological view
 Aristotle's, vii
 Plato's, vii
 Socrates', vii
Anti-chton (anti-earth), Pythagorean concept of, 74ff
Antimatter, 75, 80
Apanta, meanings of, 103
Apeiron (the boundless)
 Anaximander's primary element, 24, 35ff, 39
 Melissus' concept of, 150
Approximation(s), Pythagorean theory of, 67ff
Archaic period, 12
Arche, 24n.16, 50
 Anaxagoras', 198, 203
 Anaximander's, 36ff, 169
 Anaximenes, 45, 169
 Heraclitus' concept of, 106, 133
 Milesians' concept of, 50, 159
 Parmenides' concept of, 136
 proti (firs principle and element), 24ff
 Thales', 30ff, 32, 169
 see also Primary substance, substratum, source, Presocratics' concept of
Ariston (the best)
 Heraclitus' concept of, 116
 Xenophanes' concept of, 93
Arithmetic
 Egyptian, 17
 Greek, 17
 Pythagorean, 62, 64ff
Arrow of time, 163f

Art, 83
classical, 15, 198
Geometric, 259
Greek, 12ff
lyrical, 10
Asia Minor, 7, 257ff, 258n.2
Astrology, Babylonian, 17ff
Astronomy, 22
Anaximander's concept of, 40
Babylonian, 16ff
Egyptian, 16ff
Greek, 16ff
Pythagorean, 62, 64
Atomic theory
Aristotle's view of, 236
atomists', 26, 155, 162, 188, 211ff, 218, 233ff
Avenarius' view of, 240
Avogadro's concept of, 239
Bacon's view of, 237
Basso's view of, 238
Berthelot's view of, 239
Berzelius' concept of, 239
Bohm's concept of, 234
Bohr's, 23, 243ff
Boltzmann's view of, 240
Boyle's concept of, 238
Cannizaro's concept of, 239
Dalton's, 187ff, 238ff
Democritus', 238ff, 251
Dumas' view of, 239
Einstein's experimental proof of, 241
Epicurus' concept of, 234ff
Feynman's concept of, 233
Galilei's view of, 237
Gassendi's view of, 236
Gay Lussac's concept of, 239
Greek concept of, 18
Heisenberg's concept of, 234
history of, 234ff
Jordan's concept of, 234
Kant's view of, 236
Leibniz's view of, 236
Lémery's view of, 238
Lord Kelvin's view of, 240
Lucretius' concept of, 236
Mach's view of, 240ff, 241n.77
Magnier's concept of, 237
Maxwell's view of, 240
Mendelejew's concept of, 239
Meyer's concept of, 239
Newton's view of, 237
Ostwald's view of, 239, 241n.77
Perrin's experimental proof of, 241
Planck's view of, 240
Plato's concept of, 235ff
Poincarè's view of, 240
Popper's concept of, 234
roots of, 211
Rutherford's, 23
Sennert's concept of, 237ff
Thomson's, 241
Weizsäcker's concept of, 234
Atomists, vii, 26, 151
A-tomos (indivisible), 26, 213
Atom(s), atomists' concept of, 26, 127, 159ff, 214ff, 249
Attraction
atomists' concept of, 211
Empedocles' concept of, 172ff, 177, 212
Auto-kinesis, Ionian natural philosophers' concept of, 169, 171

B

Balance
Anaximander's concept of dynamic, 41, 43
Empedocles' concept of, 183
Heraclitus' concept of dynamic, 105, 127, 251
Barbarian, 7
Becoming
Greek concept of, 18, 101, 126
Heraclitus' concept of, vii, 107ff, 120, 127, 133, 169
modern physics' concept of, 129
Presocratics' concept of, viii, 25
Sophists' concept of, 125
Being, 2, 5, 126
Anaxagoras' concept of, 198
Aristotle's concept of, vi, 161
atomists' concept of, 26, 212ff, 214
Empedocles' concept of, 170, 171f
existential concept of, 145
Greek concept of, 5, 18, 101
Heraclitus' concept of, 111, 133
Melissus' concept of, 150ff
modern physics' concept of, 129
Parmenides' attributes of, 141ff, 164ff
Parmenides' concept of, vii, 26f, 135ff, 138ff, 145ff, 148, 160ff, 249
Plato's concept of, 160
predicative concept of, 145ff
Presocratics' concept of, viii, 25
see also Einai (being)
Big Bang, 191ff, 191n.45
Big Crunch, 195

Biology, Democritus' concept of, 221ff
Xenophanes', 86ff
Black holes, 183
Boundless, Anaximander's concept of, 35ff

C
Calculus
differential, 156
integral, 156
Catharsis, Empedocles' concept of, 177ff
Causality
Anaximander's concept of, 42
Milesians' concept of, 50
modern physics' concept of, 244
Plato's concept of, 235
Cause(s)
Anaximander's concept of sufficient 40
natural, 8
physical, 20
supernatural, 20
Change, 26
Anaxagoras' concept of, 199
Anaximenes' concept of, 45, 49
Democritus' concept of, 26, 133
Empedocles' concept of, 171f
Heraclitus' concept of, vii, 25, 103ff, 131, 169
Milesians' concept of, 24
modern sciences' concept of, 131
Parmenides' concept of, vii, 25, 133ff, 163
Presocratics' concept of, 15, 25
Zeno's, 133ff
Chaos
Hesiod's concept of, 11ff, 36
modern physics' concept of, 98, 131
Presocratics' concept of, 1
Chemical formula(s)
Democritus' concept of molecular, 239
Democritus' concept of stereochemical, 239
Democritus' concept of structural, 239
Empedocles' concept of empirical, 187
Empedocles' concept of molecular, 187
modern physics' concept of empirical, 187
modern physics' concept of molecular, 187
Chemistry, 22, 130ff, 186ff, 237ff, 240
Democritus' concept of, 215ff, 238ff
Empedocles' concept of, 186ff
City-state(s), *see Polis*
Civilization(s), 18
Assyrian, 16
Babylonian, 16
Egyptian, 16
European, 7
Greek, 95
Hittite, 16
Israelite, 16
Mycenaean, 6
Persian, 16
Phoenician, 16
Summerian, 16
Clepsydra (water-clock), Empedocles', 183, 183n.28
Cogito ergo sum, 140
Cognition, Empedocles' concept of, 176ff
Comedy, 198
Community
Heraclitus' concept of, 119ff
Pythagorean, 55
Xenophanes' concept of, 92f
Condensation, Anaximenes' concept of, 46
Conjecture(s), 23
Consciousness, Greek, 7f
modern physics' concept of, 253
Penrose's concept of, 207
Pythagorean concept of, 24
Continuity
Greek concept of, 18
Melissus' concept of, 150
modern physics' concept of, 128
Zeno's concept of, 127
Cosmic cycle, Empedocles' concept of, 171, 172ff
Cosmios (decorous), 6
Pythagorean concept of, 57, 59
Cosmogony(ies)
Anaxagoras' concept of, 201
Anaximander's, 39ff
Anaximenes', 48f
Democritus' concept of, 219ff
Empedocles', 189ff
first 10ff
Hesiod's, 10ff
Pythagorean, 72ff
Cosmology
Anaxagoras', 201
Anaximander's, 40ff
Anaximenes', 48
Democritus' concept of, 220ff
Empedocles', 170f, 173ff
Greek, 133
Heraclitus', 118
Parmenides, 138, 143ff, 162ff
Pythagorean, 74ff
Thales', 30f
Xenophanes', 86f, 87

Cosmos
Anaxagoras' concept of, 200
Anaximander's concept of, 43
Aristotle's concept of, 204
Democritus' concept of, 249
Empedocles' concept of, 26, 195
Greek concept of, 5ff, 7, 18, 19
Heidegger's view of, viii, 251
Heraclitus' concept of, 25ff, 118ff, 249
Hesiod's concept of, 12
meanings of, 103
modern physics' concept of, 253
natural philosophers' concept of, 185
Parmenides' concept of, 25f, 144
Presocratics' concept of, 22, 251
Pythagorean concept of, 56, 59, 62, 73
Thales' concept of, 33
Creation
Anaxagoras' concept of, 202
Empedocles' concept of, 175ff
Hesiod's concept of, 11ff
Presocratics' concept of, 1
Critical spirit, 3, 50, 98
philosophical, vii
Presocratics' viiff, 21, 27, 248
see also Thought, critical
Culture(s)
Eastern, 4f
Egyptian, 8
Indian, 8
Mycenaean, 7

D

Daemon (spiritual, guardian powers)
Empedocles' concept of, 179ff
Heraclitus' concept of, 120, 121
Death
Democritus' concept of, 222
Heraclitus' concept of, 106n9, 117ff
Parmenides' concept of, 144
Dedekind 'cuts', 79
Democracy, 13, 15
Democritus' concept of, 233f
Greek concept of, 254
Demythification 10
Anaximander's, 43
Hesiod's, 12
Thales', 29, 31ff
Determinism
Anaxagoras' concept of, 203f
Anaximander's concept of, 42
Empedocles' concept of, 184
Greek, 9
Milesians' concept of, 50
modern physics' concept of, 244
Thales', 31, 33
Dialectic, Heraclitus', 125
Dike
Heraclitus' concept of, 119, 120
meanings of, 103
Discreteness
Greek concept of, 18
modern physics' concept of, 128
Zeno's concept of, 127
Divine
Anaximander's concept of, 37f
Anaximenes' concept of, 47ff
Empedocles' concept of, 180f
Greek concept of, 6, 8f
Heraclitus' concept of, 106ff, 117, 121ff, 122
Milesians' concept of, 50
Pythagorean concept of, 56, 59
Thales' concept of, 32f
Xenophanes' concept of, 25, 90
Divinity
Chinese, 9
Greek, 9
Hindu, 9
Xenophanes' concept of, 88ff
Divisibility
Anaxagoras' concept of infinite 199
atomists' concept of, 213ff, 246f
Parmenides' concept of, 211ff
Dogma, 8
Dokos (guesses, opinion), Xenophanes' concept of, 93
Doxa (mere opinion), 94, 99
meanings of, 137
Parmenides' concept of, 142ff, 145f, 148ff
Dualism, Presocratics', 1
matter's, 242ff
wave-particle, 243ff
Dynamei on (that which is potentially), Aristotle's concept of, 204

E

Earth, *see Gaia* (earth)
Eclipse, lunar, 16
solar, 16ff, 30
Education, Pythagorean concept of, 61
Effect, photoelectric, 128, 242
Compton, 128
Eidenai (seeing with the mind's eye, knowing), Parmenides' concept of, 137

Eimarmeni (destiny), 8
  Heraclitus' concept of, 119
Einai (being), 137
  Greek concept of, 108n.15, 136
  meanings of, 137
  see also Being
Élan vital, 205
Eleatic School, 137, 150ff
Electromagnetism, 127, 240
Elegy, 14
Element(s)
  Boyle's concept of, 238
  Empedocles' concept of, vii, 26, 170ff
  Plato's concept of, 182
Empiricism
  classical, 183
  sophists', 249
Energeia on (that which is actually), Aristotle's concept of, 204
Enlightenment, 95, 238
En (one), 126
Entelecheia, Aristotle's concept of, 147n.21
Entelechy, 205
Entropy, 163, 240
Enylon eidos (form involved in matter), 76
Episteme (scientifi knowledge), 99
Epitaph, Pericles', 208
Ereignis (event), Heidegger's concept of, 164
Erinyes (avenging deities and restorers of order), Heraclitus' concept of, 114, 119, 120
Eris
  Heraclitus' concept of, 107
  meanings of, 103
Eros, 11ff
  see also Love
Essentia (essence), 160
Ethics
  Democritus' concept of, 228ff
  European, 3
  Heraclitus' concept of, 119ff
  Pythagorean, 59, 60f
  Xenophanes' concept of, 91ff
Eutaxia (harmonious orderliness), Descartes' concept of, 204
Evolution, 203
  Anaxagoras' concept, 203
  Anaximander's concept of, 39, 41
  Darwin's concept of, 205
  Milesians' concept of, 24, 50
  Presocratics' concept of, 251
  Wallace's concept of, 205
  see also Natural selection
Existence, Parmenides' concept of, 146
Existentia (existence), 160
Existential, 149n25
Experience, 26
  Parmenides', 162, 165
Experiment(s)
  by Democritus, 217ff
  by Presocratics, 20ff, 27
  by Pythagoreans, 68ff, 80
  in modern science, 27
  two-holes, 244

F
Fire
  Empedocles' concept of, 170ff, 174ff, 183, 192
  Heraclitus' concept of, 109ff, 119, 122
Flux
  Heraclitus' concept of, 25
  modern physics' concept of, 129, 130, 163
  Plato's concept of, 182
Force(s), 43
  carriers, 245ff
  Empedocles' theory of, vii, 171ff, 180, 184, 185, 212
  Greek concept of, 18
  Heraclitus' concept of, 118ff
  Hesiod's concept of, 11ff
  modern physics' concept of, 12, 185ff
  Presocratics' concept of, 1
Fragmentation, 2
  modern physics' view of, 252ff
Freedom, Greek concept of, 208, 254

G
Gaia (earth)
  Anaximander's concept of, suspended in space, 40f, 42
  Empedocles' concept of, 170ff, 174ff
  Hesiod's concept of, 11ff
  Plato's concept of, 182
  Pythagorean concept of, 74
Genesis (coming-to-be), Aristotle's concept of, 161
Geometrical algebra, Pythagorean concept of, 66, 67ff
Geometric period, 7, 10ff
Geometry
  Egyptian, 17
  Euclid's, 65ff
  Greek, 7ff, 17
  Pythagorean, 62, 64ff, 79
  theoretical, 30

Gignesthai (becoming), Greek concept of, 108n.15
Gnomon, 16
God(s)
  Heraclitus' concept of, 122
  Olympian, 9
  Xenophanes' one, 88ff, 94
Golden Age, 198
Grand unifie theory (GUT), 79
Gymnastic(s), Pythagorean concept of, 57, 59

H
Harmonia, 81
Harmonic(s), Pythagorean concept of, 64
Harmony
  Greek concept of, 254
  Heraclitus' concept of, 106, 133
  Kepler's concept of, 70, 75, 78
  Leibniz's concept of, 70
  modern sciences' concept of, 70ff
  Pythagorean concept of, 56ff, 59ff, 68ff, 70ff, 78, 81, 83
  Xenophanes' concept of, 94
Health, Pythagorean concept of, 70
Hellene (Greek), 7, 259
Heroism, Greek, 95
Heteron, meanings of, 103
Heterotita (otherness), Heraclitus' concept of, 109
Holism, 22
  modern physics' concept of, 98, 101ff, 245f, 252ff
  Presocratics' concept of, viiiff, 1, 251ff
  Pythagorean concept of, 75
  see also Unity
Homoeomeroi, Anaxagoras' concept of, 26, 199f
Homogeneity, 8
  Anaximander's concept of, 41
  Empedocles' concept of, 190ff
  Presocratics' concept of, 251
  Pythagorean concept of, 59f
Homo-genoi (congeners), Heraclitus' concept of, 117
Homologia (agreement, concession), Heraclitus' concept of, 115ff, 120
Homo-logos (in unison), Heraclitus' concept of, 113ff, 122
Hoti estin (the thing 'which is'), Parmenides' concept of, 160
Hybris (insolence; wanton violence), 8, 9, 22
  Anaximander's concept of, 38
  Heraclitus' concept of, 114, 120
Hydor (water)
  Empedocles' concept of, 170ff, 175f
  Plato's concept of, 182
  Thales' primary element, 24, 30ff
  Xenophanes' concept of, 90ff
Hygiene, Pythagorean concept of, 57, 59

I
Iambus, 14
Idea, eidos
  Aristotle's concept of, 161f, 204
  Plato's theory of, 63, 75ff, 159ff, 164f, 204
Idealism
  Aristotle's, 249
  Plato's, 249
  Socrates', 249
Identity, Heraclitus' concept of, 103, 108ff, 133
Idios, meanings of, 103
Idiosynkrasia (idiosyncrasy), Parmenides' concept of, 144n.11
Iliad, 10
Imagination, Greek, 9
Immortality, 17
  Pythagorean concept of, 58ff
Incommensurability, Pythagorean concept of, 66
Individualism, Sophists', 249
Infinit
  Anaximander's concept of, 36, 41ff
  Pythagorean concept of, 79ff
Interaction
  Heraclitus' concept of dynamic, 133
  modern physics' concept of, 128ff, 133
Internal set theory (IST), 157
  Presocratics' concept of, 1
Interrelation, Presocratics' concept of, 251
Intuition
  Democritus' concept of, 223
  Greek, 13f
  Presocratics', vii, 1, 21, 27
  in scientifi thinking, 12ff, 253
Ionian Enchantment, vii, ix
Ionian school, 49ff
Ionia, vii, 10ff, 19, 23, 258n.2
Irrational, 53
  Pythagoras' concept of, 51
Isomerism
  stereo, 239
  structural, 239
Isotropy
  Anaximander's concept of, 40, 42
  Democritus' concept of, 217
  Empedocles' concept of, 190ff
  Presocratics' concept of, 251

J

Justice
- Anaximander's concept of, 38ff, 42
- Heraclitus' concept of, 114, 119, 121
- Presocratics' concept of, 8

K

Kalon (good and beautiful), Democritus' concept of, 230

Knowledge, 23, 209
- Democritus' concept of, 213ff, 225ff
- Empedocles' concept of, 176ff, 182ff
- empirical, 165
- Greek concept of, 2, 101
- Heraclitus' concept of, 108f, 115ff
- modern sciences' concept of, 185
- Presocratics' concept of, 25
- Thales' concept of, 33
- Xenophanes' concept of, 92ff, 96ff

Kouros, 15

Krasis (blending; temperament), Parmenides' concept of, 144, 144n.11

Kykeon, 107

L

Language, 1
- Democritus' concept of, 227f
- Greek, 7, 8ff
- Heraclitus', 103
- modern physics', 132ff
- Parmenides, 139n.8, 143n.10

Lanthano (not to be seen), 147

Law(s)
- Anaximenes' concept of natural, 48ff
- Empedocles' concept of natural, 183
- of gravitation, 83
- Heraclitus' concept of, 113ff, 119ff, 121
- natural, 50
- Presocratics' concept of natural, 251
- Pythagorean concept of natural, 69
- thermodynamics' second, 163f

Learning, Democritus' concept of, 231f
- Pindar's idea of, vi

Legein, meanings of, 111, 149

Lethe (concealment), Heraclitus' concept of, 111

Light
- Empedocles' concept of, 186
- modern physics' theory of, 127ff, 185f

Limit(s)
- Empedocles' concept of human, 182
- Xenophanes concept of human, 25, 93, 94

Literature, Greek, 13

Litho (forget), 147

Logical principle
- of contradiction, 139
- of exclusive third, 139
- of identity, 140
- of *tertium non datur*, 132ff

Logic
- Aristotle's categories of, 161f
- classical, 132ff
- Greek, 13
- Heraclitus', 125
- modern physics', 132ff
- Parmenides', 169

Logos
- Aristotle's concept of, 164
- Heraclitus' concept of, 25, 102ff, 110ff, 111ff, 122, 130, 133
- meanings of, 103
- Parmenides' concept of, 164
- Pythagorean concept of, 66, 83
- Xenophanes' concept of, 25, 92ff

Love, 26, 170f, 172ff, 175ff, 178ff, 194ff
- Empedocles' concept of, 26, 170f, 172ff, 178ff, 194ff
- Plato's concept of, 182
- *see also Eros*

M

Macrocosm
- Anaximenes' concept of, 47
- Empedocles' concept of, 176, 184
- Heraclitus' concept of, 117ff
- modern physics' concept of, 184ff
- Pythagorean concept of, 59, 60f, 70

Materialism
- Democritus', 77
- dialectical, 125

Mathemata (sciences), Pythagorean, 57ff, 64ff

Mathematici, Pythagorean, 57ff

Mathematics, 77
- Babylonian, 64
- Democritus', 226ff
- Egyptian, 64
- Euclidean, 63
- Greek, 63ff
- Pythagorean, 63ff, 77ff
- Thales', 30

Mathematization
- modern physics', 82
- Pythagoras', vii, 77f, 251

Matter
- Anaxagoras' concept of, vii, 26, 198ff
- atomists' concept of, 127, 212ff
- dualistic nature of, 242ff

Empedocles' concept of, 187ff
Empedocles' notion of extracelestial, 194
Greek concept of, 18
Heraclitus' concept of, 25, 251
Hesiod's concept of, 11ff
Lenin's view of, 241
modern physics' concept of, 12, 22, 80, 128ff
Pythagorean concept of, 58

Measure
Democritus' concept of due, 230f
Eastern concept of, 81n.85
Greek concept of, 254
Heraclitus' concept of, 109ff, 113ff, 119, 129, 133
natural philosophers' concept of, 50
Presocratics' concept of, 251ff
Pythagorean concept of, 81ff, 83
Xenophanes' concept of, 91

Measurement, 27

Medicine
Democritus' concept of, 221ff
Greek, 49

Metempsychosis
Orphic concept of, 56
Pythagorean concept of, 56

Meteorology
Anaximander's concept of, 41f
Anaximenes' concept of, 48f

Methexis (participation), Pythagorean concept of, 56ff, 59, 62

Method
atomists' deductive, 213ff
atomists' inductive, 213ff
deductive, 20, 76, 162n.52, 233
inductive, 20, 76n52, 213ff
Parmenides' deductive', 214
Zeno's dialectic, 151

Metron (due measure), *see* Measure

Microcosm
Anaximenes' concept of, 46ff
Empedocles' concept of, 176, 184
Heraclitus' concept of, 117ff
modern physics' concept of, 184ff
Pythagorean concept of, 57, 59, 60f, 70

Mimeisthai (reproduce), Pythagorean concept of, 57

Mimesis (imitation), Pythagorean concept of, 59, 62

Mind
Anaxagoras' concept of, vii, 15, 198ff, 203ff, 207ff, 212
atomists' concept of, 211
Greek, 12
Hume's concept of, 206
modern physics' concept of, 253
Penrose's concept of, 207, 248f

Mixing
Empedocles' concept of, 171, 186ff
Parmenides' concept of, 143ff

Moira (fate), 8
Heraclitus' concept of, 120

Monochord, Pythagoras' use of, 68

Monody, 14

Motion
Democritus' concept of, 26, 214ff
Greek concept of, 18, 26
Heraclitus' concept of, 25, 107, 130, 169
Milesians' concept of, 24
Parmenides' concept of, 25, 211ff
Presocratics' concept of, 15, 26

Movement, *see* Motion

Music, Pythagorean concept of, 57, 59, 62ff, 68ff, 80

Mystic(ism), 17
Anaxagoras' concept of, 197, 203
Eastern, 126ff
Greek, 56
Newton's, 167n.2
Orphic, 9, 56
Pythagorean, 55, 76
Western, 125

Mythical cycles
of Argonauts, 10
of Argos, 10
of Thebes, 10
of Troy, 10

Mythology, Greek, 6

Myth(s), 4, 9ff, 18, 94, 95
cosmogonic, 10
homeric, 11
Thales' concept of, 33
theogonic, 10

N

Naming, Parmenides' concept of, 165

Natural phenomena
Democritus' concept of, 221
Presocratics' concept of, 251

Natural selection
Darwin's theory of, 193f, 205
Empedocles' concept of, 175ff, 193f
Presocratics' concept of, 251
Wallace' theory of, 205
see also Evolution

Nature
  Greek, 4ff, 7f
  Milesians' concept of, 50
Necessity
  Anaximander's concept of, 38ff, 40, 41, 42
  Democritus' concept of, 216, 218ff, 227
  Heraclitus' concept of, 114, 121
  Leucippus' concept of, 209
  Parmenides' concept of, 144
Neoplatonism, 76
Neo-positivism, 97, 98
Neutrino, Pauli's hypothesis of, 23
Nihilism, Sophists', 249
Nirvana, 127
Noein (apprehend, conceive, presume, think)
  Aristotle's concept of, 161f, 164
  Parmenides' concept of, 137, 148ff, 161f, 164
Nomos, meanings of, 103
Non-being
  atomists' concept of, 212ff
  Parmenides' concept of, 138ff, 141ff, 146ff
Nonstandard analysis, 156ff
Number
  Aristotle's concept of, 76
  Greek concept of, 18, 50
  Pythagorean concept of, 24, 66ff, 68ff, 71ff, 76

O

Observation
  Democritus' concept of, 213
  Presocratics', viii, 27, 251
Odyssey, 10
Oedipus, 148n.23
One
  Heraclitus' concept of, 118f, 133
  modern sciences' concept of, 83
  Pythagorean concept of, 73
  Thales' concept of, 33
  Xenophanes' concept of god, 88ff
Oneness, *see* Unity
Ontical, meaning of, 149n.25
Ontological, meaning of, 149n.25
Ontology
  atomists' concept of, 211, 214f
  existential, 126
  Parmenides', 136, 160ff, 214, 242
  Plato's, 159ff
Ontos onta, Plato's concept of, 160
Opposite(s)
  Anaximander's concept of, 37ff, 41, 42
  Anaximenes' concept of, 46
  Empedocles', concept of, 169, 183
  Heraclitus' concept of, 15, 25ff, 125, 133
  Parmenides' concept of, 143ff
  Presocratics' concept of, 251
  Pythagorean theory of, 71ff, 80
Orao, (see; look), 5
Order
  Greek concept of, 10, 254
  Empedocles' concept of, 192
  Heraclitus' concept of, 114, 131
  modern physics' concept of, 131
  Parmenides' concept of, 143
  Presocratics' concept of, vii, 9, 19, 251
  Pythagorean concept of, 57
  Thales' concept of, 33
Origin, cosmic, 3ff, 8
Orphism, 9, 17
Ousia, Aristotle's concept of, 161

P

Panta (all), 126
  Heraclitus' concept of, 133
  Hesiod's concept of, 11
Paradox(es)
  EPR, 98, 131n.60, 245f
  Zeno's, 67ff, 151ff
  Zeno's 'Achilles and the tortoise', 154
  Zeno's 'dichotomy', 153
  Zeno's 'flyin arrow', 154
  Zeno's 'millet seed', 153f
  Zeno's plurality, 151ff
  Zeno's space, 152ff
  Zeno's 'stadium', 155
Particle(s)
  modern physics' concept of, 12
  Presocratics' concept of, 1
Periodicity, 79
Personality, Greek, 12ff, 14
Phenomenalism, Sophists', 249
Philo-sophos(i) (lover(s) of wisdom)
  Heraclitus' concept of, 116, 122
  Pythagoras' concept of, 56
Philosophy
  Anaxagoras', 198
  Aristotelian, 76, 249ff
  Attic, 249ff
  Chinese, 13
  Eastern, 126
  European, 3ff, 135
  Greek, 2, 13ff, 19, 29, 29n.1, 249
  Ionian, 9
  natural, 20
  Platonic, 76, 249ff
  Presocratic, viii, 1, 10, 19
  Pythagorean, 56ff, 62ff, 81ff

Phrin (mind), Empedocles' concept of, 180
Phrontis (care), Empedocles' concept of, 180
Physicoi (natural philosophers), 25, 26, 29, 30, 47ff, 101, 103, 144, 169, 185
Physics
  classical, 244
  contemporary, 126
  Greek, 7ff
Physiology, 302ff
Physis
  Greek concept of, 6ff, 126, 171, 174
  Parmenides' concept of, 147, 148, 150, 160ff, 164
  Plato's concept of, 160
Planck's constant, 80, 128n.48
Plurality
  Democritus' concept of, 26
  Eleatics' concept of, 212
  Empedocles' concept of, 171
  Milesians' concept of, 50
  modern sciences' concept of, 83
  Presocratics' concept of, 251
Pneo (breath), 47
Pneuma, Anaximenes' concept of, 47f, 49
Poetry
  choric, 14
  elegiac, 4, 14
  epic, 4, 9ff, 13ff, 16, 259
  lyric, 4, 10ff, 13ff
Polis, 7ff, 60n24, 97f
Polos, 16
Polyhedron(s), Plato's concept of, 182
Positivism, Sophists', 249
Primary substance, substratum, source, Presocratics' concept of, 20f, 24ff
Principle(s)
  Anaxagoras' eutaxiological, 203ff, 203n11, 206ff
  Anaxagoras' teleological, 200ff, 203ff, 203n.11, 212
  Anaximander's concept of reciprocity, 42
  Aristotle's four, 204n.13
  Aristotle's immanent teleological, 204f
  atomists' teleological, 211
  Bergson's teleological, 205
  of complementarity, 244
  of contradiction, 139ff
  Darwin's concept of the teleological, 205
  dialectic materialism's concept of the teleological, 205
  Driesch's teleological, 205
  Hegel's teleological, 205
  Kant's teleological, 205
  kinetic, 26
  Leibniz's teleological, 205
  Monod's concept of, 205
  Newton's eutaxiological', 204
  Plato's transcendental teleological, 204f, 235
  Russell's concept of the teleological, 205
  Tennant's concept of the teleological, 205
  of uncertainty, 244ff, 246
Probability, 244
Process
  Anaximenes' concept of, 46
  chemical, 109
  cyclic, 131
  Heraclitus' concept of, 25, 105ff, 126, 129ff, 131, 251
  modern sciences' concept of, 129ff, 131
Progress
  Democritus' concept of, 227f
  Xenophanes' concept of, 25, 89ff
Prologue, Parmenides', 138
Pro-phanes (made manifest), Parmenides' concept of, 146
Protista (first) Hesiod's concept of, 11ff
Psyche (soul), *see* Soul
Psyho (breath, blow), 47
Purification(s
  Empedocles', 168f, 177ff, 180f
  Pythagoras' concept of, 56ff, 59f, 62
Pyr, meanings of, 103
Pythagorism, 17

Q

Quantum theory, 23, 42, 80, 98ff, 131ff, 163, 242ff, 246ff, 252
  Bohr's, 243ff
  Born's, 243ff
  Copenhagen interpretation of, 244ff
  Dirac's relativistic, 75, 128, 157, 243
  Heisenberg's, 243ff
  Jordan's, 243
  Schrödinger's, 243ff

R

Radiation
  black-body, 23, 242
  cosmological microwave background, 189
Radioactivity, 241
Rarefaction, Anaximenes' concept of, 46
Rationalism
  Anaxagoras' concept of, 203
  Popper's critical, 99

Rational spirit
  Anaximander's, 43
  European, 259
  Presocratics', viiff, 1
  Pythagorean concept of, 51, 58, 81f
Ratio(s), Pythagorean concept of, 65, 66
Reality
  Einstein's concept of, 163
  Greek concept of, 5, 254
  Heraclitus' concept of, 25
  modern sciences' concept of, 253
  Parmenides' concept of, 162ff, 169
  Presocratics' concept of, 25, 27, 251
  Weyl's concept of, 163
Reason
  Greek, 9, 13, 19, 94
  Leucippus concept of, 209
  Parmenides' concept of, 162
  Presocratics', 27
  Xenophanes' concept of, 92ff
Reductio ad absurdum, applied by Aristotle, 67n.40, 213
Refutationism, 23, 98
Reincarnation, Empedocles' concept of, 179ff
Relativism, Sophists', 249
Relativity
  general theory of, 23, 79f, 128, 163, 183
  special theory of, 128, 252
Religion, 4, 7ff, 18
  Einstein's concept of, 253
  Pythagorean concept of, 24, 62, 63
Renaissance, 20, 50, 77, 125ff, 250, 251
Reproduction, Empedocles' concept of, 188f
Repulsion
  atomists' concept of, 211
  Empedocles' concept of, 172ff, 212
Res cogitans, 133
Res extensa, 133
River image, Heraclitus', 108f
Röntgen-rays, 241
Root(s), Empedocles' concept of, 26, 159ff, 180, 187ff

S
Salvation, Pythagorean concept of, 55ff
Samadhi, 127
Schrödinger's cat, 131n.60, 245f
Science
  Greek, 13, 16, 19, 23ff
  Eastern, 17
  European, 3ff, 76
  modern, 23, 84
  Presocratics', viii, 1, 19ff
  Pythagorean, 55ff, 59, 62ff
  Thales', 32
Script, Mycenaean B, 7
Sea, 16
  Greek, 5
Seed(s)
  Anaxagoras' concept of, 26, 199f
Seeming
  Parmenides' concept of, 138ff, 143ff, 148ff, 163
  Xenophanes' concept of, 25
Sensation
  Democritus' concept of, 222ff
  Empedocles' concept of, 176ff, 182ff
Sense(s), 25
  Democritus' concept of, 222ff
Separation, Empedocles' concept of, 171f, 186ff
Seven Wise Men, 8, 29
Shamanism, 17, 54n.9
Silence, Pythagorean, 58
Silloi, 85
Sima (grave), Heraclitus' concept of, 117
Skepticism, Sophists', 249
Society
  Greek concept of, 6ff, 97ff
  Pythagorean concept of, 60ff
Soma (body), Heraclitus' concept of, 117
Sophia (wisdom)
  Democritus' concept of, 231f
  Heraclitus' concept of, 115ff
Sophist(s), 125, 249ff
Soul, 47
  Anaxagoras' concept of, 200
  Anaximenes' concept of, 47f
  Democritus' concept of, 222, 232n.47
  Empedocles' concept of, 177ff
  Heraclitus' concept of, 116ff
  Pythagorean concept of', 58ff, 59ff, 70
  Thales' concept of, 32f
Space
  Anaximander's concept of isotropic, 40, 42
  Greek concept of, 18
Spacetime, four-dimensional, 163
Speaking, saying, Parmenides' concept of, 140, 148ff
Sperma(ta), seed(s), Anaxagoras' concept of, 159, 199
Spheros, Empedocles' concept of, 190ff
Stability, Presocratics' concept of, vii
Standard Model, 245ff
Stoichiometry, 215

Stoichion ('letter' of the alphabet; element), 215
Stoic(s), 125
Strife, Empedocles' concept of, 26, 170f, 172ff, 175ff, 178ff, 191ff
  Heraclitus' concept of, 107, 110, 121
Syllogi (collection; intellectual activity), 149n.24
  Plato's concept of, 182
Syllogism (thought), Heraclitus' concept of, 113
Syllogizomai, meanings of, 149n.24
Symbola (secret codes), Pythagorean, 57ff
Symbolon (one half of a cooperative whole), Empedocles' concept of, 188f
Symbol(s), modern sciences', 132
Symmetry
  Anaximander's concept of, 41ff
  Eastern concept of', 78n.75
  Empedocles' concept of, 183, 190, 192
  gauge, 128
  Milesians' concept of, 50
  modern physics' concept of, 78ff, 78n.77, 80
  Plato's concept of, 78, 245
  Presocratics' concept of, 1, 8, 251
  Pythagorean concept of, 78
Synthesis, of contradictory concepts, 128

T
Tao, 126
Tauton, meanings of, 103
Tautotita (maintenance of identity), Heraclitus' concept of, 109
Telos (limit; fina cause), 147n.21
  Anaxagoras' concept of, 203
  Bacon's concept of, 204
  Darwin's concept of, 205
Testament, New, 16
  Old, 6, 8ff, 16, 47, 91
Thea (view), 5
Theogony(ies)
  Akousilaus', 11
  Epimenides', 11
  Hesiod's, 10ff, 13
  Musaius', 11
  Orpheus', 11
  Pherekydes', 11
Theologian(s), 11, 12
Theology, 76
  Democritus' concept of, 227ff
  Pythagorean, 61ff
  Thales' concept of, 32
  Xenophanes' negative, 24, 88ff
Theorein (observing), 6
Theoria, 5, 57
  Anaxagoras' concept of, 208
Theor(y)ies
  Anaxagoras' intelligent-design, 251
  Anaxagoras' unified 208
  dualistic, 26
  of everything (TOE), 128, 191n.45
  of gases, 240
  monistic, 26, 159, 169, 198
  pluralistic, 26, 170
  Presocratics', 23
  scientific 21
  unified 43f, 128
  Xenophanes' anthropocentric, 89f
Thinker(s), Greek, 1
  Milesian, vii
  Presocratic, 20, 27
Thought, critical, 21, 53
  Aristotle's concept of, 161f
  Democritus' concept of, 225
  Empedocles' concept of, 182
  Greek, 1f, 16, 19, 49
  Hellenistic, 125
  Heraclitus', 105, 113, 127
  Hesiod's, 11ff
  Hume's concept of, 206
  mythological, 3
  Parmenides' concept of, 140ff
  Presocratics', 2f, 4, 15, 27, 250f
  rationalistic tradition, 3
  Western, vii, 10
Ti estin ('what' a thing actually 'is'), Plato's concept of, 160
Time
  Greek concept of, 18, 38n.6ff
  Empedocles' concept of, 193f
  modern physics' concept of, 193f
  Pythagorean concept of, 80
Tragedy, Attic, 15, 16, 25, 198, 208
Tragic element, 85
  Heraclitus', 122ff
  Xenophanes', 93ff
Transcendentalism
  Aristotle's, 249
  Plato's, 249
  Socrates', 249
Transmigration, of soul, 17
Truth, 3, 21, 23
  Aristotle's concept of, 160ff, 164
  Democritus's concept of, 23, 209, 214ff
  Empedocles' concept of, 182ff

Greek concept of, 18, 126
Heraclitus' concept of, 23, 25ff, 136
Hesiod's concept of, 11ff
Parmenides' concept of, 25ff, 138ff, 145f, 147ff, 160ff, 164ff
Plato's concept of, 160
Presocratics' concept of, viii, 10
Protagoras' concept of, 125
Sophists' concept of, 249f
Xenophanes' concept of, vii, 23, 25ff, 98ff
Tyche (chance)
Anaxagoras' concept of, 203
Democritus' concept of, 218ff
Empedocles' concept of, 171, 174f
Tyranny, 6

U
Unificatio of, 43f, 128
corpuscular and wave theory of light, 128
electromagnetic and weak interactions, 128
forces in physics and chemistry, 128
kinetic mechanics and electrodynamics, 128
mechanics and the theory of heat, 128
quantum mechanics and the special theory of relativity, 128
Unity, 195f
Empedocles' concept of, 195f
Heraclitus' concept of, 103, 106, 109, 112ff, 125, 127
Presocratics' concept of, vii, 19, 251
Pythagorean concept of, 81
Thales' concept of, 31
Xenophanes' concept of, 87f
see also Holism
Universe
Empedocles' oscillating, 183, 195
expanding, 189f

V
Verificationism 98
Vision, Greek, 95
Void
atomists' concept of, 26, 212, 214, 216f, 249
Empedocles' concept of, 171
Melissus' concept of, 151, 211
modern physics' concept of, 246f
Parmenides' concept of, 25, 141ff, 211
Pythagorean concept of, 73, 79

W
War(s)
Heraclitus' concept of, 25, 107
Persian, 15
Way of life
Democritus' concept of, 232ff
Pythagorean, 55ff
Wheeler-DeWitt equation, 163
Wholeness, *see* Holism
Wiener Kreis, 97
Wisdom, 209
Heraclitus' concept of, 115ff
Wonder, 3, 5, 24
World
Empedocles' concept of, 169
Heraclitus' concept of, 118ff
World map, Anaximander's, 36

X
Xynon (common, impartial), Heraclitus' concept of, 107, 107n.10, 112ff

Z
Zen Buddhism, 75
Zero, 64n.36

The Time: Chronological Table

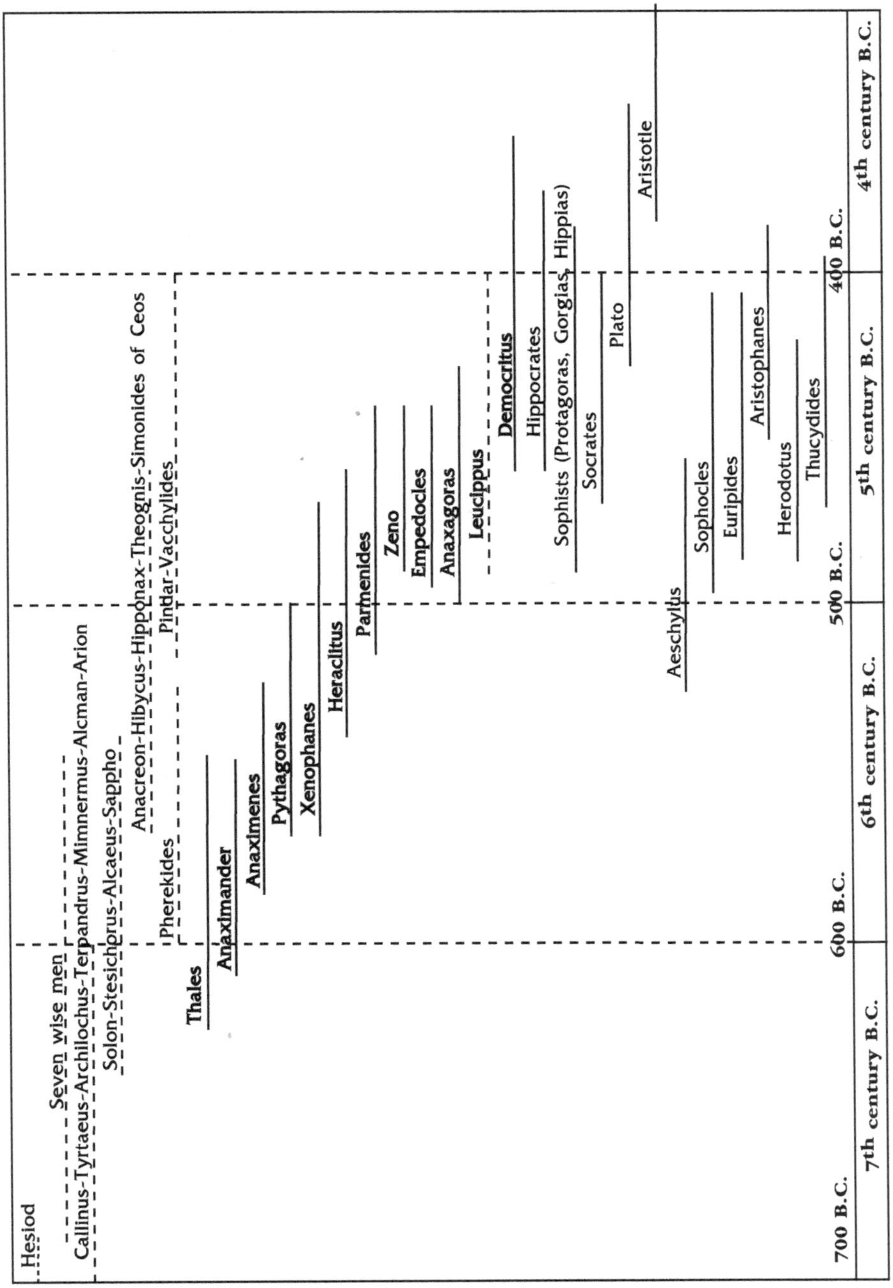

The Space:The Greek Civilization during the Presocratic Period

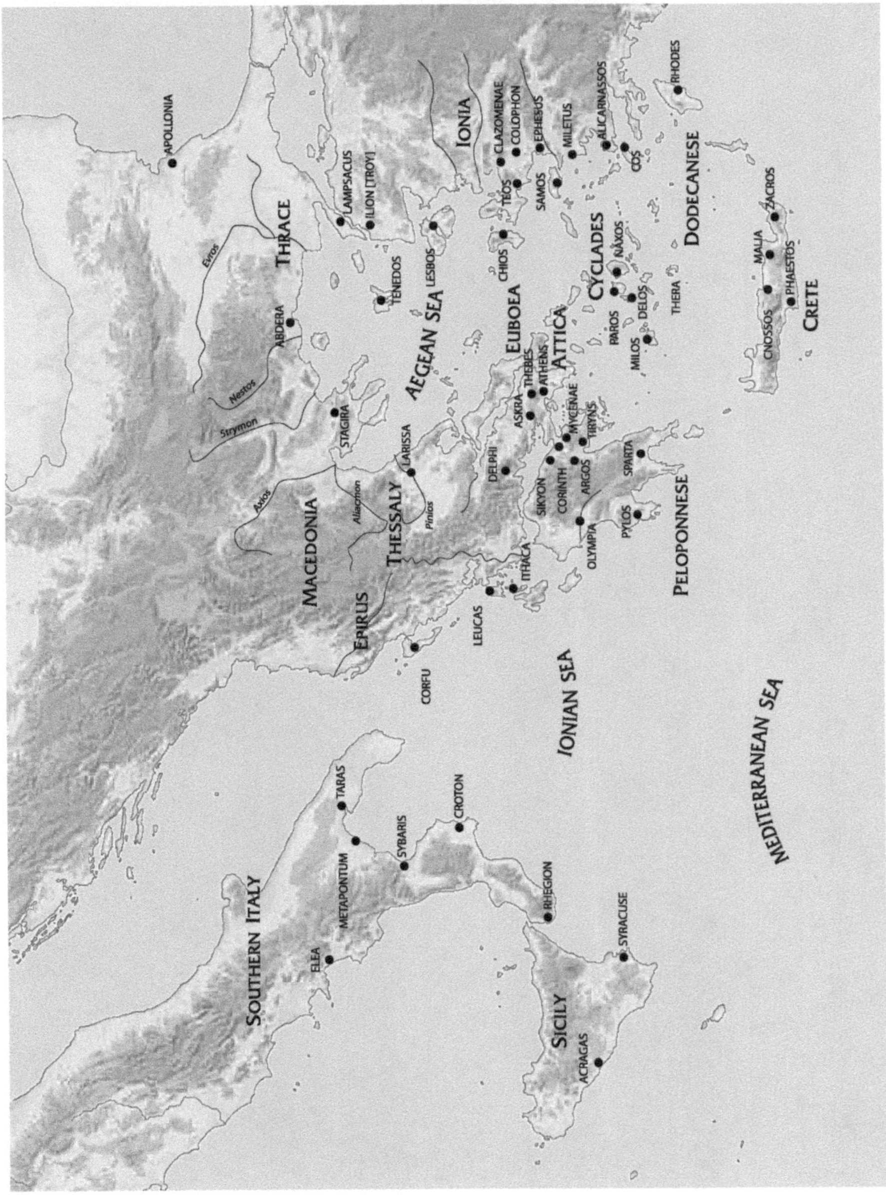